ADVANCES IN MATERIALS RESEARCH

Series Editor-in-Chief: Y. Kawazoe

Series Editors: M. Hasegawa A. Inoue N. Kobayashi T. Sakurai L. Wille

The series Advances in Materials Research reports in a systematic and comprehensive way on the latest progress in basic materials sciences. It contains both theoretically and experimentally oriented texts written by leading experts in the field. Advances in Materials Research is a continuation of the series Research Institute of Tohoku University (RITU).

Please view available titles in *Advances in Materials Research* on series homepage http://www.springer.com/series/3940

Kazuo Nakajima
Noritaka Usami
Editors

Crystal Growth of Si for Solar Cells

With 159 Figures

Springer

Professor Kazuo Nakajima
Professor Dr. Noritaka Usami
Tohoku University, Institute for Materials Research
Katahira 2-1-1, Aoba-ku, Sendai 980-8577, Japan
E-mail: nakasisc@imr.tohoku.ac.jp, usa@imr.tohoku.ac.jp

Advances in Materials Research ISSN 1435-1889

ISBN 978-3-642-26075-9 e-ISBN 978-3-642-02044-5

DOI 10.1007/978-3-642-02044-5

Softcover reprint of the hardcover 1st edition 2009

Typesetting: Data prepared by SPI Kolam using a Springer LaTeX macro package
Cover concept: eStudio Calmar Steinen
Cover production: SPI Kolam

SPIN: 12443024 57/3180/SPI
Printed on acid-free paper

9 8 7 6 5 4 3 2 1

springer.com

Preface

This book, a continuation of the series "Advances in Materials Research," is intended to provide the general basis of the science and technology of crystal growth of silicon for solar cells. In the face of the destruction of the global environment, the degradation of world-wide natural resources and the exhaustion of energy sources in the twenty-first century, we all have a sincere desire for a better/safer world in the future.

In these days, we strongly believe that it is important for us to rapidly develop a new environment-friendly clean energy conversion system using solar energy as the ultimate natural energy source. For instance, most of our natural resources and energy sources will be exhausted within the next 100 years. Specifically, the consumption of oil, natural gas, and uranium is a serious problem.

Solar energy is the only ultimate natural energy source. Although 30% of total solar energy is reflected at the earth's surface, 70% of total solar energy can be available for us to utilize. The available solar energy amounts to several thousand times larger than the world's energy consumption in 2000 of about 9,000 Mtoe (M ton oil equivalent). To manage 10% of the world's energy consumption at 2050 by solar energy, we must manufacture 40 GW solar cells per year continuously for 40 years. The required silicon feedstock is about 400,000 ton per year. We believe that this is an attainable target, since it can be realized by increasing the world production of silicon feedstock by 12 times as much as the present production at 2005. To accomplish this target of using solar energy for 10% of the world's energy consumption at 2050, we must develop several key materials to establish a clean energy cycle, taking into consideration the lifespan of various materials.

Among the various materials, silicon, which accounts for more than 90% of solar cells today, is undoubtedly the key, especially if we consider a large-scale deployment. To accelerate the deployment of photovoltaic technology by development of high-efficiency crystalline silicon solar cells, the Institute for Materials Research (IMR), Tohoku University, organized a unique domestic workshop in 2004 and 2005 to discuss the approach from the view point of materials science by putting emphasis on crystal growth. Based on the success of the domestic workshop, the first international workshop on "Science and Technology of Crystalline Si Solar Cells (CSSC)" was held at IMR on 2–3 October 2006, coorganized by IMR and Japan Society for the Promotion of Science (JSPS), No. 161 Committee on "Science and Technology of Crystal Growth." The international workshop has been continued as a forum to provide an opportunity for scientists/engineers from universities, government institutes, and industry to meet and discuss on the latest achievements and challenges in crystalline silicon solar cells from the viewpoint of materials science. CSSC-2 was held in Xiamen, China, on 7–9 December 2007, and CSSC-3 was held in Trondheim, Norway, on 3–5 June 2009. In addition, IMR and JSPS No. 161 Committee organized a special symposium on "Solar Cells and Clean Energy Technology" in the 4th Asian Conference on Crystal Growth and Crystal Growth Technology (CGCT-4), held in Sendai, Japan, on 21–24 May 2008. Most of topics in this book are based on the discussions during these symposia. The editors acknowledge all the participants in CSSC and CGCT-4 for fruitful discussions.

There are several books on the general aspects of solar cells. However, emphasis is mostly placed on the device physics and little attention has been paid for crystal growth technologies. This is partly due to the misunderstanding that there is no room for further improvement of crystals for solar cells. However, this is not true even for multicrystalline Si wafers, whose macroscopic properties could be altered by manipulating their microstructures with the aid of "crystal growth." The fundamental knowledge obtained through this book is believed to contribute to future developments of novel crystal growth technologies for solar cell materials.

We thank JSPS No. 161 Committee of the "Science and Technology of Crystal Growth" for supporting this book. We also wish to thank Prof. T. Fukuda, the former chairman of the No. 161 committee, Prof. Y. Kawazoe, the Series Editor-in-Chief, for their support and continuous encouragement. We are grateful to Y. Maeda for her formatting assistance. Dr. C. Ascheron and his team from Springer are gratefully acknowledged for the good collaboration.

Sendai
July 2009

Kazuo Nakajima
Noritaka Usami

Contents

10 Vapor Phase Epitaxy

Mustapha Lemiti

11 Thin-Film Poly-Si Formed by Flash Lamp Annealing

Keisuke Ohdaira

12 Polycrystalline Silicon Thin-Films Formed by the Aluminum-Induced Layer Exchange (ALILE) Process

Stefan Gall

13 Thermochemical and Kinetic Databases for the Solar Cell Silicon Materials

Kai Tang, Eivind J. Øvrelid, Gabriella Tranell, and Merete Tangstad

Contributors

Thorvald Engh
Norwegian University of Science and Technology
N-7491 Trondheim
Norway

Alain Fave
Institut des Nanotechnologies de Lyon (INL), UMR 5270
INSA de Lyon, Blaise Pascal
7 avenue Jean Capelle
69621 Villeurbanne Cedex
France

Kozo Fujiwara
Institute for Materials Research
Tohoku University, 2-1-1, Katahira
Aoba-ku, Sendai 980-8577
Japan

Stefan Gall
Helmholtz Zentrum Berlin für Materialien und Energie GmbH
Institute Silicon Photovoltaics
Kekulestr. 5, D-12489 Berlin
Germany

Astrid Gutjahr
ECN, P.O. Box 1
1755 ZG Petten
The Netherlands

Giso Hahn
University of Konstanz,
Jacob-Burckhardt-Str. 29
78464 Konstanz
Germany

Chao-Kuan Hsieh
Sino-America Silicon Products Inc.
No. 8. Industrial East Road 2.
Science-Based Industrial Park
Hsinchu, Taiwan, R.O.C.

Wen-Chin Hsu
Sino-America Silicon Products Inc.
No. 8. Industrial East Road 2.
Science-Based Industrial Park
Hsinchu, Taiwan, R.O.C.

Koichi Kakimoto
Research Institute for Applied Mechanics, Kyushu University
6-1, Kasuga-koen, Kasuga 816-8580
Japan

Kazuhiko Kuribayashi
Department of Space Biology & Microgravity Sciences,
The Institute of Space & Astronautical Science (ISAS)
Japan Aerospace Exploration Agency (JAXA) 3-1-1 Yoshinodai
Sagamihara 229-8510
Japan

Kentaro Kutsukake
Institute for Materials Research
Tohoku University, 2-1-1
Katahira, Aoba-ku
Sendai 980-8577
Japan

Chung-Wen Lan
Photovoltaics Technology Center
ITRI, 195 Chung Hsing Rd.
Sec.4 Chu Tung, Hsin Chu
Taiwan 310, R.O.C.

Mustapha Lemiti
Institut des Nanotechnologies de
Lyon (INL), UMR 5270
INSA de Lyon, Blaise Pascal
7 avenue Jean Capelle
69621 Villeurbanne Cedex
France

Anke Luedge
Leibniz Institut für
Kristallzuechtung Max-Born-Str. 2
D-12489 Berlin
Germany

Kosuke Nagashio
Department of Materials
Engineering, Graduate School
of Engineering
The University of Tokyo
7-3-1, Hongo, Bunkyo-ku
Tokyo 113-8656
Japan

Kazuo Nakajima
Institute for Materials Research
Tohoku University, 2-1-1
Katahira, Aoba-ku
Sendai 980-8577
Japan

Keisuke Ohdaira
School of Materials Science
Japan Advanced Institute of Science
and Technology (JAIST), 1-1,
Asahidai, Nomi, Ishikawa 923-1292
Japan

Eivind J. Øvrelid
SINTEF Materials and
Chemistry, Alfred Getz vei 2
NO-7465 Trondheim
Norway

Helge Riemann
Leibniz Institut für
Kristallzuechtung Max-Born-Str. 2
D-12489 Berlin
Germany

Axel Schönecker
RGS Development B.V.
P.O. Box 40, 1724ZG Oudkarpsel
The Netherlands

Kai Tang
SINTEF Materials and
Chemistry, Alfred Getz vei 2
NO-7465 Trondheim
Norway

Merete Tangstad
Norwegian University of Science and
Technology, N-7491 Trondheim
Norway

Gabriella Tranell
Norwegian University of Science and
Technology, N-7491 Trondheim
Norway

Noritaka Usami
Institute for Materials Research
Tohoku University
2-1-1, Katahira, Aoba-ku
Sendai 980-8577
Japan

1

Feedstock

Eivind J. Øvrelid*, Kai Tang, Thorvald Engh, and Merete Tangstad

Abstract. This chapter will give a short introduction to the well-described processes for the production of metallurgical silicon and solar grade silicon by the Siemens process. Among the new methods for the production of solar grade silicon, the upgraded metallurgical silicon is a good alternative to replace the feedstock produced by the Siemens process. Many of the new methods consist of several steps. The most common refining processes are described in this chapter. The intention of this chapter is to give an overview and understanding of the new processes that are emerging and to give some tools to optimize and develop new methods for production of solar grade silicon.

1.1 Introduction

The global energy consumption is predicted to increase dramatically every year. Higher energy prices and public awareness for the global warming problem have opened up the market for solar cells. Today, the majority of solar cells are made of Si. Experts believe that it will take at least a decade before other PV technology based on other materials can be competitive. The growth in the PV industry has, however, caused a lack of solar grade Si (SoG-Si), i.e., Si with the required chemical purity for PV applications, resulting in increased prices for such material. Presently, the shortage of low-cost SoG-Si is the main factor preventing environmentally friendly solar energy from becoming a giant in the energy market.

1.1.1 Main Supply Route Today

Scrap, a rejected and nonprime material from the semiconductor was the main supply route in the early days of PV. Due to the fast growth of the market, scrap is not sufficient and the main source today is nonprime polysilicon, deliberately produced by operating the conventional Siemens process with more economical parameters (e.g., faster production rates, lower energy consumption, higher impurity levels). New processes are emerging along different routes: Modification of the chemical path to SoG-Si, or refining of metallurgical Si based on metallurgical refining steps.

1.1.2 Impurities

The level of impurities tolerated in the solar cell process is becoming less stringent with the development of the solar cells and the understanding of the effect of each element. The impurities can be divided into three main categories:

(1) Dopants that affect the resistivity of the Si:P, B, (As, Sb, and Al).
(2) Light elements like O, C, and N will form inclusions above the solubility limits and is found in feedstock as SiO_2, SiC, and Si_3N_4. These impurities can create problems in the wafering process and during solidification. The consequence can be structure loss in the CZ process and instability in the MC process.
(3) Metals reduce the lifetime and the cell efficiency. Examples are Fe, Ti, Cu, Cr, Ti, and Al. At too high-levels, metals will also give problems in the solidification processes.

1.2 Metallurgical Si

Metallurgical Si is produced by reducing quartz by C in an electric arc furnace. Consumable graphite electrodes are used to supply the necessary energy for the reaction. The overall reaction in an idealized form can schematically be written:

$$SiO_2 + 2C + Energy = Si + 2CO(g) \qquad (1.1)$$

However, the scenario is more complex. The process can be looked upon as a semicontinuous counter current process. Excavation of furnaces after cool-down, have given a good understanding of the process. In Fig. 1.1, the inner zone of an electric arc furnace is shown just before stoking of the furnace.

The charge mix, added at the top of the furnace, will move down through the furnace pot. The gases that form in different zones of the furnace will flow towards the top of the furnace. In the zone around the electrode, the temperature is high and SiO(g) and CO(g) will form and travel to the top of the furnace. The SiO is supposed to react with the C feed into the furnace, to produce SiC in the prereaction zone.

$$2SiO(g) + C(s) = SiC(s) + CO(g) \qquad (1.2)$$

Both quartz and SiC will descend down to the reaction zone, that is, the cavity around the electrodes. Here,SiO will be produced according to the two

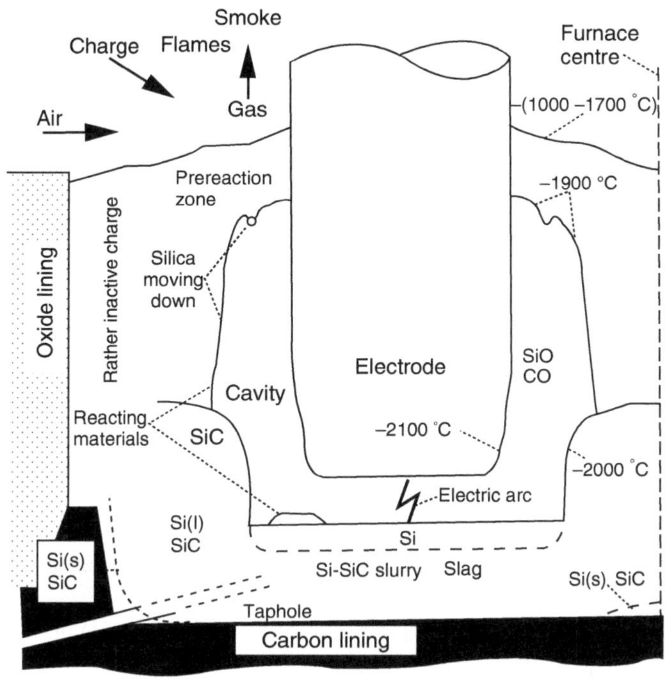

Fig. 1.1. The inner zone around one electrode in a Si electric arc furnace [1]

reactions:

$$SiO_2(l) + Si(l) = SiO(g) \tag{1.3}$$

and

$$SiC(s) + 2SiO_2 = 3SiO(g) + CO(g) \tag{1.4}$$

The metal will be produced according to reaction

$$SiC(s) + SiO(g) = Si(l) + CO(g) \tag{1.5}$$

The book of Schei et al. [1] can be referred for a more detailed description of the carbothermic reduction process. The best description of practical aspects is given by Andresen [2] who relates the theory, thermodynamics and kinetics to the operation of the furnace. A thermodynamics analysis of the carbothermic reduction of SiO_2 was also given by Tuset [3].

For the solar industry, impurity control and alternative raw materials are of great interest. In the normal metallurgical silicon production furnace, silica is charged into the form of quartz lumps. The carbonaceous materials have many forms like charcoal coke, coal, and wood chips. The charge is carefully composed and the raw materials are selected to give best possible yield with respect to Si and to avoid unwanted impurities. Also, the refractory materials and the consumable electrodes are source of impurities. The Si from the arc furnace will be saturated with C at the high-temperature in the furnace and SiC particles are precipitated as the metal cools down.

1.3 The Siemens Process

The extreme purity Si required in the photovoltaic or electronics applications is obtained by converting metallurgical-grade Si into silanes, which are then distilled and decomposed into pure Si. An overview of the most common pathways to polysilicon production, by the chemical means is given in Fig. 1.2 [4].

Today the majority of polysilicon is either produced from trichlorosilane (TCS) or monosilane. In the TCS synthesis, Si fines in the 40 μm range are reacted with HCl in a fluidized bed reactor. The TCS can be purified by fractional distillation. It is possible to reduce the electric active elements including B and phosphorus down to 1 ppm [5]. Silicontetrachloride (STC) is produced as a byproduct in the Siemens process, and can be hydrogenated back to TCS. The Ethyl process is an alternative method to produce SiH_4 based on the byproduct SiF_4 from production of fertilizers (alternatively $SiF_4 + AlMH_4 = SiH_4 + AlMF_4$).

The most common way to deposit the Si is the Siemens type of deposition, where the Si is deposited on a Si filament in a bell jar reactor. For deposition of Si from TCS, typically a temperature of 1100°C is applied on the seed rods. The deposition is carried out inpresence of hydrogen, and the reaction

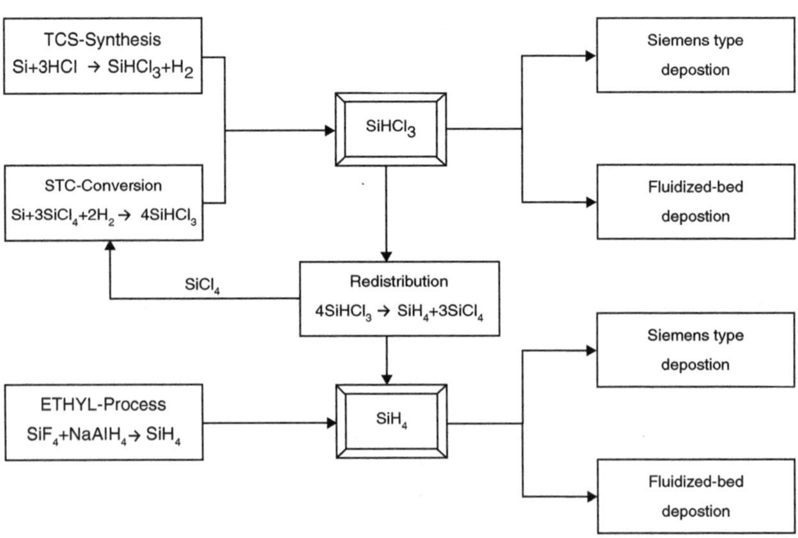

Fig. 1.2. Pathways for polysilicon production [4]

is given by:

$$H_2 + HSiCl_3 = Si + 3HCl \tag{1.6}$$

Due to the large temperature gradients in the reactor, other reactions also occur. Silicontetrachloride is one of the major by-products, according to the following reaction:

$$HCl + HSiCl_3 = SiCl_4 + H_2 \tag{1.7}$$

The decomposition processes are affected by the parameters, such as temperature, gas flow rate, and TCS/hydrogen ratio. Details are given by Mazumder [6].

For deposition of Si from monosilane, the reaction is given by:

$$SiH_4(g) = Si(s) + 2H_2(g) \tag{1.8}$$

Several processes for alternative decomposition of silanes are being explored. Examples are production of pure Si by reaction of $SiCl_4$ with Zn (Hycore, Natsume) and vapor to liquid extraction (Tokuyama).

1.4 Refining of Si for the PV Applications

If we consider metallurgical Si as a starting point for solar grade Si, all other processes can be looked upon as refining process to remove certain impurities. We have several routes or multistep processes where each step removes an impurity or reduces the total level of impurities down to a lower level. In this section, the most usual refining processes for removal of impurities from Si are sketched.

1.4.1 Removal of Boron by Oxidation

B is one of the common impurities in the metallurgical grade Si. Since B is difficult to remove by either of directional solidification or evaporation, the oxidative refining processes are frequently used to remove B from liquid Si.

The oxygen needed for the refining reactions in oxidative refining can be introduced either

(1) As a gas in the form of air or steam at the metal surface, or by gas-blowing through a lance, a nozzle, or plug in the bottom of a refining vessel or
(2) In the form of SiO_2 (in slag or in oxy-fluoride mixture) as an oxidizing agent

Gas-blowing is usually combined with the addition of some kind of slag-forming additives (e.g., CaO, MgO, and fluorides) that may also act as oxidizing agents. The removal of B from molten Si by a steam-added plasma can also be regarded as an oxidative refining process.

A diagram for the Gibbs energy change of oxide formation with temperature (Ellingham's diagram) is shown in Fig. 1.3. This diagram shows the

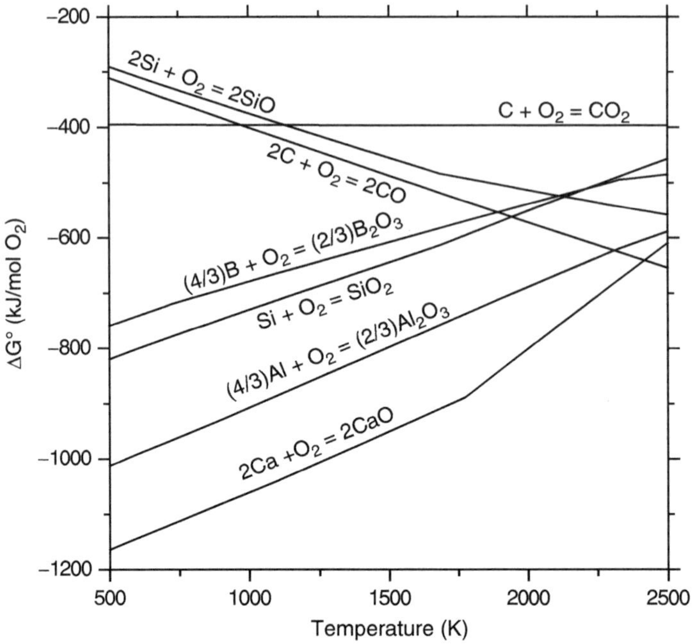

Fig. 1.3. The Gibbs energy change of oxides formation with temperature (Ellingham's diagram)

temperature dependence of the Gibbs energy change for an element reacting with 1 mol of oxygen gas at 1 bar pressure to form oxides. The reaction with the lowest Gibbs energy is the most stable. From the equilibrium point of view, the oxidation of Si is more favorable than the oxidation of B. SiO vapor will be converted to a gas species at high-temperatures, which makes the situation more complicated.

In refining Si for solar cell applications, the dissolved elements will be at ppb (part per billion) levels. Transport of impurities from the metal to the reacting surface/interface will, therefore, be the rate limiting step in the refining process. When oxygen comes in contact with Si, a boundary-layer of oxide film forms at the metal surface as a result of the reaction between oxygen and Si. The oxygen potential in the metal phase is then defined by the following reaction:

$$\frac{1}{2}\left(\mathrm{SiO_2}\right) = \frac{1}{2}\mathrm{Si} + \underline{\mathrm{O}} \tag{1.9}$$

The reaction of removal of B from Si by using basic fluxes can be expressed:

$$\underline{\mathrm{B}} + \frac{3}{4}(\mathrm{SiO_2}) = \frac{3}{4}\mathrm{Si} + (\mathrm{BO_{1.5}}) \tag{1.10}$$

The B dissolved in the liquid Si reacts with the silica in the slag. Boron gets oxidized and enters the slag, while the silica gets reduced and enters the liquid melt. If we use [%B] for the concentration by mass of B in the metal, and (%B) for the concentration of B by mass in the slag, we can write, for the equilibrium constant:

$$K = \frac{a_{\mathrm{Si}}^{3/4} a_{\mathrm{BO}_{1.5}}}{a_{\mathrm{SiO}_2}^{3/4} a_{\mathrm{B}}} = k_{x\to\%} \frac{(\%\mathrm{B})\, \gamma_{\mathrm{BO}_{1.5}}}{[\%\mathrm{B}]\, f_{\mathrm{B}}} \left(\frac{a_{\mathrm{Si}}}{a_{\mathrm{SiO}_2}} \right)^{3/4}, \tag{1.11}$$

where $\gamma_{\mathrm{B_2O_3}}$ and f_{B} are, respectively, the activity coefficients of $\mathrm{BO}_{1.5}$ and B in slag and metal phases. $k_{x\to\%}$ is the coefficient for molar fraction to the mass percentage. The distribution coefficient of B is given as the ratio between the B content in the slag and the B content in the metal:

$$L_{\mathrm{B}} = \frac{(\%\mathrm{B})}{[\%\mathrm{B}]} = \frac{K f_{\mathrm{B}}}{\gamma_{\mathrm{BO}_{1.5}} k_{x\to\%}} \left(\frac{a_{\mathrm{B_2O_3}}}{a_{\mathrm{Si}}} \right)^{3/4}, \tag{1.12}$$

where square brackets denote B dissolved in the metal and parentheses denotes B in the slag. This distribution coefficient is a measure of the possibility for removal of B from Si. Hence, for a given slag at a fixed temperature, the distribution coefficient is constant.

Suzuki et al. [7] tested a number of different slag systems for the removal of B under varying temperatures and in different atmospheres. They found that optimal distribution coefficients (L_{B}) exist for various fluxes (Fig. 1.4 and Table 1.1). The maximum value obtained by Suzuki et al. was approximately 1.7 at 1,500°C. They operated with a initial B content of 30–90 mass ppm and the experimental time varied between 1.8 and 10.8 ks. The activity of Si was set to be 1. The experiments were conducted under CO atmosphere at 1,500°C. Similar experiments were performed by Tanahashi [8]. Some of the results are given in Table 1.1.

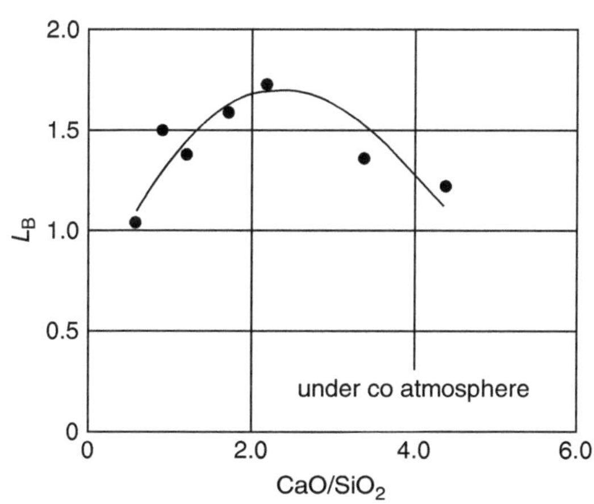

Fig. 1.4. Experimentally determined relation between L_{B} and CaO/$\mathrm{SiO_2}$ ratio for the CaO-30% $\mathrm{CaF_2}$–$\mathrm{SiO_2}$ flux [7]

Table 1.1. Values of L_B for different systems at 1,500°C

System	CaO/SiO_2	L_B	Reference
CaO-10%MgO–SiO_2(–CaF_2)	∼0.9	1.7	Suzuki [7]
Ca-10%BaO–SiO_2(–CaF_2)	∼1	1.7	Suzuki [7]
CaO-30%CaF_2–SiO_2	∼2.2	1.7	Suzuki [7]
6%$NaO_{0,5}$-22%CaO–72%SiO_2	∼3.3	3.5	Tanahashi [8]

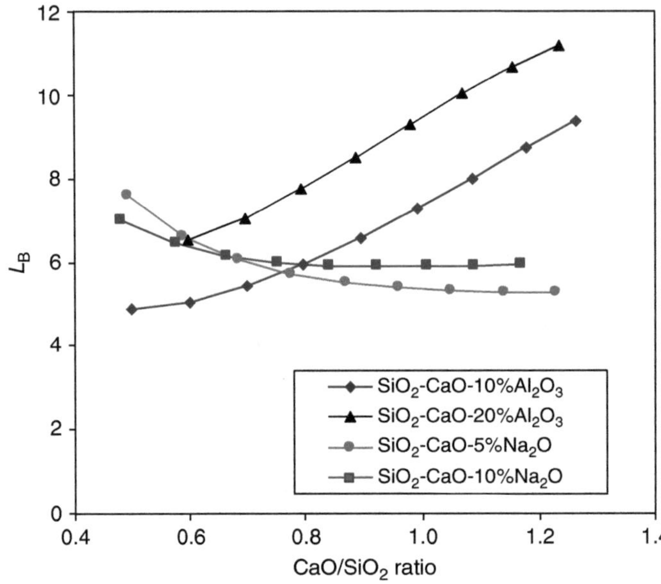

Fig. 1.5. Calculated boron distribution coefficient between SiO_2–CaO–Al_2O_3 or SiO_2–CaO–Na_2O slags and Si melt at 1,550°C

The equilibrium B distribution coefficients between the SiO_2–CaO–Al_2O_3 or SiO_2–CaO–Na_2O slags and liquid Si melt at 1,823 K were also simulated using the new assessed thermochemical databank (see Chap. 13) together with the FACT oxide thermodynamic database [9]. Figure 1.5 shows the calculation results. The SiO_2–CaO–Al_2O_3 slag seems work better than the SiO_2–CaO–Na_2O slag. The calculated L_B values are approximately two times higher than the experimental values, which indicate that the reaction kinetic barriers may play important roles in the refining processes. This will be discussed in detail in the following section.

1.4.1.1 Slag Refining in a Ladle

When a dissolved element is refined by slag treatment, as illustrated in Fig. 1.6, it goes through following five steps:

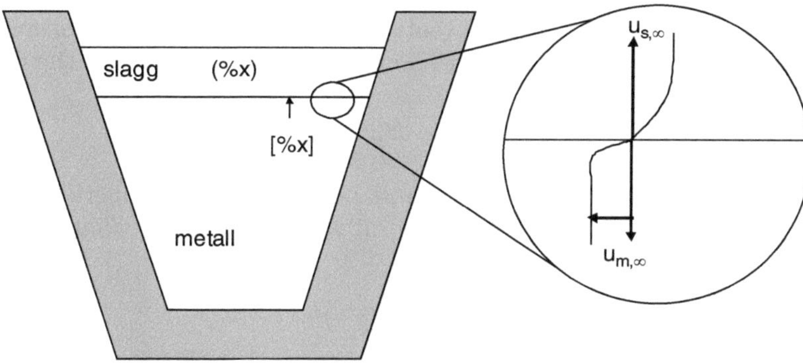

Fig. 1.6. Schematic picture of profile of impurity in metal and in slag

1. The impurity element must be transferred from the bulk metallic phase to the metal boundary layer, $[\%X]_b \rightarrow [\%X]_\delta$
2. The impurity element must diffuse through the metal boundary layer, $[\%X]_\delta \rightarrow [\%X]_i$
3. The metal is oxidized at the interphase between metal and slag, $[\%X]_i \rightarrow (\%X)_i$
4. The impurity element diffuses through the slag boundary layer, $(\%X)_i \rightarrow (\%X)_\delta$
5. The impurity element is transferred from the slag boundary layer to slag bulk phase, $(\%X)_\delta \rightarrow (\%X)_b$

Steps 1 and 5 depend on stirring and mixing in the metal and slag. Stirring is often done by gas bubbling or by mechanical devices to increase the mass transfer in the bulk phases. Hence, slag properties, such as viscosity are important. A high-viscosity will give low velocities in the slag and hence, a low mass transfer of the impurity element. High-viscosity also lowers the diffusivity of the impurity element.

Steps 2 and 4 depend on the mass transfer coefficient in the metal, k, and in the slag, k_s, respectively. With the assumptions that steps 1, 5 and 3 are must faster than step 2, the final concentration can be calculated.

A serious difficulty with refining metal by extraction to a second (slag) phase is the problem of mixing in the slag phase. Often, the slag phase is relatively viscous, so that it is difficult to mix the impurity element throughout the slag. Nevertheless, here we assume that the slag phase is also "completely mixed". A more detailed description can be found in the book, written by Engh [10].

Accumulation rate of x in the slag = removal rate through melt boundary layer:

$$-M\frac{\mathrm{d}\,(\%x)}{\mathrm{d}t} = k_t \rho A_s \left([\%x] - [\%x]_e\right), \tag{1.13}$$

where $[\%x]_{\mathrm{e}}$ is the hypothetical concentration of the metal in equilibrium with the actual concentration of the slag $(\%x)$, so that $[\%x] - [\%x]_{\mathrm{e}}$ is the driving force.

$$[\%x]_{\mathrm{e}} = \frac{\gamma_x (\%x)}{K f_x}. \tag{1.14}$$

To integrate (1.13) over time, $[\%x]_{\mathrm{e}}$ must be replaced by a function of $[\%x]$. $[\%x]_{\mathrm{e}}$ may be obtained in terms of $[\%x]$, if we consider that whatever leaves the melt enters the slag:

$$M \left([\%x]_{\mathrm{in}} - [\%x]\right) = M_{\mathrm{s}} (\%x), \tag{1.15}$$

where M, is the amount of slag. Here, we assume that the slag originally did not contain any component x. Thus, the above equation together with (1.14) gives

$$[\%x]_{\mathrm{e}} = \frac{\gamma_x}{f_x K} \frac{M}{M_{\mathrm{s}}} \left([\%x]_{\mathrm{in}} - [\%x]\right). \tag{1.16}$$

The driving force in (1.13) becomes

$$[\%x] - [\%x]_{\mathrm{e}} = [\%x] \left(1 + \frac{\gamma_x M}{K f_x M_{\mathrm{s}}}\right) - \frac{\gamma_x M}{K f_x M_{\mathrm{s}}} [\%x]_{\mathrm{in}}. \tag{1.17}$$

The lowest value of $[\%x]$ attainable is when the driving force given by (1.17) becomes zero. Then

$$[\%x] = [\%x]_{\infty} = \frac{M [\%x]_{\mathrm{in}}}{M + K f_x M_{\mathrm{s}} / \gamma_x}. \tag{1.18}$$

$[\%x]_{\infty}$ is the value of $[\%x]$, when equilibrium between slag and melt is finally reached (at time $t \to \infty$). In terms of $[\%x]_{\infty}$, we may write for the driving force:

$$[\%x] - [\%x]_{\mathrm{e}} = \left(1 + \frac{\gamma_x M}{K f_x M_{\mathrm{s}}}\right) \left([\%x] - [\%x]_{\infty}\right). \tag{1.19}$$

This equation, introduced into (1.13), gives

$$\int_{[\%x]_{\mathrm{in}}}^{[\%x]} \frac{\mathrm{d}[\%x]}{[\%x] - [\%x]_{\infty}} = - \int_{0}^{t} \frac{k_t \rho A_{\mathrm{s}}}{M} \left(1 + \frac{\gamma_x M}{K f_x M_{\mathrm{s}}}\right) \mathrm{d}t. \tag{1.20}$$

This gives, on integration and assuming that k_t, y_x, f_x and so on do not change with time equation (1.21). In Fig. 1.7, we see that $[\%x]$ drops exponentially down to the value $[\%x]_{\infty}$ given by the equilibrium between x in the melt and in the slag.

Here, we have assumed that resistance is in the melt boundary layer. However, resistance in the slag boundary layer can also be taken into account. Then k_t is replaced by a total mass transfer coefficient that is obtained by summing the two resistances.

$$\frac{[\%x] - [\%x]_{\infty}}{[\%x]_{\mathrm{in}} - [\%x]_{\infty}} = \exp\left\{ -\frac{k_t \rho A_{\mathrm{s}} t}{M} \left(1 + \frac{\gamma_x M}{K f_x M_{\mathrm{s}}}\right) \right\}. \tag{1.21}$$

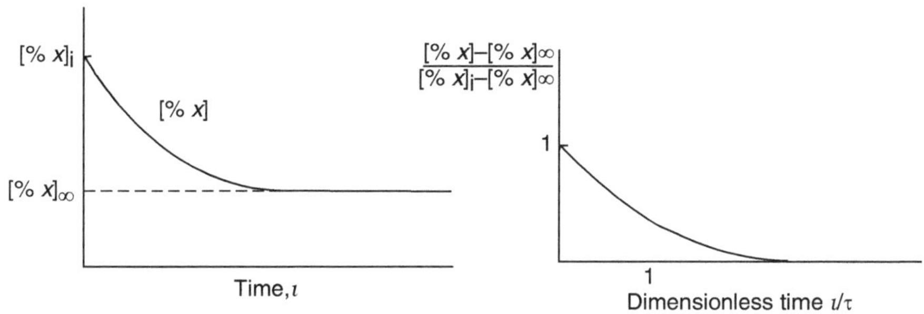

Fig. 1.7. The removal of an element x from metal to slag [10]

1.4.1.2 Slag Properties

Refining by slag treatment is dependent on several parameters, such as reaction kinetics, diffusion of impurities, partition ratios, etc. Meanwhile, these parameters are dependent on the type of slag and its thermophysical and thermochemical properties. For example, viscosity, density, and interfacial properties affect the separation of slag from the metal and the duration of slag refining.

1.4.2 Removal of Boron by Reaction with Water Vapor

B is less volatile than Si, however, B may react with oxygen and hydrogen at elevated temperatures to form the volatile species BHO, BO, and BH_2. To increase the temperature at the bath surface, plasma heating has been employed in laboratory experiments. The reacting species calculated from thermodynamic data are given by Alemany et al. [11] and shown in Fig. 1.8. One problem is that Si will also be oxidized by the water vapor, giving some loss of Si, both as SiO and SiO_2.

1.4.3 Removal of Phosphorous by Vacuum Treatment

As soon as Si is molten, P will start to evaporate from the free surface of a Si melt. The thermodynamics of P evaporation shows that for high P levels, P_2 is the dominating species to evaporate, but for low-concentrations, P(g) will be the dominating species. P_3(g) and P_4(g) are always negligibly low in the gas phase [12]. The partial pressures as a function of P concentration are presented in Fig. 1.9.

Miki et al. [12] gave a relationship between the yield of Si and the attained phosphorus concentration during vacuum treatment at 1823 K, see Fig. 1.10. This result is based on the assumptions that (1) there is no resistance in the melt boundary layer, (2) a perfect vacuum is maintained.

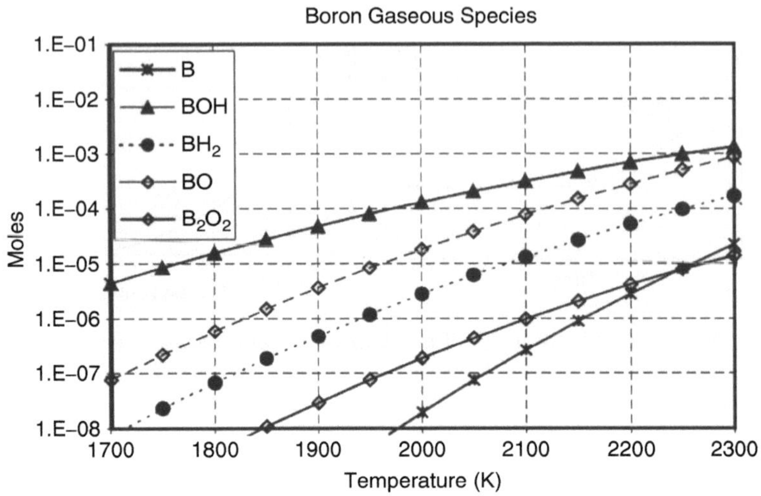

Fig. 1.8. Main B species reaction with water vapor, the initial concentration ratio of B/Si = 10^{-4}. The total pressure is 1 bar [11]

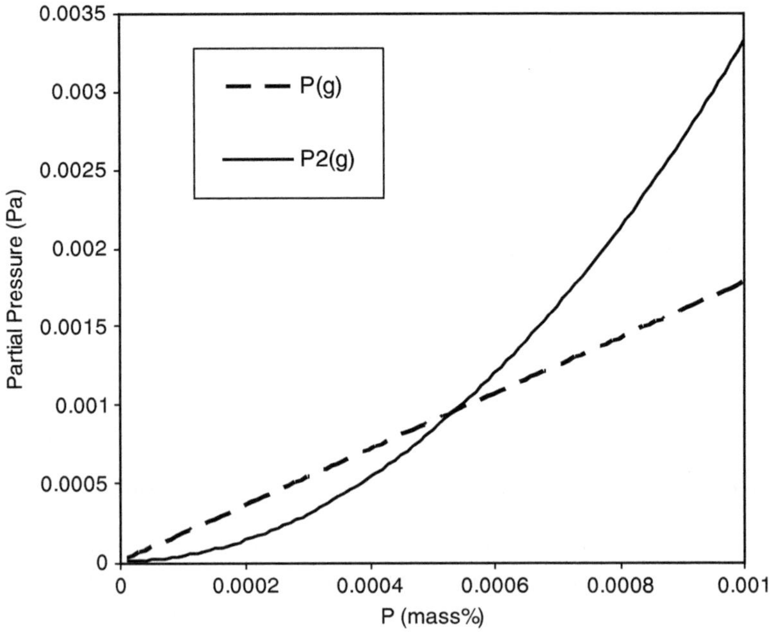

Fig. 1.9. Partial pressure of P species as a function of concentration

The evaporation of P can also be described by the rate equation, if P_2(g) species is neglected,

$$\frac{\mathrm{d}X_\mathrm{P}}{\mathrm{d}t} = -\frac{A}{V}kX_\mathrm{P}^0. \tag{1.22}$$

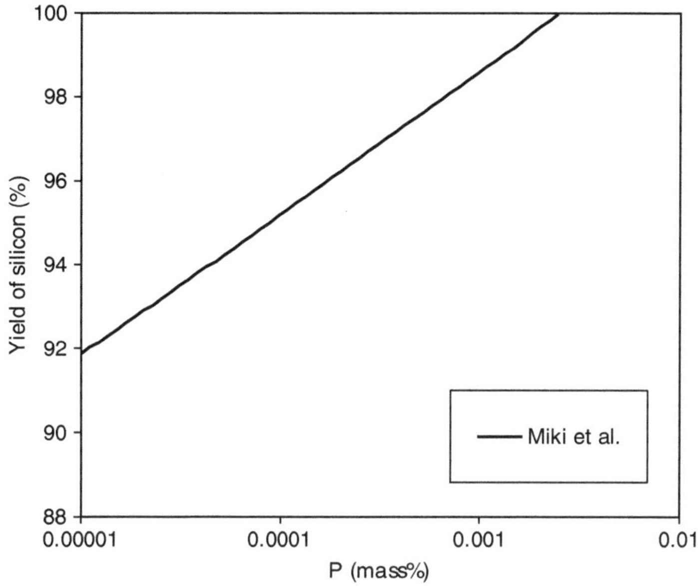

Fig. 1.10. The equilibrium relationship between the yield of Si and the attained phosphorus concentration during the vacuum treatment at 1,823 K

Then

$$\frac{X_{\mathrm{P}}}{X_{\mathrm{P}}^{0}} = \exp\left(-\frac{A}{V}kt\right), \tag{1.23}$$

where k is the so-called specific evaporation constant. For P in liquid Si, it was reported that $k = 1 \times 10^{-4}\,\mathrm{cm\,s^{-1}}$ at temperatures near the melting point of silicon. Using the experimental data reported by Suzuki et al. [7] we evaluated the value of Ak/V. The following equation, taking the temperature into account, can be used to determine the phosphorus content in Si melt for the vacuum treatment:

$$X_{\mathrm{P}} = X_{\mathrm{P}}^{0} \exp\left[(-1.53 \times 10^{-4} - 3 \times 10^{-7}T)t\right]. \tag{1.24}$$

The calculated phosphorus removal rates as a function of evaporation time at different temperatures are shown in Fig. 1.11. The experimental results of Suzuki et al. [7] are included in the same figure for comparison.

1.4.4 Refining by Solidification

As a part of solar grade Si production, it is necessary to cast the material after refining. A substantial refining effect can be attained, if planar front is achieved during the solidification. The solubility of many impurities is higher in liquid Si than in solid Si. In directional solidification with a planar front, there will be a clearly defined interface observed between solid and liquid Si.

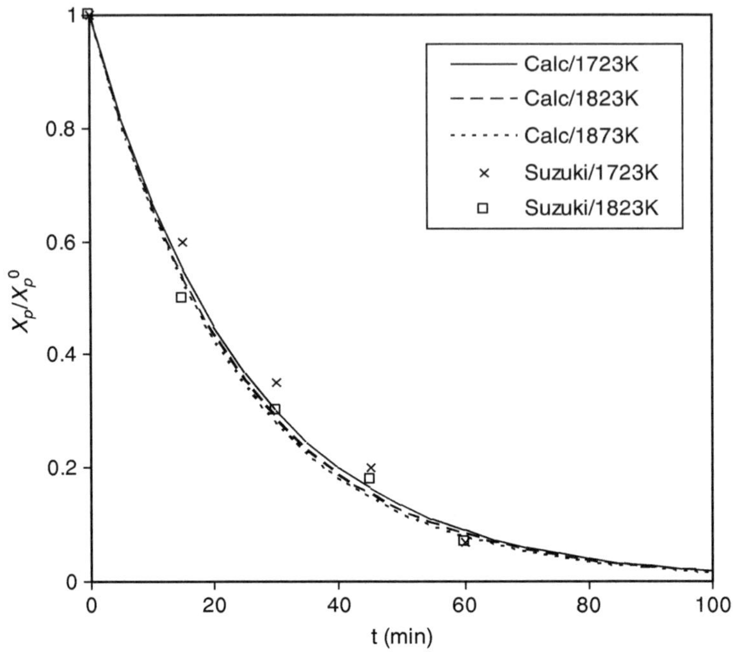

Fig. 1.11. P removal rate as a function of evaporation time at different temperatures

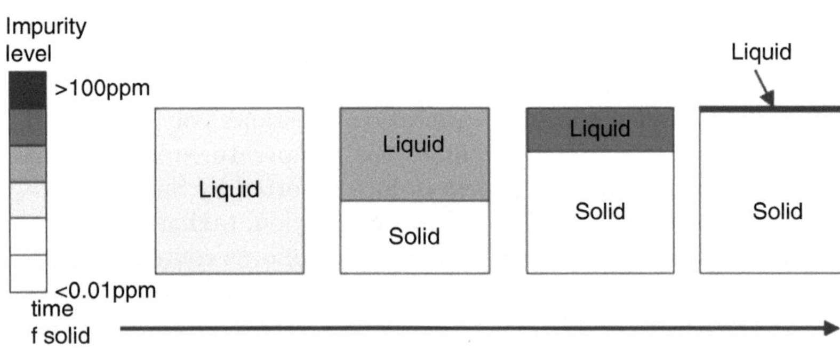

Fig. 1.12. Refining by directional solidification for $k << 1$. The *dark color* indicates level of impurity (*white* is <0.01 ppm and *black* is >100 ppm)

At the solid/liquid interface, the concentration of the impurities in the solidified Si will be in equilibrium with the molten metal. The ratio between the two concentrations is defined as the partition coefficient, given by $k = C_s/C_l$, where C_s and C_l are the concentrations of solid and liquid. For k values less than 1, there will be a refining effect. The principle of refining by solidification is illustrated in Fig. 1.12.

Assuming that there is thermodynamic equilibrium at the solid/liquid interface, no diffusion of impurities in solid, and complete mixing in the liquid,

Table 1.2. Segregation of impurities in Si where $k << 1$

C/C_0	1	2	4	8	16
f_s	0	0.5	0.75	0.875	0.19375
$1 - f_s$	1	0.5	0.25	0.125	0.0625

we can, for small values of k (for $k << 1$), derive a simple expression for the concentration as a function of fraction solid.

Imagine that you start with 1 ppm in liquid, after 50% solidification, the concentration x is doubled; after 75%, it doubled again, and so on. The reason is the relatively much lower solubility in solid than in liquid and, therefore, the total amount of solute in the liquid remains practically constant, while the volume of the liquid is reduced, as solidification proceeds.

From Table 1.2, one may easily derive (1.25)

$$\frac{C}{C_{\text{in}}} = \frac{1}{f_s}, \tag{1.25}$$

where C is the concentration of the liquid at a given fraction solid, f_s, and C_{in} is the concentration of the Si when solidification begins. Note that the segregation pattern in (1.25) is independent of k, under the given assumptions.

A more general equation is the well-known Scheil's equation:

$$\frac{C}{C_{\text{in}}} = (1 - f_s)^{k-1}. \tag{1.26}$$

Obviously, Scheil's equation (predicting an infinite solute concentration when the solid fraction becomes 1) is invalid above a certain system-specific liquid concentration, where the remaining liquid solidifies by some eutectic reaction.

In the industrial process, we want to cut-off the high-concentration area and use the remaining material. The average concentration of the remaining material for a given cut-off fraction is given by integration of the Scheil equation. Replacing f_s with $1-f$, where f is fraction remaining liquid, gives for the solid:

$$\frac{C_s}{C_{\text{in}}} = kf^{k-1}. \tag{1.27}$$

Integrating (1.27) from $f = 1$ to f gives the refining ratio [12], where $\bar{C}_s$ is the mean relative impurity content of the solid, C_{in} is the initial concentration of the metal to be refined.

$$\frac{\bar{C}_s}{C_{\text{in}}} = \frac{k}{1-f}\int_f^1 f^{k-1}\mathrm{d}f = \frac{1-f^k}{1-f}. \tag{1.28}$$

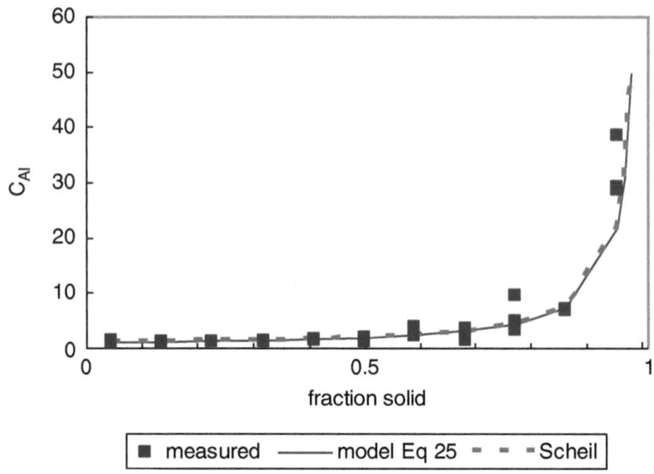

Fig. 1.13. Segregation pattern in 12 kg ingot with $C_0 = 8$

For double solidification, the mean relative impurity content in the cut-off fraction becomes the initial concentration in the second solidification process, giving:

$$\frac{\bar{C}_{s2}}{C_{in}} = \frac{\bar{C}_{s1}}{C_{in}} \frac{\bar{C}_{s2}}{\bar{C}_{s1}} = \left(\frac{1-f^k}{1-f}\right)^2 . \tag{1.29}$$

The value for the segregation coefficient of Al in the Czochralski process has been reported to be $k_{Al} = 0.002$. We wanted to check if this value is valid for our 12 kg lab-scale furnace. An ingot was made with 8 ppm Al, and the concentration in the ingot was determined from the resistivity. The results are shown in Fig. 1.13 and calculations showed that a value of $k_{Al} = 0.003$ fitted better to the measured values. Using (1.29) for cut-off at $f_s = 0.9$ and one time directional solidification, we can see that the difference in $\bar{C}_s$ for $k = 0.002$ and $k = 0.003$ is $\bar{C}_s(0.002) = 0.04$ ppm and $\bar{C}_s(0.003) = 0.06$, respectively. The refining ratio for one and two times directional solidification of Al is shown in Fig. 1.14.

1.4.4.1 Factors that Reduces the Refining Efficiency and the Stability of Crystallization

From our example, we could see that for our lab-scale furnace with only one impurity element, the segregation follows closely to the ideal case. The difference between the ideal case measured in the Czochralski material, where the stirring and resulting transport of solutes from the interface is good, and our casting furnace with relatively low melt velocities ($V \sim 1.5$–$2\,\mathrm{mm\,s^{-1}}$ Meese et al. [20]) is due to the build up of impurities during the solidification due to low transport of solute from the interface to bulk and due to recirculation

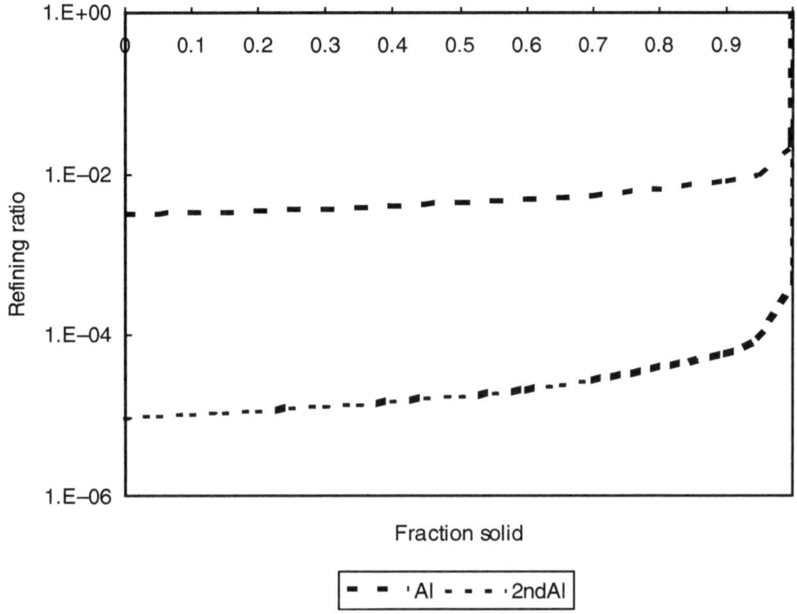

Fig. 1.14. Refining ratio of Al for 1 or 2 times directional solidification assuming $k = 0.002$ and 10% cut-off

zones in the melt. The recirculation zones will act as independent zones in the furnace with reduced exchange of heat and impurities, giving macrosegregation and possible breakdown of the solidification front. For a larger furnace, this picture is even more complicated.

Another problem for high-levels of impurities is possible breakdown of the solidification front. Figure 1.15 shows a cross section of a 12 kg ingot, where the Si contained both Al and Fe in the 100 ppm range (total). Examination of the ingot showed that after the breakdown, small particles (AlFeSi) could be found in the grain boundaries. Analysis of the material by chemical methods showed that the composition after the breakdown was in the same range as that of initial composition, C_{in}.

1.4.5 Solvent Refining

During solidification of a Al–Si melt, the first phase that precipitated is solid Si. Most of the impurity elements have higher solubility in the liquid alloy than in solid Si and can, therefore, effectively be removed by the solvent refining method. The Si will of course be saturated with Al, which has to be removed at a later stage.

The segregation coefficient of an impurity can be expressed as

$$k_{\mathrm{i}} = \frac{x_{\mathrm{i}}^{\mathrm{S}}}{x_{\mathrm{i}}^{\mathrm{L}}} = \frac{\gamma_{\mathrm{i}}^{\mathrm{L}}}{\gamma_{\mathrm{i}}^{\mathrm{S}}} \exp\left(\frac{-\Delta G_i^0}{RT}\right), \tag{1.30}$$

Fig. 1.15. Picture that shows breakdown of the solidification front in MC Si

where x_i^S and x_i^L are the molar fractions of impurity in solid and liquid phase, respectively. γ_i^S and γ_i^L are the corresponding activity coefficients. ΔG_i^0 refers the Gibbs energy of fusion of impurity. The principle of selecting solvent is such that it can increase the solid activity coefficient and/or reduce liquid activity coefficient. The choice of solvent also greatly depends on the terminal solubility, which is decided by careful examination of the solvus line of the equilibrium phase diagram. It is desirable that the solute precipitates retrogradely.

Over the past decades, several papers were published regarding purification of Si using molten aluminium as solvent [13–16]. Yoshikawa and Morita [17] have estimated the segregation coefficient for various elements between solid Si and the Si–Al melt (Table 1.3).

We also used the new assessed thermodynamic database (see Chap. 13) to evaluate the possible candidates for the solvent refining of B in Si. Figure 1.16 shows the calculated B and P segregation coefficients in Si–Al melts, which are close to the experimental values. The model also predicts that the removal of B by Si–Zn melt is even better than the Si–Al melt.

1.4.6 Removal of Impurities by Leaching

Solidified metallurgical Si consists of relatively pure Si with most of the impurities present as additional phases between these crystals. The impurities can be removed by crushing and subsequent etching, giving Si grains at approximately 2 mm. If 5% Ca is added to the molten Si, the impurities will be found

Table 1.3. Segregation ratio between solid Si and Si–Al melt

Element	Segregation ratio between solid Si and Si–Al melt			$k_{Eq} = x_s/x_l$
Temperature	1,073°K	1,273°K	1,473°K	
Iron	1.7 E-11	5.9 E-9	3.0 E-7	6.4 E-6
Titanium	3.8 E-9	1.6 E-7	9.6 E-7	2.0 E-6
Chromium	4.9 E-10	2.5 E-8	2.5 E-7	1.1 E-5
Manganese	3.4 E-10	4.5 E-8	9.9 E-7	1.3 E-5
Nickel	1.3 E-9	1.6 E-7	4.5 E-6	1.3 E-4
Copper	9.2 E-8	4.4 E-6	2.5 E-5	4.0 E-4
Zinc	2.2 E-9	1.2 E-7	2.1 E-6	1.0 E-5
Gallium	2.1 E-4	8.9 E-4	2.4 E-3	8.0 E-3
Indium	1.1 E-5	4.9 E-5	1.5 E-4	4.0 E-4
Antimony	3.4 E-3	3.7 E-3	8.2 E-3	2.3 E-2
Lead	9.7 E-5	2.9 E-4	1.0 E-3	2.0 E-13
Bismuth	1.3 E-6	2.1 E-5	1.7 E-4	7.0 E-4

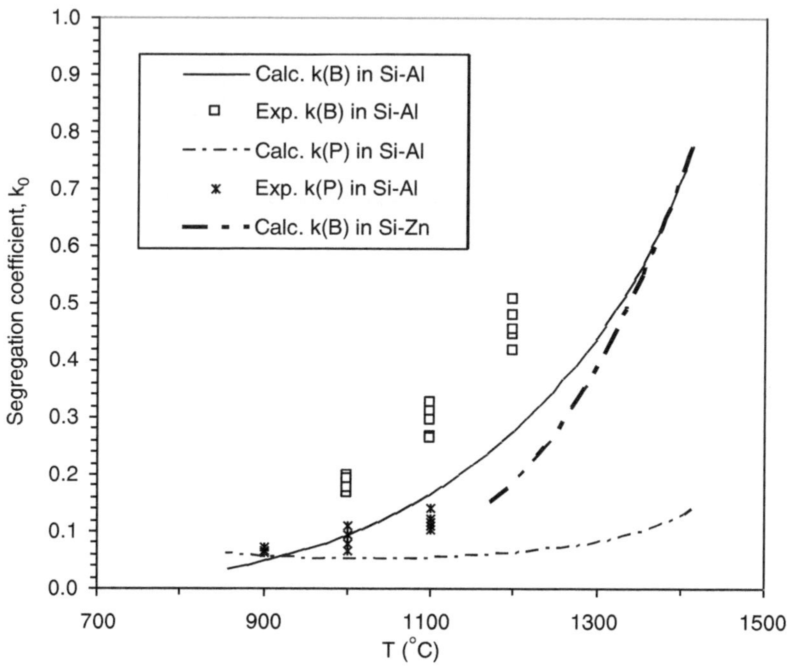

Fig. 1.16. The segregation coefficients of B and P in the Si–Al and Si–Zn melts

as small grains in the calcium disilicide phase, $CaSi_2$. If 5% Ca is added to the alloy before casting, the impurities will be precipitated into the $CaSi_2$ phase that forms between the Si grains as seen in Fig. 1.17.

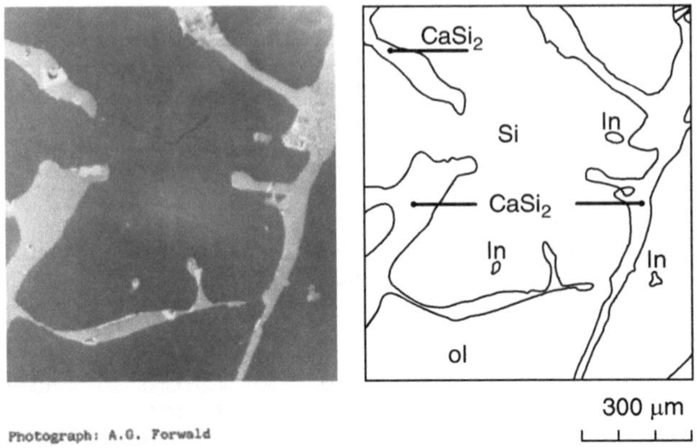

Fig. 1.17. The leaching alloy seen in an electron microscope [18]

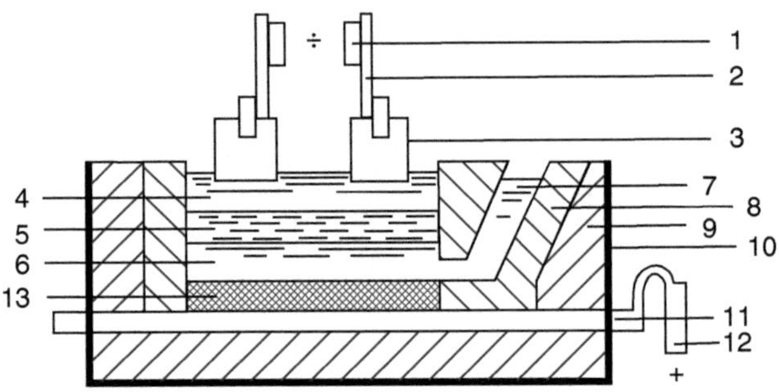

Fig. 1.18. Cell for three-layer refining: 1-cathode bus bar, 2-cathode hanger, 3-graphite cathode, 4-refininged metal, 5-electrolyte, 6-anode alloy, 7-forewell, 8-refractory lining (magnesia), 9-refractory lining (chamotte), 10-steel shell, 11-adode connection, 12-anode buss bar, 13-C bottom (ref]

1.4.7 Electrolysis/Electrochemical Purification

This route has been investigated only on a lab-scale [21,22]. This could be a viable process in the long term for low-cost and with low-energy consumption. The principle of the three-layer electrorefining of Si in molten oxides is shown in Fig. 1.18. The anode is a mixture of Cu and Si (the Cu makes it denser), and the electrolyte is a mixture of oxides that is less dense than the alloy. The purified Si is the cathode.

1.4.8 Removal of Inclusions by Settling

Settling can be modeled by Stokes law, in which the force on particles due to gravity and the difference in density of SiC and Si is balanced with the drag on the particles as shown schematically in Fig. 1.19.

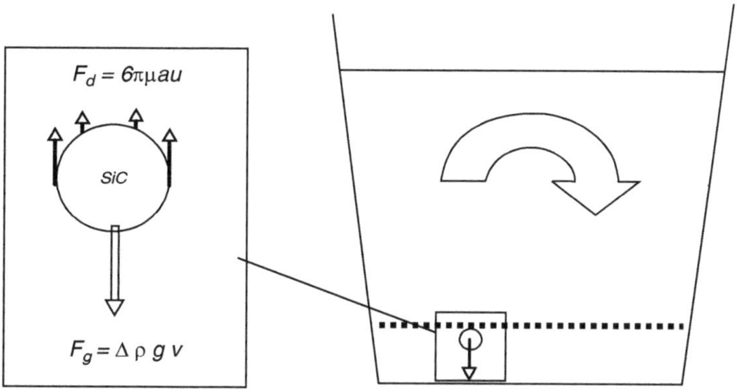

Fig. 1.19. Forces that acts on a particle (*left*) and a schematic drawing of the settling reactor (*right*). (V is the volume of the particle.)

The resulting equation for the settling velocity in the bulk melt for small particles ($Re_\mathrm{p} < 2$):

$$u_\mathrm{r} = \frac{2\Delta\rho g a^2}{\rho\nu 9} \tag{1.31}$$

where $\Delta\rho = 770\,\mathrm{kg\,m^{-3}}$ is the difference in density between Si and SiC, a is the particle radius, g is the acceleration due to gravity, and ν is the kinetic viscosity, $\nu = \mu/\rho$.

Using the settling velocity as mass transfer coefficient and assuming complete mixing in the reactor, the refining efficiency for spherical particles is given by:

$$\frac{C}{C_0} = \exp\left(-\frac{Au_\mathrm{r}}{V}t\right). \tag{1.32}$$

The term A is the area of the bottom of the reactor and V is the volume of the reactor, C_0 is initial concentration of particles, C is the concentration at a given time t. Here, we have used the approximation that the settling velocity at the bottom is the same as in the bulk melt. In Fig. 1.20, the refining efficiency for different settling times is presented as a function of particle size. We can see from the figure that after 1 h, 10 µm particles are removed down to 15% of the initial concentration, while almost all 20 µm are removed. These numbers were used to design a pilot scale reactor. Experiments were performed with SOLSILC material, initially containing 700 ppmw of C. After the settling process, the concentration of C in liquid Si was less than 50 ppm [19].

1.4.9 Removal of Inclusions by Filtration

Inclusions can also be removed by filtration. The minimum level for removal of impurities like N, O, and C is down to the equilibrium level at the filtration temperature between the dissolved element and the corresponding inclusions

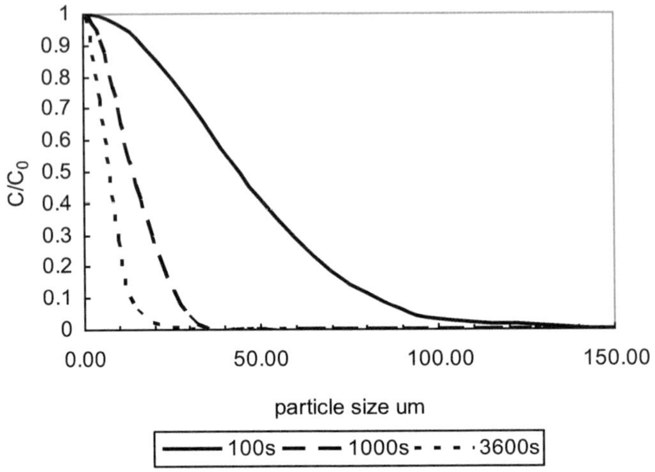

Fig. 1.20. Theoretical settling curves for SiC in Si and the ratio $A/V = 1.8\,m^{-1}$

Fig. 1.21. Pictures from cross section of graphite filter after filtration of Si. *Left*: Electron microscope. *Right*: Light microscope. *Gray* particles are SiC

SiO_2, SiC, and Si_3N_4 that are going to be removed [23]. Usually, filters have less than 100% filtration efficiency, meaning that some of the particles will escape through the filter. The principle and theory is the same as for other metals, but great care has to be taken to find filter materials that do not contaminate the Si. The pictures in Fig. 1.21 shows how the particles stick to the walls of the filter.

References

1. A. Schei, J. Tuset, H. Tveit, *Production of High Silicon Alloys* (Tapir Forlag, Trondheim, Norway, 1998)
2. B. Andresen, in *The Silicon Process*, ed. by H.A. Øye et al. Silicon for the Chemical Industry VIII (Tapir Forlag, Trondheim, 2006) p. 35

3. J.K. Tuseth, in *Thermodynamics of the Carbothermic Silicon Process*, ed. by H.A. Øye et al. Silicon for the Chemical Industry VIII (Tapir Forlag, Trondheim, 2006) p. 23
4. K. Hesse, D. Pãtzold, in *Survey over the TCS process*, ed. by H.A. Øye et al. Silicon for the Chemical Industry VIII, (Tapir Forlag, Trondheim, 2006) p. 157
5. W.C. O'Mara, R.B. Herring, L.P. Hunt (eds.), *Handbook of Semiconductor Silicon Technology* (Noyes, New Jersey, 1990)
6. B. Mazumder, *Silicon and Its Compounds* (Science, Enfield, 2000)
7. K. Suzuki, T. Kumaga, N. Sano, ISIJ Int. **32**, 630 (1992)
8. M. Tanahashi et al. J. Mining Mater. Process. Inst. Jpn. **118**, 497 (2002)
9. C. Bale et al., CALPHAD **26**, 189 (2002)
10. T.A. Engh, *Principles of Metal Refining* (Oxford Science, Oxford, 2002)
11. C. Alemany et al., Sol. Energy Mater. **72**, 41 (2002)
12. T. Miki, K. Morita, N. Sano, Metall. Mater. Trans. B **27**, 937 (1996)
13. J.L. Gumaste et al., Sol. Energy Mater. **16**, 289 (1987)
14. R.K. Dawless et al., J. Cryst. Growth **89**, 68 (1988)
15. T. Yoshikawa, K. Morita, Metall. Mater. Trans. B **36**, 731 (2005)
16. T. Yoshikawa, K. Morita, Sci. Technol. Adv. Mater. **4**, 531 (2003)
17. T. Yoshikawa, K. Morita, Refining of silicon during its solidification from a Si-Al melt. The 4th Asian Conference on Crystal Growth and Crytsal Technology. Sendai, Japan, 2008
18. T. Shimpo, T. Yoshikawa, K. Morita, Metall. Mater. Trans. B **35B**, 277 (2004)
19. E. Øvrelid et al., *Silicon for the Chemical Industry WIII* (Tapir Forlag, Trondheim, 2006)
20. E. A. Meese et al. High Accuracy Predictive Furnace Model For Directional Growth Of Multicrystalline Silicon, in proceedings of the 22th european photvotaic solar energy conference, (Milano, 2007) p. 793
21. K. Yasuda, et al. Electrochimica Acta, **53**, 106 (2007)
22. G.M. Haarberg et al. Electrorefining of metallurgical silicon in molten chloride and fluoride electrolytes, in the 3rd International Workshop on Science and Technology of Crystalline Si Solar Cells, (Trondheim, Norway 2009)
23. A. Ciftja et al. Removal of SiC and Si_3N_4 particles from silicon scrap by foam filters, in Recycling and Waste Processing, TMS annual meeting Orlando, 67 (2007)

2

Czochralski Silicon Crystal Growth for Photovoltaic Applications

Chung-Wen Lan*, Chao-Kuan Hsieh, and Wen-Chin Hsu

Abstract. The fast growing photovoltaic market is mainly based on crystalline silicon. The strong demand on silicon requires wafer manufacturers to produce high-quality material through high productivity processes with low-cost. Due to the higher energy conversion efficiency of single crystalline silicon (sc-Si), the Czochralski (Cz) pulling remains the key technology in photovoltaics. However, when compared with the multicrystalline silicon (mc-Si) production by the directional solidification, the current Cz technology is still more costly, due to the lower throughput and more energy consumption. Therefore, to retain the competition of sc-Si in the PV market, high efficient Cz ingot pulling is needed. In this chapter, we discuss some important issues in the Cz sc-Si production. Special focuses will be on the hot-zone design and multiple charges. The implementation of these concepts has led to significant cost reduction and yield improvement for both 6 in. and 8 in.-diameter solar-grade silicon in production. Some comments for the future development are also given.

2.1 Introduction

Photovoltaics (PV) is solar electric power that converts sunlight to electricity. Although the research and development of PV technology have been over 50 years, the solar market growth was slow until recent years due to the supports of various incentive programs, particularly in Japan and Europe. As a result, over the last 5 years, the markets have grown by an average of more than 35%. In 2007 alone, about 4 GWp PV modules were produced [1]. Among them, approximately 45% of the silicon was grown by the Czochralski (Cz) method, i.e., the same method for producing silicon for semiconductor industry. However, the module price drops in a rate of 20% per year, in which about 50% of the cost is attributed to the ingot and wafer production [2,3]. Therefore, how to reduce the production cost and increase the yield, without sacrificing much the quality, remains the key issue for crystal growers. Meanwhile, the low-cost and high-productivity of multicrystalline silicon (mc-Si) by casting has been increasing the market share, due to its higher throughput and lower energy consumption. This also imposes a great challenge on Cz silicon producers to improve their process and lower the cost.

On one hand, the production per hour (PPH) is about 7 kg $h^{-1}$for a current advanced GT 450 directional solidification (DS) furnace for a period of 50 h in production (with a growth rate of 2 mm h^{-1}). On the other hand, for a Cz puller with 100 kg charge and 30 h growth time (with a pulling rate of 50 mm h^{-1}), the PPH is about 2 kg h^{-1}. On the other hand, the energy consumption for a DS furnace is about 10 kWh kg^{-1}, but for a Cz puller it is more than 30 kWh kg^{-1}. In terms of main power productivity, it could be 5–10 times more for the Cz growth depending on the level of automation. Therefore, to keep the Cz sc-Si growth competitive in the PV market, in addition to the significant improvement in the production efficiency, the high silicon quality is also crucial. Indeed, the improvement of 1% in solar conversion efficiency, which strongly relies on the crystal quality, could reduce the cell production cost about 7%, and the sc-Si production cost needs to be compensated by the cell production cost.

Efforts to improve Cz silicon growth usually focus on the hot-zone designs [4–11], multiple charges [12], crucible developments for lower cost and longer lifetime [13,14], and growth ambient control to reduce argon consumption and improve the yield [7,11]. At the same time, increasing charge, ingot diameter and length are also very useful to the throughput, in terms of PPH. Although using continuous casting is particularly effective on PPH, little effort has been paid until recently by SUMCO using the electromagnetic continuous casting (EMCC). In general, the PPH with the Cz process is typically about 1.5 kg h^{-1}, and for the mc-Si process, such as the heat exchanger method (HEM) [15] is about 3.5 kg h^{-1} with energy and material costs being much lower [15,16]. The silicon charge up to 650 kg has been reported recently, but the largest charge available for DS furnaces in the market for DS is about 450 kg. Therefore, to be competitive with the mc-Si process, significant process improvement, especially on high-speed pulling and multiple charges, for Cz growth is very important. The joint effort of Siemens Solar Industries (SSI) and Northwest Energy Efficiency Alliance (NEEA) is a successful story that the power reduction was more than 40% [7]. With multiple charges, a PPH up to 1.7 kg h^{-1} was achieved [12]. In 2001, we also launched a similar project for high-efficient solar ingot growth, mainly for 5.5 in.-diameter solar ingots [11]. The project ended in 2003, and the improvement was better than that reported by SSI/NEEA. The reduction of power and argon consumptions was 65% and 53%, respectively. In addition, the pulling speed was increased 44%, and the yield with multiple charges was increased up to 31%. With the initial success, at the end of 2003, the effort was shifted to 8 in.-diameter solar ingots using the same type of furnaces. With an improved hot-zone, a PPH up to 2 kg h^{-1} could be achieved for single charge. In this chapter, we briefly discuss the typical approaches toward the efficient solar ingot growth. Some comments for the future development are given as well.

2.2 Hot-Zone Design

Most of the hot-zone designs have been focused on the improvement in ingot quality for Cz silicon growth [4–7, 10]. However, for PV applications, the cost of ingot pulling is one of the major concerns, while the specifications for ingot quality are much flexible. Therefore, an efficient hot-zone for solar ingot growth is usually focused on the lower power consumption and the higher ingot pulling speed. With an effective hot-zone, the usage of the consumables, such as argon and graphite components (mainly the heater and susceptor), could also be significantly reduced. Computer simulation has been a useful tool for the hot-zone design, and several computer packages are available [17–19]. The calculated results are reliable if thermal properties are reasonably accurate [20]. We have been using STHAMAS [18] as a simulation tool for hot-zone designs, and the time for experimental trial-and-error has been significantly reduced. STHAMAS is a finite-volume based simulation package for axisymmetric growth systems developed by Muller's group in Germany. Melt and gas convections, as well as the face-to-face radiative heat transfer, are considered in this model. Figure 2.1 illustrates the approach for hot-zone design used in our research project. To ensure the reliability of the simulated result, benchmark comparison is very important. The inconsistency requires careful judgment, and sometimes tuning thermal data is inevitable because some simplifications are made in the simulation. This is particularly true for the bottom insulation. In reality, with four electrodes there, it is far away from axisymmetric.

For example, the hot-zone of a Kayex CG6000 puller, as shown in Fig. 2.2a has the crystal diameter of 5.5 in. and quartz crucible of 16 in.. The hot-zone could be divided into six major components: (1) radiation shield (or cone); (2) top side insulation; (3) additional side insulation; (4) bottom insulation; (5) top-wall insulation; and (6) argon venting. The default hot-zone from Kayex did not include the cone and the top-wall insulation. The room for the bottom and side-wall insulations was still large. The temperature distribution calculated by STHAMAS is also shown on the left hand side of Fig. 2.2a,

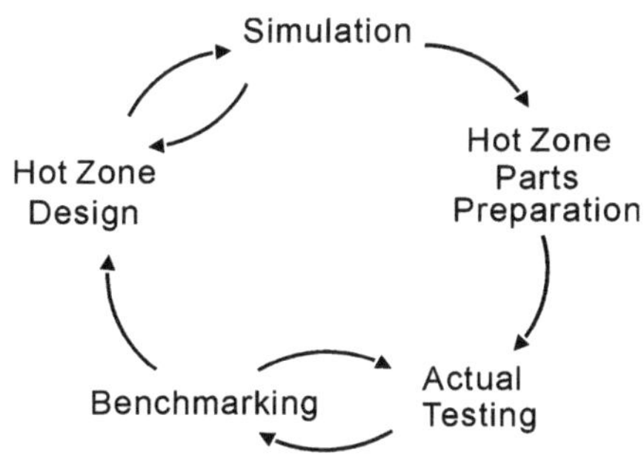

Fig. 2.1. A typical hot-zone design approach; the preparation of hot-zone parts could take several months

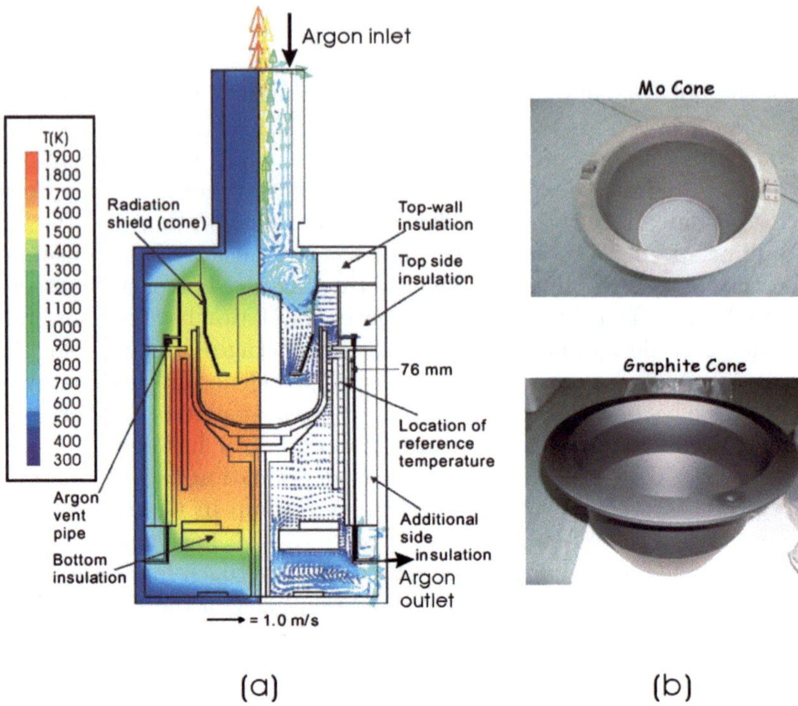

Fig. 2.2. (**a**) Schematic representation of the hot-zone (six major components) for a Kayex CG6000 puller using 16 in. crucible; typical thermal (*left*) and argon flow fields (*right*) are shown; (**b**) photographs for the molybdenum (*top*) and the graphite (*bottom*) cones

while on the right hand side, the argon flow field is shown. Both molybdenum (Mo) and graphite with coating have been considered for the cone material. Figure 2.2b shows the photographs of both cones. The lifetime of both cones is long, but the Mo cone is more robust and can be cleaned easily by sand blast.

2.2.1 Power and Growth Speed

The main purpose of using the cone is to block the thermal radiation from the melt to crystal, so that the crystal can be cooler and pulled faster. Such an idea can be easily understood from the energy balance at the growth interface:

$$k_S G_S - k_L G_L = \rho_S \Delta H V, \tag{2.1}$$

where k_S and k_L are the thermal conductivities of the crystal and melt, while G_S and G_L are the thermal gradients in the solid and melt, respectively; ρ_S is the density, ΔH heat of fusion, and V the pulling speed. Obviously, as the thermal gradient G_S in the crystal is increased, the growth rate V can also

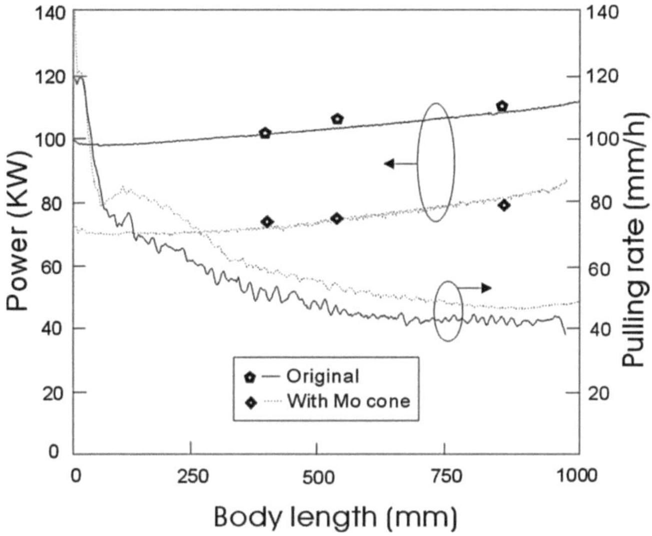

Fig. 2.3. History of power consumption and growth speed with and without molybdenum cone; the simulated data in power consumptions are included (*diamond symbols*) for comparison

be increased. The cone is used not only to block the thermal radiation, but also to reflect the radiation back to the melt. Hence, the cone material needs to have a high-reflectivity and can be operated at high-temperature, such as molybdenum. Graphite is also often used because of the low-cost. To enhance the reflectivity of the graphite cone, SiC or Pyrolitic Carbon (PC) coating can be used. Figure 2.3 shows the comparison of the power and pulling speed for the growth of 5.5 in.-diameter silicon without and with the Mo cone; other insulation options were not installed in this case. As shown, with the Mo cone, the power consumption was greatly reduced (from about 105 to 75 kW) and the growth speed was significantly increased. The average growth rate in the body (more than 50 mm h^{-1}) was increased as well; the growth rate at the body length of 200 mm was up to 76 mm h^{-1} and the crystal quality and RRV (radial resistance variation) were still better than the original one. The better RRV for the wafer was due to the smaller interface concavity. The power consumptions predicted by STHAMAS were quite satisfactory for both cases. In addition to being as a radiation shield, the cone can be designed as a thermal insulator as well. In such a case, a composite cone is preferred.

To further reduce the power consumption, effective insulation is necessary. Through a few tests, we have found that the bottom insulation plays a critical role. With good bottom insulation, the power consumption could be further reduced to about 50 kW during the body growth. Figure 2.4 shows the improvements of several hot-zone designs on the power consumptions and growth speeds. The standard design (STD) refers to the original hot-zones,

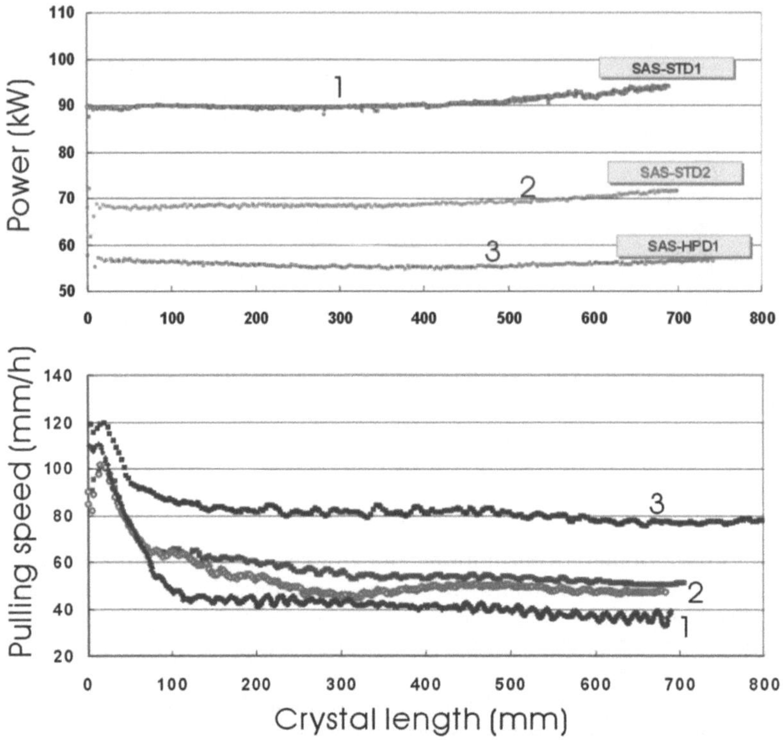

Fig. 2.4. Power (*top*) and pulling speed (*bottom*) histories for different hot-zone designs; the standard design (STD) and high-performance design (HPD) refer to the original default design and the computer-added one

and the high-performance design (HPD) refers to the design screened by computer simulations. As shown, an average growth speed of 80 $\mathrm{mm\,h^{-1}}$ can be achieved. Adding a cooling ring to cool down the crystal can further increase the growth speed, while reducing the interface concavity.

2.2.2 Interface Shape and Thermal Stress

Besides the lower power consumption and the faster pulling speed, an important benefit from the hot-zone design is the small interface deformation. Using the cone is an effective way to reduce the interface concavity. According to simulation, it is found that the cone shape and cone material are the two major factors for the interface control. In general, the cone edge needs to be close to the melt/crystal/gas trijunction line as much as possible [11]. Nevertheless, lowering the cone position increases the risk of melt splashing. The simulation result also reveals that the smaller interface concavity gives a smaller von Miss thermal stress, which is usually concentrated at the trijunction line [11, 21].

In addition to the pulling stage, a proper hot-zone design is also useful for melting and cooling down. For example, during melting down, the crucible position is the lowest, and the radial heating could generate significant thermal gradients and, thus, thermal stress in the crucible. This increases the risk of crucible breakage. The bottom heater is the most effective way to reduce such thermal gradients. However, to save power consumption, the bottom heater is usually turned off during the pulling stage. The power ratio control of both heaters is not easy, but again computer simulation is very useful to optimize the setting. According to our experience, as reported in [11], the additional side and top insulations are less effective. Since the graphite felt is not cheap, using an effective insulation is necessary to reduce the cost.

2.2.3 Argon Consumption and Graphite Degradation

During silicon growth, oxygen is dissolved from the crucible into the melt, forming silicon monoxide (SiO). Carrying SiO away from the melt by argon flow is very important in practice as too much deposition of SiO particles on cooler surfaces, such as the chamber wall, crucible inner wall, and ingot surface, could cause problems when the particles fall down to the melt. The structure integrity is lost, i.e., having dislocations, if the particles get into the growing crystal; then, the growth needs to be restarted and PPH or growth yield is significantly affected. Therefore, the argon flow rate and its path are important to avoid the catastrophe. Further, with the cone, it is observed that argon flow consumption could be significantly reduced. The major reason is that the flow space between the cone and the melt is small and the argon flow across the melt surface becomes faster. Accordingly, the removal of SiO from the melt surface is more effective. The argon flow rate could be reduced from 60 to 15 slpm in our study. This corresponds to 27 cf per kg of Si; the original was 93 cf per kg. The argon flow path is shown by the vector field on the right hand side of Fig. 2.2a. With the cone, near the melt surface, no flow circulation is found, which is believed to be useful in minimizing the falling of SiO particles from the upper cooler surfaces. Nevertheless, the upper part of the cone still has significant SiO deposition. With the top-wall insulation or a composite cone, the deposition is reduced and its position is higher due to the increase of cone temperature.

Besides the argon consumption, the original argon flow path has to run through the graphite heater and SiO reacts with graphite forming silicon carbide. This deteriorates the heater and shortens heater's lifetime. The same is true for graphic susceptor. By redirecting argon to the side insulation, it is found that the heater lifetime, as well as that of the graphite susceptor, could be significantly elongated. Over the project period, heater lifetime is increased to 4,600 h from the original lifetime of 3,000 h. The lifetime of the graphite crucible (susceptor) is increased to 1,500 h, which is almost double that of the original one (880 h). Figure 2.5 shows the photographs of the heater before

Fig. 2.5. Graphite heaters after growth: (**a**) without redirected argon flow; (**b**) with redirected argon flow

(Fig. 2.5a) and after (Fig. 2.5b) the venting pipe was installed. As shown, as argon is redirected to the side wall, the deposition of SiO on the heater is significantly reduced after crystal growth.

In addition to the cleaner heater, with the cone and the redirected argon flow, the grown crystal surface is found very shining, without any surface oxidation. Figure 2.6 shows the photographs of 5.5 in.- (Fig. 2.6a) and 8 in.-diameter solar ingots. The left two ingots were grown without using the cone. As shown, the colorful oxidation rings are clear. Furthermore, with the cone, the melt leftover in the crucible could be significantly reduced because silicon remains molten near the end of the growth due to the less radiation heat loss. As a result, the wastage of material is reduced significantly. For the same amount of charge, the pulled crystal is, thus, longer.

2.2.4 Yield Enhancement

The yield for dislocation-free growth is the most important factor for PPH. In addition to cost reduction and pulling speed enhancement, a successful pulling without losing structure integrity, i.e., dislocation-free, is extremely important. Once a dislocation is generated during growth, the growth needs to be restarted over, which significantly reduces the production yield. It is believed that particles are responsible for the formation of the dislocations. Therefore, beside the quality of polysilicon raw material, particles due to crucible erosion and SiO deposits peeling off from the top surfaces need to be carefully prevented. This could be done by using a coated crucible and an effective argon venting design. Reducing the argon flow resistance and preventing flow recirculation are found to be useful. The lower crucible temperature due to the lower heating power also reduces silica erosion and particle generation.

(a)

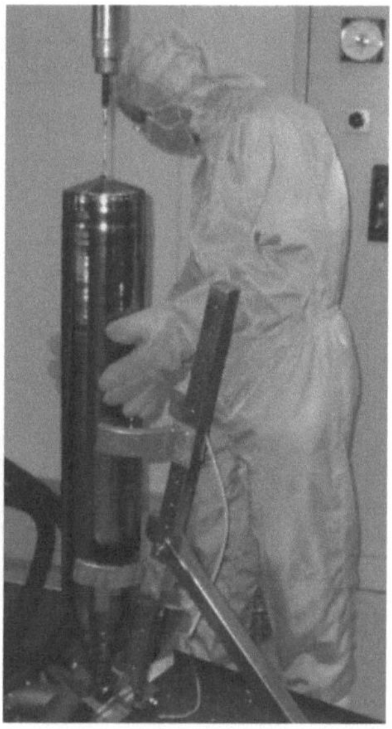
(b)

Fig. 2.6. Photographs of grown crystals: (**a**) 5.5-in. diameter; (**b**) 8 in.-diameter; the *left* two ingots in (**a**) were grown from the hot-zone without the radiation shield, so that the oxidation rings are clear on the surface

2.3 Continuous Charge

2.3.1 Multiple Charges

As mentioned, the PPH for 6–8 in.-diameter ingot growth is usually below 2 kg h^{-1}, while HEM mc-Si casting can easily reach 3–4 kg h^{-1}. The main reason is that a batch charge for HEM is up to 240–300 kg, while a typical charge for a Cz puller, such as Kayex 6000 for 6–8 in. ingot growth, is much less than 100 kg. Therefore, to increase PPH, one has to either increase the charge by using a larger crucible or use continuous charging. Since the traditional crystal pulling is a batch process, the chamber for holding the crystal has a limited length. Therefore, continuous charging is difficult. An alternative is to have multiple charges. In other words, after a crystal is pulled out and removed, without cooling down, the new material can be fed from the feeder. In SSI, this kind of feeder has been used, and the PPH for 6 in.-diameter ingots has been pushed to 1.7 kg h^{-1} [12], and ten crystals could be pulled out from one silica crucible. A similar design has also been adopted for semiconductor applications. Figure 2.7 shows a typical Kayex 150 puller with the installation of the feeder and the recharge tank. Typically, the feeder requires granular or small-chunk polysilicon. This kind of multiple charges can save the time for cooling down and, thus, improve PPH. Since several crystals can be pulled

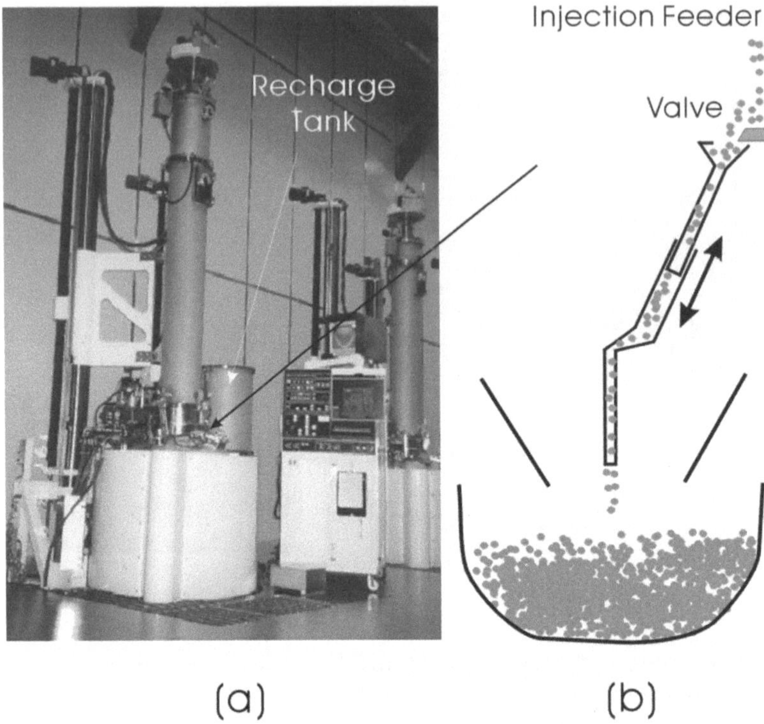

Fig. 2.7. Recharge system for a Kayex 150 puller: (**a**) photograph showing a recharge tank; (**b**) the feeder design (*top*) and the feeding of granular silicon (*bottom*) (Curtsy of Taisil Electronic Materials Inc., Taiwan)

out from one crucible, the crucible cost per kg silicon is greatly reduced. Typically, five crystals (100–150 kg each) can be pulled out from a 22 in.-diameter crucible for semiconductor applications. If such a puller is used for solar ingot with a proper hot-zone design for a high pulling speed, a PPH comparable to HEM casting can be anticipated.

For semiconductor applications, the problems of using multiple charges are metal contaminations during recharge and impurity accumulation in the crucible. The former problem can be resolved by a careful design of the feeder and the material used. However, for impurity accumulation, one has to use high-purity granular polysilicon. Fortunately, for solar-grade silicon, both problems are greatly relieved because of the lower purity requirement. The improvement of PPH further reduces the specific energy consumption (kWh/kg). In SSI, with multiple charges, overall specific energy saving is over 51% [7]. In SAS, before the project, the electricity consumption was 59.1 kWh kg^{-1}. After the hot-zone optimization and using multiple charges, the electricity consumption was reduced to only 17.4 kWh kg^{-1} for 5.5 in.-diameter ingots. For 8-in.

silicon, this figure is even lower due to the high PPH; the power consumption during growth needs to be below 65 kW.

2.3.2 Coated Crucible

The quartz crucible is a major cost for solar-grade ingot production, because no cheaper crucibles are available for PV applications. The production of quartz crucibles is quite tedious. The outer layer of the quartz crucible, the so-called bubble layer, needs to be less dense to make the crucible a good insulator, while the inner layer has to be prepared with a denser and stronger amorphous structure. During crystal growth at high-temperature, the devitrification of silica forms crystobalite and it is often detached from the crucible surface and cause dislocations in the growing crystal. For multiple charges, due to the long time operation, the situation is even worse. Therefore, a special coating on the inner crucible surface is necessary. Typically, a dense barium silicate coated layer is used [13]. In this GE patent, barium oxide is used as a good devitrification promoter that helps forming a uniform dissolution layer on the crucible inner surface during crystal growth. Hence, particle generation is reduced. There are also other methods to improve the crucible's inner surface for reducing erosion and particle generation [14]. Of course, if other materials, such as Si_3N_4-coated carbon [22] can be used or the crucible could be reused, and the cost could be further reduced. Again, carbon contamination remains a great concern.

2.3.3 Large Size and Continuous Growth

As discussed in (2.1), the increase in the crystal thermal gradient can increase the pulling rate. However, the gradient (G_s) needs to be kept constant over the entire interface during the whole growth period, and cannot be too high for thermal-stress reason. In addition, crystal quality usually decreases with the increasing cooling rate ($G_S V$). Therefore, the highest pulling rate is usually below 100 mm h^{-1} for Cz silicon, and about 30 mm h^{-1} for HEM mc-Si casting. Moreover, thermal instability often occurs at the edge of the interface due to the larger radial cooling. Hence, as a general rule, to ensure thermal stability and a planar interface, the ratio of volume to surface must be as large as possible [23]. Of course, the availability of large and reliable silica crucibles and furnaces at an affordable price has to be considered. The cost of graphite heaters and elements cannot be ignored as well. In addition, from the investor point of view, the higher capital investment somehow discourages the use of large pullers for cheaper products. Therefore, large pullers, such as Kayex 150 or higher models, have not yet been used for solar ingot production.

However, the cost for equipments and parts increases rapidly with the size. Therefore, without increasing the diameter for an economic ingot production, a continuous process should be considered seriously in the near future. In fact, among the available crystal growth techniques for PV applications,

Cz and HEM are the only batch processes. Others, such as edge-defined film-fed growth (EFG), are continuous. Semicontinuous Cz, i.e., using multiple charges, has been widely used. Recently, the electromagnetic continuous pulling (EMCP) process [24] combining the advantages of cold crucible melting (no crucible consumption and low-pollution) with those of continuous casting has been proposed for PV silicon. Similarly, ingot growth using a square die with a continuous feeding could be a feasible process. The floating-zone technique can also be useful. In fact, Topsil's PV-FZ silicon ingots are available in the market; the solar cell conversion efficiency could be greater than 22% easily. Thanks to the high thermal gradients and crucible free of the FZ technique, its PPH can be very high. However, the pulling of mc-Si by the Cz method could be an interesting consideration; the effort for seeding and keeping structure integrity is removed. Otherwise, the use of a long crucible is a simple way to increase PPH. The same is true for HEM. Furthermore, the less puller stations for the same amount of ingot production also imply the lower personal cost.

2.4 Crystal Quality Improvement

Finally, since crystal quality is directly related to solar cell efficiency, it has to be taken into account for the overall PV cost. In general, the sc-Si ingot is pulled in the <100> direction and this is preferred in the alkaline texturing for the inverted pyramids during solar cell fabrication. Also, because (111) slip planes are oblique to the growth direction, dislocations generated can propagated outward to the crystal surface leading to a dislocation-free ingot. Such an advantage allows the dislocation-free ingot to be pulled at an extremely high pulling rate. One of the major indicators for the solar-cell performance of a wafer is the lifetime of minority-carriers. The control of oxygen and carbon contents has been found important to improve the minority-carrier lifetime. Especially, for N-type silicon, oxygen precipitates usually reduce the minority-carrier lifetime due to the formation of the interface traps at the oxide surfaces [25]. Transition metals are detrimental as well [26]. Boron doped P-type silicon has been found to have a severe light-induced degradation due to a recombination-active boron/oxygen complex, as a result of the reaction involving substitutional boron atoms and interstitial oxygen dimers [27]. Although a special 200°C annealing treatment has been proposed to resolve this problem recently [28], the low-oxygen ingots are still preferred in applications. The oxygen content in the grown crystal is affected by many factors. However, crucible temperature is believed to be an important one. With a lower crucible temperature, the oxygen dissolution into the melt is less. Based on the hot-zone simulation [11], we observed that the crucible temperature in contact with the melt decreases with the decreasing power consumption. Therefore, an energy-saving design is favored for the lower oxygen content. In our hot-zone design, the oxygen content has been reduced to about 17 ppma in the head and 12 ppma in the tail for 5.5 in.-diameter ingots. The average minority-carrier lifetime is longer than 60 μs.

Similar to oxygen, carbon is also detrimental. In Cz silicon growth, carbon is released mainly from the graphite heater and susceptor, as well as raw polysilicon. Therefore, the control of argon flow path is important to reduce carbon contamination. In fact, after redirecting the argon flow, even with a flow rate of 15 slpm, our hot-zone always keeps the carbon content below 0.03 ppma.

In addition to oxygen and carbon, transition metals and types A and B swirls have also negative effect on the minority-carrier lifetime [29]. Therefore, controlling the growth at a high V/G ratio for getting a vacancy-rich ingot [30,31] is helpful. Still, near the crystal shoulder, due to the higher cooling rate there, getting vacancy-rich is difficult, unless an exceptional high pulling rate is used. Accordingly, the oxidation-induced-stacking-fault ring (OISF-ring), which a visible boundary roughly separating the vacancy-rich (inside) and the self-interstitial-rich regions (outside), is often observed in the solar ingot near the shoulder. The lifetime of the minority carriers is also shorter there. Again, with the improved growth speed using a high-performance hot-zone, the grown ingot can be almost vacancy rich, and the portion of interstitial rich is significantly reduced. However, there is always a trade-off on the high growth rate. Because the cooling rate, i.e., VG_S, is proportional to the pulling rate, and the minority-carrier lifetime usually decreases with the increasing cooling rate [29], finding an optimal V/G is necessary for crystal quality.

2.5 Conclusions and Comments

With the dramatic increase of the demand on silicon solar ingots since 2003, the price of the solar silicon has increased rapidly in conjunction with the soaring price of raw polysilicon. Such a demand is believed to continue for another 2 or 3 years. Indeed, this should be an incentive for silicon producers to switch their interest from IC to PV industries. However, it is also a historical trend that the price of solar cells will continue to decrease, and this implies that the production of solar ingots has to face the same challenge of cost reduction. Hot-zone design has been found extremely useful in this aspect. Besides the faster growth and less power consumption, the cost for consumables, such as argon and graphite components, is significant reduced. Crystal quality can also be improved due to the less oxygen and carbon contents, as well as the vacancy-rich growth at the high growth rate. However, there is always a limitation for the hot-zone to improve. Therefore, using multiple or continuous charges is inevitable for production. This includes a better design of the feeder to reduce contamination and preparing small chuck or granular silicon. In addition, a long-life silica crucible or its substitute is required; a reusable crucible is preferred.

Meanwhile, the high-productivity DS mc-Si casting appears as a strong competitor to Cz ingot pulling. As a result, the market share of sc-Si in the PV industry continues to be deprived by mc-Si. Although the higher mechanical strength of the single crystals is believed to be a unique advantage for thinner solar wafers, say 100 μm or less, in the future, the slicing technology continues to improve for better yield. Therefore, the share shrinkage of single crystal silicon seems to be inevitable. Furthermore, the requirement of a dislocation-free growth remains a critical factor for the yield of a single crystal, even though its solidification speed can be several times faster than that for DS. This situation could be even worse with low-grade raw polysilicon. Therefore, pulling polycrystalline silicon or continuous casting may not be an unrealistic idea for Cz crystal growers to consider in the future. Again, the grain size, defect, and contamination control, as that in DS, remains a great challenge for Cz mc-Si to have a comparable solar cell efficiency as its single crystals.

Acknowledgments

The highly-efficient hot-zone design project was sponsored by the Ministry of Economics through the Strategic Technology Development Program. The simulation tool STHAMAS provided by Prof. G. Müller is highly appreciated. We are also grateful for the generous support from SAS, especially from the President Doris Hsu.

References

1. P.D. Maycock (ed.), in *PV News*, PV Energy Systems (Warrenton, VA, March 2004)
2. T. Surek, J. Cryst. Growth **275**, 292 (2005)
3. K.E. Knapp, T.L. Jester, Home Power, Dec. 2000/Jan. 2001, pp. 42–26
4. I. Yamashita, K. Shimizu, Y. Banba, Y. Shimanuki, A. Higuchi, H. Furuya, US Patent 4,981,549, Jan. 1, 1991
5. K. Takano, I. Fusegewa, H. Yamagishi, US Patent 5,361,721, Nov. 8, 1994
6. T. Tsukada, M. Hozawa, J. Chem. Eng. Jpn. **23**, 164 (1991)
7. G. Mihalik, B. Fickett. Energy efficiency opportunities in silicon ingot manufacturing. Semiconductor Fabtech. 10th edn. ICG Publishing, 191–195 (1999)
8. P. Sabhapathy, M.E. Sacudean, J. Cryst. Growth **97**, 125 (1989)
9. P. Sabhapathy, Report on "Computer simulation of flow and heat transfer in CG6000 during 144 mm diameter crystal growth with 16-in hot zone". It is an internal report at Kayex Inc., July 1998
10. E. Dornberger, W.V. Ammon, J. Electrochem. Soc. **143**, 1648 (1996)
11. L.Y. Huang, P.C. Lee, C.K. Hsieh, W.C. Chuck, C.W. Lan, J. Cryst. Growth **261**, 433 (2004)
12. B. Fickelt, G. Mihalik, J. Cryst. Growth **211**, 372 (2000)
13. R.L. Hansen, L.E. Drafall, R.M. McCutchan, J.D. Holder, L.A. Allen, R.D. Shelley, US Patent No. 5,976,247, Nov. 1999

14. Y. Ohama, S. Mizuno, US Patent No. 6,886,342, May 2005
15. C.P. Khattak, H. Schmid, 26th IEEE PVSC Conference, Anaheim, CA, 1997
16. J.M. Kim, Y.K. Kim, Sol. Energy Mater. Sol. Cells **81**, 217 (2004)
17. T. Sinno, E. Dornberger, W. von Ammon, R.A. Brown, F. Dupret, Mater. Sci. Eng. **28**, 149 (2000)
18. J. Fainberg, Ph.D. Thesis, University of Erlangen-Nürnberg, 1999
19. C.W. Lan, Chem. Eng. Sci. **59**, 1437 (2004)
20. E. Dornberger, E. Tomzig, A. Seidl, S. Schmitt, H.-J. Leister, Ch. Schmitt, G. Müller, J. Cryst. Growth **180**, 461 (1997)
21. D. Bornside, T. Kinney, R.A. Brown, G. Kim, Int. J. Numer. Methods Eng. **30**, 133 (1990)
22. T. Saito, A. Shimura, S. Ichikawa, Sol. Energy Mater. **9**, 337 (1983)
23. F. Ferrazza, Sol. Energy Mater. Sol. Cells **72**, 77 (2002)
24. F. Durand, Sol. Energy Mater. Sol. Cells **72**, 125 (2002)
25. J.R. Davis Jr., A. Rohatgi, R.H. Hopkins, P.D. Blais, P. Rai-Choudhury, J.R. McCormick, H.C. Mollenkopf, IEEE Trans. Electron. Devices **ED-27**, 677 (1980)
26. J.M. Hwang, D.K. Schroder, J. Appl. Phys. **59**, 2487 (1986)
27. J. Schmidt, K. Bothe, Phys. Rev. B **69**, 024107 (2004)
28. B. Lim, S. Hermann, K. Bothe, J. Schmidt, R. Brendel, Appl. Phys. Lett. **93**, 162102 (2008)
29. T.F. Ciszek, T.H. Wang, J. Cryst. Growth **237–239**, 1685 (2002)
30. V.V. Voronkov, J. Cryst. Growth **59**, 625 (1982)
31. L.I. Huang, P.C. Lee, C.K. Hsieh, W.C. Hsu, C.W. Lan, J. Cryst. Growth **266**, 132 (2004)

3

Floating Zone Crystal Growth

Helge Riemann* and Anke Luedge

Abstract. This chapter outlines one of the two practically important bulk crystal growth methods for silicon, the crucible-less floating zone (FZ) technique, which cannot be evaluated without comparing it to the other one, the Czochralski (CZ) method. The main advantage of FZ silicon is the high purity and the resulting high electrical and structural material quality. Although, till now, FZ silicon for solar cells is mainly a matter of R&D and not of cell production for terrestrial utilization, it has a big potential for future applications, because efficiency and long-term stability of FZ silicon cells are considerably higher than that of CZ silicon cells.

Presently, the main problem with FZ solar cells is not only the high price of the special feed material for FZ, boosted by the currently unbalanced situation of the feed stock market, but also the limited crystal cross section of FZ crystals, which does not fit the standard cell formats. New concepts to overcome these difficulties are to grow crystals directly with the desired square cross section of the solar wafer, the use of cheaper feed material like granular silicon, or pulling feed rods after a downgraded CZ technique from low-price raw silicon like upgraded metallurgical grade (UMG) silicon with the benefit of further purification by segregation.

3.1 The FZ Method: Its Strengths and Weaknesses

Originally, zone melting was invented to ultra-purify semiconductors and some other congruently melting substances by the segregation effect. A rod of the treated material is moved in a horizontal boat through a narrow heating zone. Correspondingly, a molten zone is generated in the boat where almost all impurities are collected because of their generally higher solubility in the liquid state. For low crystallization rates, the ratio of impurity concentrations in melt and solid is constant and equal to the equilibrium segregation coefficient, k_0:

$$k_0 = c_s/c_l, \tag{3.1}$$

where c_s is the solubility of impurities in the solid and c_l in the melt. For almost all impurities, k_0 is smaller than 1. If the molten zone travels faster, the equilibrium is less established and the effective segregation coefficient k_{eff} raises towards 1.

In this way, the impurities are shifted with the molten zone according to their k_{eff} to the end of the rod. So, the rod part crystallized at first becomes purer. However, impurities newly introduced from the boat or the surroundings could work against this effect. Multiple zone melting improves the purity of the initially solidified fraction up to a certain theoretical limit. In practice, this limit is independent on how much impurity is newly introduced from the boat or other surroundings.

If a seed crystal is introduced at the start of the zone melting, a single crystal can be grown. This was widely applied for germanium, the first technically important semiconductor. For silicon, which became increasingly important, this method failed, since the silicon melt reacts with any boat or crucible material and the crystallized material sticks tightly at the container wall, so that both will break when cooling down.

Therefore, the vertical crucible-free floating zone (FZ) technique was developed soon [1] to grow very pure silicon single crystals without any contact of the molten zone with foreign materials. If already very pure silicon is used as starting material, the purification effect of the FZ technique is less important than the exclusion of newly introduced impurities. The state-of-the-art Siemens process provides polycrystalline feed rods of highest purity, hence, additional FZ purification runs are dispensable and, above all, would be too costly.

The modern FZ process is carried out in the growth chamber of a large FZ machine, see Fig. 3.6. The FZ setup consists of the upper and lower vertical pulling spindle carrying the polycrystalline feed rod and the thin rod-like seed crystal at their ends, respectively. Between them, the pancake-shaped inductor, a one-turn RF coil is placed for contactless inductive heating.

After pre-heating the lower end of the feed rod, it can be inductively heated by an RF current of about 2.5–3.5 MHz working frequency, until a hanging melt drop is formed and stabilized by the surface tension (Fig. 3.1a). Then, the seed is moved upwards until it touches the melt drop. After slightly back-melting, the seed is moved downwards and a thin monocrystalline neck is grown to remove all dislocations through the neck surface, after Dash [2,3]. Then, the diameter of the growing crystal can be increased (Fig. 3.1b) while rising the RF-power, lowering the pulling rate and increasing the feed rate until the final crystal diameter is achieved (Fig. 3.1c). Now, the crystal grows cylindrically with almost constant process parameters till the feed is consumed. The so-called needle-eye technique (Figs. 3.1 and 3.2a) allows the growth of large-diameter crystals without a rising zone height L_{max}, which is physically limited under gravitation by the melt density and the surface tension [4].

$$L_{max} = 2.84\sqrt{\frac{\gamma}{\rho_{melt}\, g}}. \tag{3.2}$$

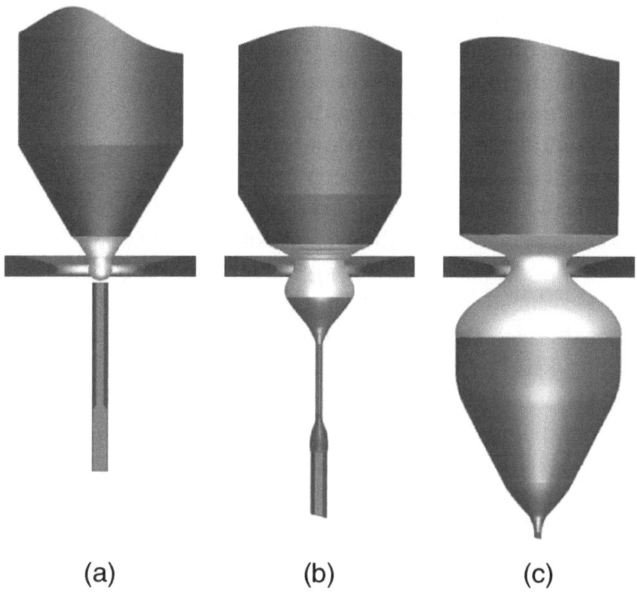

Fig. 3.1. Three phases of the FZ needle-eye process (for diameters of >30 mm) (**a**) hanging drop, (**b**) growing cone, (**c**) stable crystal growth, see also [5]

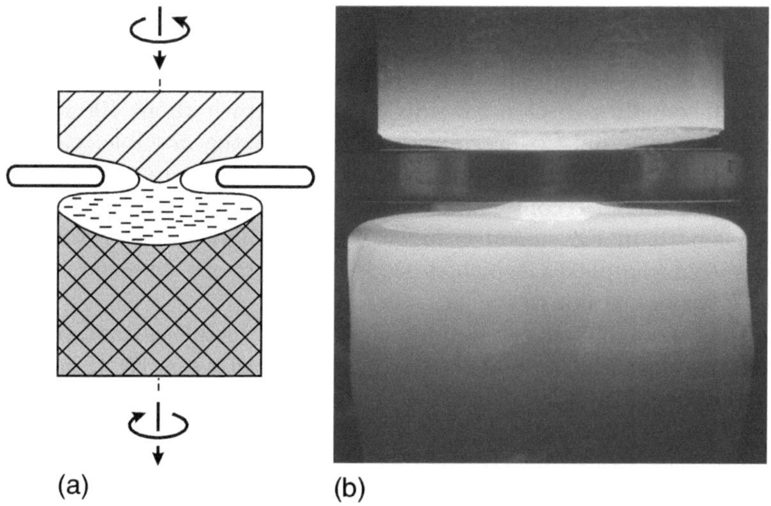

Fig. 3.2. Needle-eye technique for large FZ crystals: on bottom the growing crystal, on top the melting feed rod, in between the molten zone heated by the inductor schematic cross section (**a**) and photograph (**b**) of an FZ process (100 mm crystal diameter)

Although silicon is not the only substance, which can be grown as a monocrystal after the FZ method, its unique combination of the relatively low melt density and the high surface tension enables much bigger molten zones and crystal diameters than other semiconductors [5]. Because silicon is the most important semiconductor for electronic devices and, as well, for solar cells, always large crystal diameters have been demanded for both, Czochralski (CZ) crystals grown from the melt in a quartz crucible and FZ crystals. Today, CZ silicon crystals are produced with a diameter of more than 300 mm, whereas FZ crystals can be grown with up to 200 mm diameter [6]. Although only 5–10% of the produced semiconductor silicon wafers are FZ wafers, they are non-replaceable by the dominant CZ silicon because of their material parameters related to the higher purity.

For silicon, the generation energy for dislocations is very high. Therefore, a crystal that grows dislocation-free will hold this state as long as it is not hardly disturbed or if there are no precipitations of other phases at the crystallization front, either from impure feed material or from the growth surrounding. However, it is not possible to grow silicon single crystals having dislocations with diameters of more than about 40 mm because of the low energy for dislocation multiplication due to gliding processes driven by thermo-mechanical stress. This means, any bulk material of crystalline silicon with technically relevant dimensions, including PV silicon, is either dislocation-free single crystalline or polycrystalline.

Principally, the CZ method has a greater purification effect by segregation because all the remaining feed material is molten and the segregated impurities can be diluted, other than in the molten zone of the FZ method having a small and constant melt volume (see Fig. 3.3).

However, since liquid silicon reacts markedly with any crucible material including fused silica (SiO_2), FZ silicon can be grown about 100 times purer than CZ silicon mainly concerning, oxygen from the crucible, carbon from the graphite heater and, last but not least, transition metals (Cu, Fe, Ni, ...) acting as carrier recombination centers. In that way, the minority carrier diffusion length in FZ silicon is bigger and the light-induced generation of boron–oxygen complexes, which degrade the solar cell efficiency for boron-doped CZ wafers, is not relevant.

Another economic aspect besides saving the crucible is that the FZ process has the potential of higher maximum growth rates [7] (Fig. 3.4). The heat is effectively dissipated because the only hot parts of the FZ setup are the comparably small molten zone and the solid silicon close to the solid–liquid interfaces. Consequently, the specific energy consumption of the growth process is small compared to that of the CZ technique.

However, there are disadvantages and weaknesses of the FZ technique too: the polycrystalline silicon feed rods for the FZ production have to be well cylindrical, free of cracks, pores, precipitates (SiO_x–, SiC, Si_3N_4) and high residual stress. They are generally more expensive than those being crushed

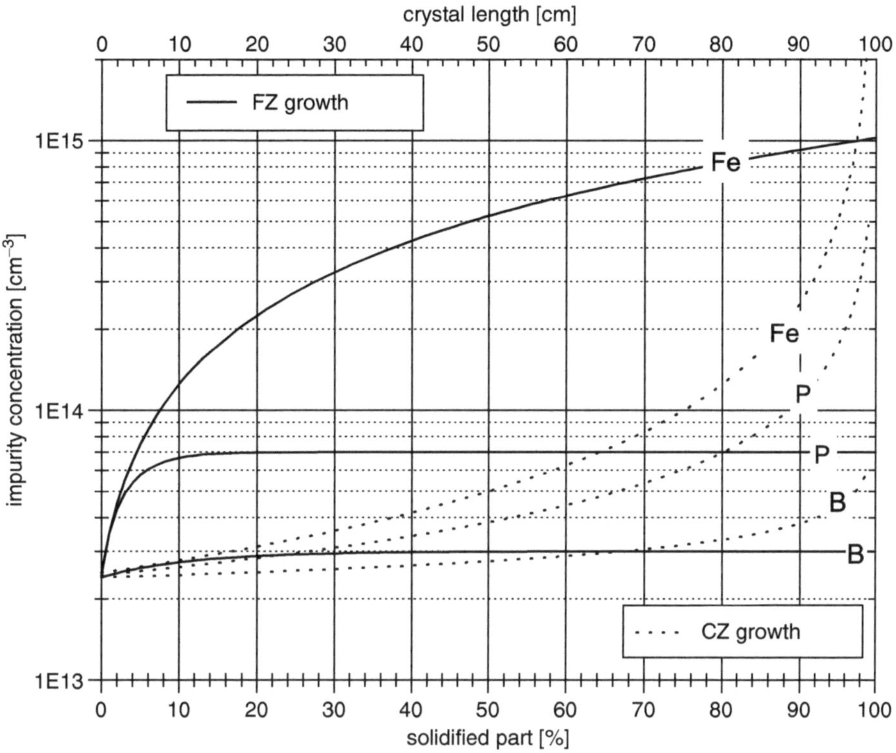

Fig. 3.3. Segregation of different impurities in FZ and CZ crystals

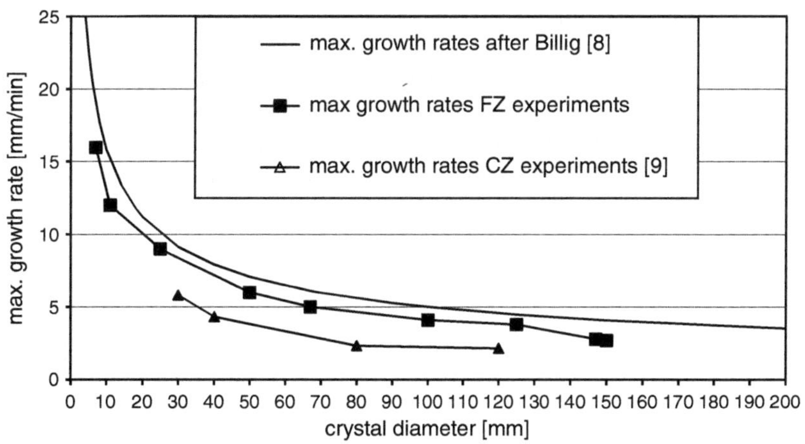

Fig. 3.4. Maximum growth rates: Calculated after Billig [8], from FZ experiments at the ICG Berlin and CZ experiments [9]

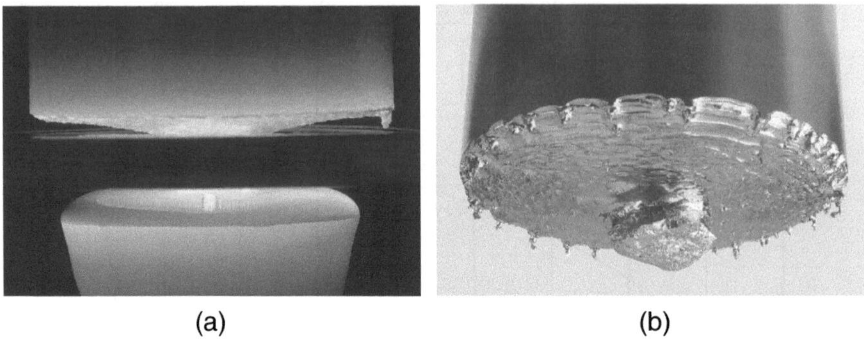

Fig. 3.5. (**a**) A risky big nose during the FZ process, (**b**) noses at the feed rod after ending the process

to fill a CZ crucible. In contrast to CZ, they need to have a diameter of at least 80–90% of that of the crystal to be grown.

More than for Cz, upscaling is limited by physical reasons. The RF threshold voltage of the inert growth atmosphere (Ar) limits the maximum crystal diameter. For rising diameters, more and more RF power is needed and finally, arcing occurs at the slit of the RF inducting coil.

A solution to overcome this situation could be an increased gas pressure, because, after the Paschen rule, the threshold voltage is proportional to this pressure, but a higher pressure means larger heat capacity of the gas, which rise the thermoelastic stress strongly. The latter is principally higher for big crystal diameters and the growing crystal tends to crack. The RF voltage would also be reduced for lower working frequencies, but in this case, due to the bigger skin depth, the melting front of the feed rod gets more and more rough forming needle-like relicts, which cannot be molten inductively and will touch the inductor later causing arcing, as shown in Fig. 3.5. Independent of the frequency, these silicon spikes or "noses" are a general problem and can occur also if the feed rod is of low structural quality having pores or fissures (Fig. 3.5b).

Adding some per mille of nitrogen to the argon protecting gas effectively reduces arcing. It lowers the impact ionisation of the Ar atoms because collisions with the dumbbell-shaped N_2 molecules are inelastic due to the low and dense rotational energy levels. However, nitrogen is continuously incorporated into the molten zone and it can form silicon nitride precipitates, if the maximum solubility of $5.4 \times 10^{15}\,\mathrm{cm}^{-3}$ is exceeded in the crystal. Therefore, the nitrogen content in the Ar gas limited below ca. 1%. Interstitially solved nitrogen in FZ silicon has some positive influences, too: A-swirls (self-interstitial clusters) as well as D-defects (voids, vacancy clusters) and some other intrinsic point defects are beneficially smaller although higher

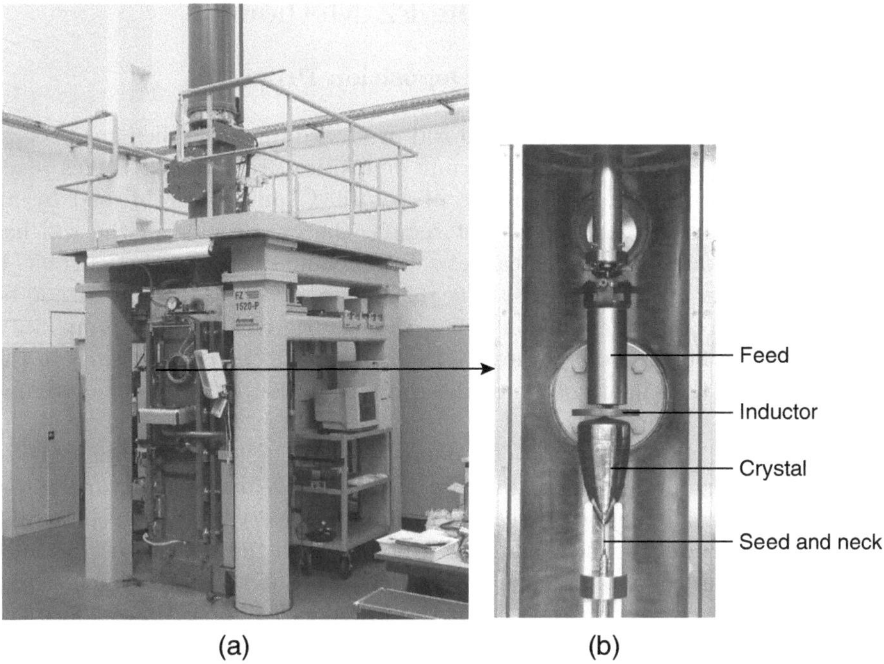

Fig. 3.6. Large FZ Puller at the Institute for Crystal Growth (ICG) Berlin. The enlargement (*right*) shows the setup in the chamber after ending the process: feed rod, inductor and crystal on the thin neck

concentrated. Additionally, dislocation gliding during high-temperature processing is markedly inhibited by the interstitially dissolved N_2. Hydrogen similarly acts against arcing and was used in the past as additive to the argon, but, as a detrimental side effect, it causes hazardous micro cracks and other defects in the crystal.

The inherent stability of the rather sophisticated FZ process is weaker than for CZ and operating needs special skills. There is always a danger for the molten zone to spill out. The process cannot be reversed if dislocations appear as it is possible for the CZ case. The FZ puller is generally more expensive than a comparable CZ machine because of the exacting mechanics and the RF power generator. The feed rod as well as the growing crystal are vertically moved by rigid shafts over a distance equal to the maximum crystal length. As a consequence, the FZ machine has to be at least four times higher than the latter, in fact ca. 9 m for growing a crystal of 2 m length (Fig. 3.6). Only short crystals can stay stable on the thin-neck, therefore, the crystal should be carefully supported by an additional elaborate device.

3.2 Silicon Feed Rods for the FZ Method

3.2.1 Siemens and Monosilane Deposition Processes

In principle, the diameter of a silicon rod produced after the Siemens process (see part I) is limited. The endothermic deposition from $SiHCl_3$ takes place at a temperature of the rod surface of >1,100°C and is established by an axial electric current. As the hottest region, the core region of the rod has the highest conductivity, which causes self-bunching of the heating current. If the rod diameter exceeds 160–180 mm, the silicon melting temperature can be reached in the core. When silicon melts, the electric conductivity jumps by a factor of about 30. Therefore, most of the current will be concentrated in the molten core. If that melt solidifies, the rod will brake by the specific volume expansion of ca. 8%.

Rods of FZ quality have to be deposited with a low rate, not only to reduce that current bunching but also to achieve a dense structure and a smooth surface. The deposition conditions must be very pure to exclude SiC, SiO_2 and other inclusions, which would not melt in the FZ process because the dwell time in the molten zone is too short, in contrast to the CZ method, where the melt can be overheated in the crucible for some time before the growth starts.

If silane SiH_4 is used instead of $SiHCl_3$, the pyrolytic deposition works at about 850°C. In principle, a bigger rod diameter should be possible in this way. However, silane is seen to be more dangerous because it is self-igniting. Only a few companies in the world are known to go this way. Generally, such rods are of high structural quality and purity and well suited for FZ growth.

3.2.2 Growth of Feed Rods

Another attempt to make feed rods for FZ crystal growth from cheap starting material is to melt solar (or lower) grade raw silicon in a quartz crucible and pull a rod of the desired diameter after the CZ method, not necessarily a single crystal. A lot of impurities can be removed by segregation and in the subsequent FZ step, almost all oxygen and most of the other impurities can be removed, too. However, the actual costs of such an approach must be carefully considered, but it is still an option to overcome the diameter limitations of the Siemens process.

3.2.3 Granular Feed Stock

Silicon can also be deposited at the surface of hot silicon particles hovering in a fluid bed by pyrolytic decomposition of SiH_4 or $SiHCl_3$. The result is granular silicon with a purity grade between very good solar and medium semiconductor quality. Compared with the Siemens process, the deposition rate is high and time-independent because of the large and almost constant total surface of the hovering grains. Therefore, a specific cost reduction is seen. Whereas, presently granular silicon is often used for the CZ growth because the crucible can be filled easily and automatically with this feed stock, a direct application for FZ growth appears difficult. However, a new forward-looking technological concept was patented by a well-known German FZ producer [10]. Here, two inductors work at two levels. In the upper level, silicon granulate is continuously fed into a special container and inductively molten. The melt flows down to the lower level forming a molten zone, where a second coil establishes the thermal conditions for the growth of an FZ-like crystal.

3.3 Doping of FZ Silicon Crystals

The main dopants, boron and phosphorus, can easily be introduced by a mass flow controlled gas stream of argon-diluted diborane (B_2H_6) or phosphine (PH_3), respectively. Both decompose at the hot melt surface to either boron or phosphorus being quickly absorbed. Because the segregation coefficients of B and P are rather large ($k = 0.8$ and 0.36, respectively), the melt is soon in equilibrium resulting in an almost homogeneous axial resistivity distribution of the crystal. For more uncommon dopants like gallium, indium, aluminium and antimony, k is small ranging from 10^{-4} to 10^{-3}. Here, axial homogeneity can be achieved by introducing a single dopant portion into the molten zone at the beginning of the FZ growth, e.g. by placing it in a small hole radially drilled in the feed rod (pill-doping).

3.4 Physical and Technical Needs and Limitations

The FZ puller is a rather heavy, high and generally sophisticated machine, as shown in Fig. 3.6. The heavy weight is due to the indispensable mechanical precision and stability of the spindle movements requiring a high stiffness of the whole construction. As already mentioned, it needs to be at least four times as high as the full pull length. Other critical parts are the high-power and low inductivity coaxial RF feed-through, the high-purity protecting and doping gas supply and, last but not least, the water-cooled and vacuum-tight inducting coil made of copper or, better, of silver, which is the "heart" of the FZ technology. The manufacturing of coils for large crystal diameters requires highly advanced material machining, welding and soldering techniques.

The RF power-generator typically works at frequencies of $\omega = 2.6$–3.2 MHz with a power of 40–80 kW, depending on the desired crystal diameter. Till now, such generators only work with an industrial electron power tube (triode), which is robust, but has a limited total operating time. The RF output parameters (ω, U, I) of the self-excited generator are defined by the tank circuit. It is essential that harmonics or other fractions of higher frequencies are widely suppressed as they can cause arcing at the coil gap.

3.5 Growth of Quadratic FZ Crystals (qFZ)

In contrast to the majority opinion, a growing FZ crystal need not necessarily has to rotate. It is clear that then the temperature field is no longer rotationally averaged and the cross section of the crystal can differ from a circular one. The temperature field caused by the induction coil as well as the surface tension dominate the cross section of the crystal [11].

Figure 3.7 shows an FZ crystal growing without rotation. At the beginning, the FZ growth started in the common way up to the steady diameter of the crystal. After that, the RF power and the feed rate were slightly lowered to reduce the volume of the melt in the zone according to the size of the square to be grown, then, the crystal rotation was stopped. Further, the crystal grew dislocation-free up to the end in a square-like shape because the inductor exhibits four slits in diagonal directions.

Such an inductor generates a temperature field of almost square symmetry. The main slit in the inductor, which forms the current loop, however, is not completely equivalent to the other three slits as seen in Fig. 3.9b. Furthermore, the surface tension surrounds the corners by minimizing the melt surface, which is not desired because the goal are quadratic wafers directly cut from the crystals. Additionally, on the straight sides, where the horizontal curvature is almost zero, growth instabilities can occur. Regarding the application for the solar industry, which would save material loss and the costs of cutting the round crystal into a square one, it is essential to get a stable

Fig. 3.7. Growing single-crystalline FZ crystal with a quasi-square shape, see also [5] (The feed rod on top of the dark coil is mostly covered by a heat shield.)

Fig. 3.8. qFZ crystal with almost quadratic cross section ($100 \times 100\,\mathrm{mm}^2$, rounded). The instabilities at the sides and the deviation from the *square shape* can be seen

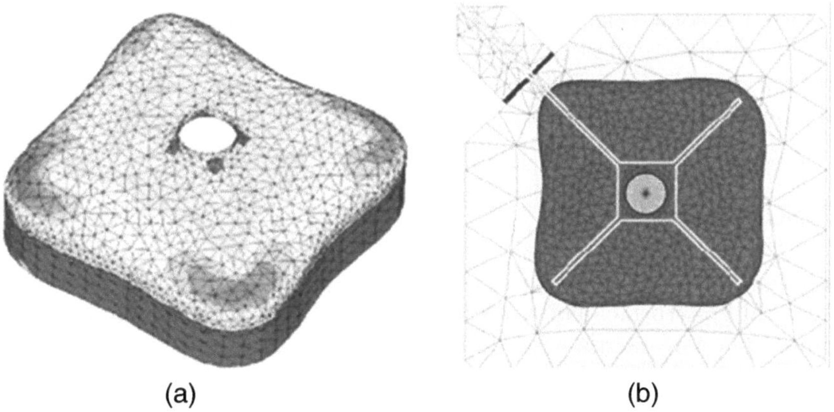

Fig. 3.9. Numerical modelling of molten zone (**a**) and inductor and calculated shape of the cross section (**b**) [12]

cross section approximating a square as good as possible. Figure 3.8 shows a single-crystalline qFZ crystal.

Numerical modelling can support process development. Also for this application, calculations were made [12] to find out, whether a symmetric cross section is possible.

Even with a rather simple inductor, the numerical calculation shows that it should be possible to form a useful crystal shape (Fig. 3.9).

3.6 Comments on the Potential of FZ Silicon for Solar Cells

In comparison to monocrystalline CZ silicon, which presently dominates the monocrystalline cell production, FZ silicon is purer, especially regarding transition metals, oxygen and carbon. The resulting higher carrier lifetime and the absence of the light-induced cell degradation enables the highest long-term efficiencies of any silicon-based solar cell. Efficiencies of more than 25% were demonstrated in the laboratory, whereas comparable CZ cells show about 22% maximum, which, for the most common boron doping, degrades below 20% after a few days of sunlight exposition.

The reasons why FZ silicon, up to now, could not be commercially launched in the terrestrial PV market are the smaller maximum crystal sizes (diameter, lengths) and the need of expensive high-quality feed rods being of limited sizes, too. Today, the biggest FZ crystal diameter on the market is 200 mm. This would fit a quadratic wafer format of $150 \times 150\,\text{mm}^2$, but growing such crystals is still extremely sophisticated and expensive, too. However, future progress in FZ growth development, as mentioned before, could make FZ solar cells competitive or even superior to CZ cells.

With FZ solar cells, the amount of silicon per Watt peak (Wp) would probably be the smallest of the crystalline silicon solar cells. Low-cost silicon feedstock development suitable for FZ is crucial for the future of FZ silicon based photovoltaic.

3.7 Summary

The purest single crystals from silicon can be grown after the crucible-free FZ crystal growth technique, presently (2008) with a maximum diameter of 200 mm. The low energy effort for the growth and no need for a crucible are the economic advantages.

The feedstock silicon for FZ growth has to be of high purity and structural perfection and is rather expensive. The development of low-cost feed rods is crucial for the commercial terrestrial solar application.

The growth of quadratic FZ crystals could reduce the material waste for solar cell production. High carrier diffusion length and low oxygen concentration enable the highest solar cell efficiencies of about 25% in laboratory scale.

References

1. H.C. Theuerer, Method of processing semiconductive materials - US Patent 3060123 v. 17.12.1952
2. W.C. Dash, J. Appl. Phys. **29**, 705 (1958)
3. W.C. Dash, J. Appl. Phys. **31**, 736 (1960)

4. W.Heywang, Z. Naturforschung, **11a** 238 (1956)
5. A. Lüdge, H. Riemann, M. Wünscher, G. Behr, W. Löser, A. Cröll, A. Muiznieks, in T. Duffar (ED) Crystal Growth Processes Based on Capillarity : Czochralski, floating zone, shaping and crucible techniques (Chapter 4), John Wiley&Sons, Chichester, England (in print)
6. W. von Ammon, in *Silicon Crystal Growth*, ed. by G. Müller, J-J. Meteois, P. Rudolph. Crystal Growth-From Fundamentals to Technology (Elsevier, Amsterdam, 2004)
7. A. Lüdge, H. Riemann, B. Hallmann, H. Wawra, L. Jensen, T.L. Larsen, A. Nielsen, ed. by Cor L. Claeys. *High Purity Silicon VII*, (Electrochemical Society, New Jersey, PV 2002–20, 2002)
8. E. Billig, Proc. R. Soc. **229**, 346 (1955)
9. H.J. Oh, et al., Electrochem. Soc. Proc. **17**, 44 (2000)
10. W. von Ammon, Process and apparatus for producing a single crystal of semiconductor material-US Patent 2003/0145781A1 Aug. 7.2003
11. A. Luedge, H. Riemann, B. Hallmann-Seiffert, A. Muiznieks, F. Schulze, ECS 317 Trans **3**(4), 61
12. A. Muiznieks, A. Rudevics, K. Lacis, H. Riemann, A. Lüdge, F.W. Schulze, B. Nacke, Square-like silicon crystal rod growth by FZ method with especially 3D shaped HF inductors. *International Scientific Colloquium Modelling for Material Processing*, Riga, 2006

4

Crystallization of Silicon by a Directional Solidification Method

Koichi Kakimoto

Abstract. This chapter introduces crystallization process of multicrystalline silicon by using a directional solidification method. Numerical analysis, which includes convective, conductive, and radiative heat transfers in the furnace is also introduced. Moreover, a model of impurity segregation is included in this chapter. A new model for three-dimensional (3D) global simulation of heat transfer in a unidirectional solidification furnace with square crucibles was also introduced.

4.1 Directional Solidification Method: Strengths and Weaknesses

Multicrystalline silicon (mc-Si) has a large demand of photovoltaics to overcome difficulty of the present green problem [1]. The directional solidification method is a key method for large-scale production of mc-Si in highly efficient solar cells. The maximum efficiency of the solar cell based on mc-Si is 18%. However, the use of commercially available wafers typically results in solar cell efficiency of about 16% in industrial solar cell processes.

There are many problems that must be solved to achieve high efficiency. Mc-Si has many dislocations and grain boundaries that are introduced during the solidification process. Moreover, Mc-Si crystals are grown in a crucible, a process that degrades purity of the crystals due to their attachment to the crucible wall. Such defects and impurity can reduce the conversion efficiency of solar cells. Therefore, we should control distributions of dislocations, grain boundaries, and impurities during the solidification process. In the recent studies, quasi single crystal or multicrystalline with large grain size have been grown by using a modified directional solidification methods [2,3].

The directional solidification process has several merits. Square-shaped crystalline silicon can be grown by using a square-shaped crucible. When round-shaped crystals grown by the Czochralski method are used as raw materials for square-shaped solar cells, a large amount of waste of silicon materials is inevitably produced. Therefore, the unidirectional solidification process has a merit concerning the feedstock problem.

4.2 Control of Crystallization Process

Control of crystallization of Mc-Si, especially cooling rate, solid–liquid interface shape, and impurity distributions, is important for obtaining solar cells with high conversion efficiency. The crystallization process is rather complicated since heat transfer system in a furnace is highly nonlinear. Due to the development of computer technology and computation techniques, numerical simulation has become a powerful tool for optimization of the directional solidification process and crystal growth process [4–12]. Since a directional solidification furnace has a nonlinear conjugated thermal system, transient simulation with global modeling is an essential tool for improvement of the directional solidification process from melting to cooling through the solidification process. The global model includes processes of radiative, conductive, and convective heat and mass transfer in a furnace. Therefore, we can quantitatively estimate the distributions of temperature in a furnace, velocity of the melt and distributions of impurities. The author developed a transient code with a global model for the directional solidification process. By using the code, the author carried out calculations to study distributions of temperature and impurities such as iron, carbon, oxygen, nitrogen, and nitride in a silicon ingot, during the directional solidification process [13–15].

The dimensions of a small unidirectional solidification furnace for producing mc-Si is shown in Fig. 4.1. The computation grid of the entire furnace is shown in the right part of Fig. 4.1. To establish a discrete system for numerical simulation using the finite difference method, the domains of all components in a unidirectional solidification furnace are subdivided into numbers of block regions, for example, as shown in the left part of Fig. 4.1, in which subdivision resulted in a total of 13 blocks. Each of these blocks is then discretized by a structured grid as shown in the right part of Fig. 4.1. Two heaters marked with numbers 12 and 13 can be recognized in the furnace. The following major assumptions are used in the present model: (1) the geometry of the furnace configuration is axisymmetric, (2) radiative transfer is modeled as diffuse-gray surface radiation, (3) melt flow in the crucible is laminar and incompressible, and (4) the effect of gas flow in the furnace is negligible.

A local view of the computation grid in the melt-crystal domain when the interface is moving upward as a function of time is shown in Fig. 4.2. In order to fit to the moving interface as a function of time, the grid cells in the melt and solidified ingot are stretched, and the grid points are moved upward as well.

Conductive heat transfer in all solid components, radiative heat exchange between all diffusive surfaces in the unidirectional solidification furnace, and the Navier–Stokes equations for the melt flow in the crucible are coupled.

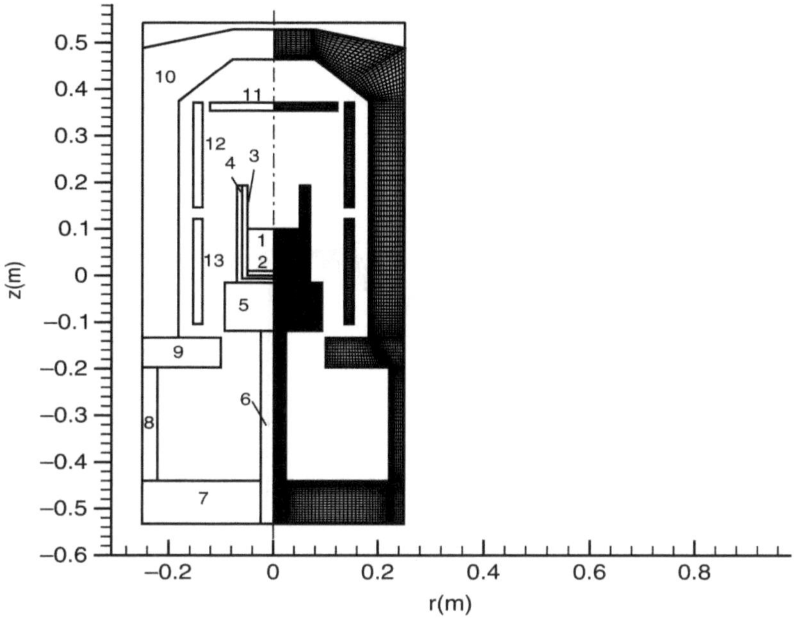

Fig. 4.1. Configuration and computation grid of a directional solidification furnace

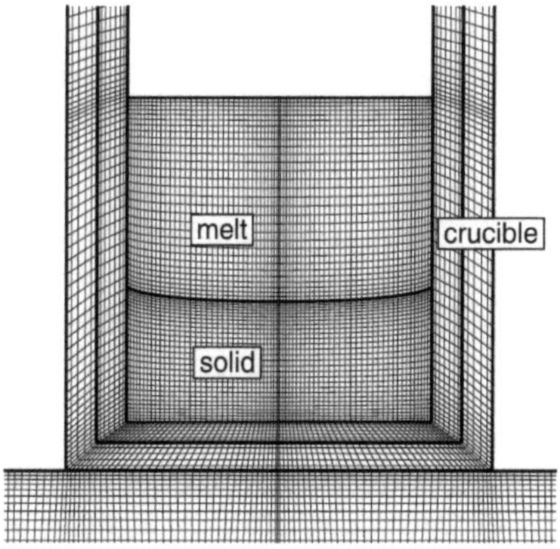

Fig. 4.2. Dynamic grid of the melt and the solid in a crucible

Then, they are solved iteratively in a transient way. Time histories of heater power, fraction solidified, and growth velocity during a unidirectional solidification process are shown in Fig. 4.3 [16]. Variation of heater power as a function of time was imposed as a process parameter.

The impurity distribution in the silicon melts and solidified silicon ingot was solved on the basis of solutions of the thermal field and melt flow in a

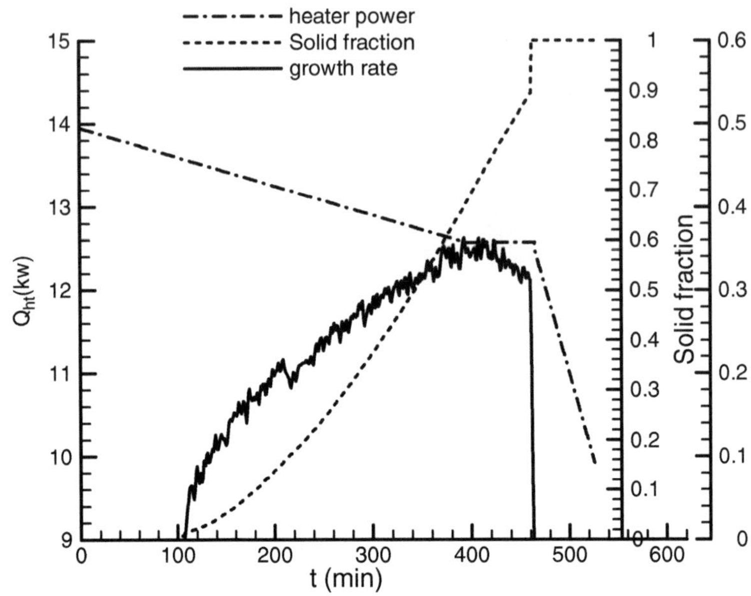

Fig. 4.3. Heater power, fraction solidified and solidification rate as a function of time during solidification process

crucible. Impurity segregation at the melt–crystal (m–c) interface was taken into account. The m–c interface shape was obtained by a dynamic interface tracking method. The global iterative procedure is described in [16].

Concentration taking into account on segregation of impurity was expressed by (4.1) at the m–c interface:

$$D_{\mathrm{m}}\frac{\partial C_{\mathrm{m}}}{\partial n}+V_{\mathrm{g}}C_{\mathrm{m}}\left(1-k_0\right)=D_{\mathrm{s}}\frac{\partial C_{\mathrm{s}}}{\partial n}, \tag{4.1}$$

where C_{m}, C_{s}, and V_{g} are the impurity concentrations of the melt and solid and the growth velocity, respectively. D_{m} and D_{s} show diffusion constants of the impurity in the melt and the crystal, respectively.

The geometry of the furnace configuration is axisymmetric. The diameter of the inside wall of the crucible and the height of the solidified ingot are 100 mm. The author carried out a transient global calculation for a fast-cooling solidification process in which the cooling rate is 0.42 k W h^{-1} for the first 1.8 h and 0.084 k W h^{-1} thereafter. The initial heater power is 13.9 kW. Duration of the solidification process is 8.1 h. The average solidification rate is about 0.21 mm min^{-1}.

The heater power was decreased at a constant rate until 400 min and then kept constant until 460 min. Then a fast cooling rate was imposed during the cooling process after the completion of solidification. Solidification was started after 108 min from the initial stage of the process, which corresponds

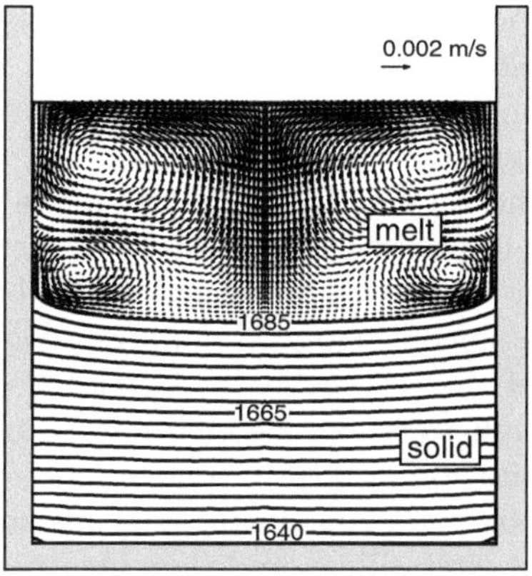

Fig. 4.4. Melt convection and distribution temperature during solidification of the ingot

to the time when heater power started to be decreased. The growth rate was increased during the period of decrease in heater power and it was decreased when the heater power was kept constant. Time response of the imposed heater power to the growth rate was calculated to be about 20 min. The fraction solidified was gradually increased and finally became constant. The whole process was completed in about 6 h.

The melt flow pattern and temperature distribution in the solidified ingot when half of the volume of melt has been solidified are shown in Fig. 4.4. Two pairs of weak vortices were formed in the melt. Flow velocity of the melt is in the order of several millimeters per second, which is almost one order of magnitude smaller than that of Czochralski crystal growth. The m-c interface is concave to the crystal in this case, since the diameter of the crucible was smaller than that of a commercially used crucible. The present calculation revealed that the temperature gradient in the solid, which is $10\,\mathrm{K\,cm^{-1}}$, is about one-third to one-fourth of the values in the Czochralski crystal growth system. Growth rate of the unidirectional solidification process is about one third to one fifth of that of the Czochralski growth system. Therefore, we can discuss about distribution of point defects in both single crystals and mc-Si based on the Voronlov's theory, which is expressed by the ratio between growth rate and temperature gradient in a growing crystal (V/G) [17].

4.3 Incorporation of Impurity in Crystals

There are several impurities in mc-Si, including carbon, oxygen, nitrogen, and iron. Carbon is one of the major impurities in silicon feedstock. When the carbon content exceeds its solubility limit in silicon, it will precipitate to form SiC particles in a directional solidification process. It has been experimentally proved that the dislocation density is a function of carbon concentration in mc-Si [18]. The SiC precipitates can cause severe ohmic shunts in solar cells [5] and result in nucleation of new grains in mc-Si. Both carbon and SiC precipitates in mc-Si can greatly reduce the conversion efficiency of solar cells. However, due to the shortage of high-quality silicon feedstock and requirement of cost reduction in photovoltaic industries, metallurgical-grade silicon (MG-Si) feedstock is a candidate for usage as feedstock to produce high-quality multicrystalline silicon ingots to maintain continued rapid growth of the photovoltaic market. Since the MG-Si feedstock contains higher level of impurities, it is important to study the characteristics of carbon segregation and SiC particle precipitation in a directional solidification process.

The author focused on modeling the primary phase of SiC particles, which are precipitated in the molten silicon and then incorporated into the solidified ingot [19]. The carbon precipitation in solid is neglected due to the small diffusion constant in a solid.

Figure 4.5 shows a phase diagram between silicon and carbon. The diagram shows how SiC particle precipitation forms in Si-melt in the C-rich domain [20], in which the solubility limit of carbon in Si-melt $C_{\mathrm{L}}(T)$ is approximated by a polynomial function,

$$C_{\mathrm{L}}(T) = 8.6250 \times 10^{-4} T^2 - 2.7643T + 2222.9. \qquad (4.2)$$

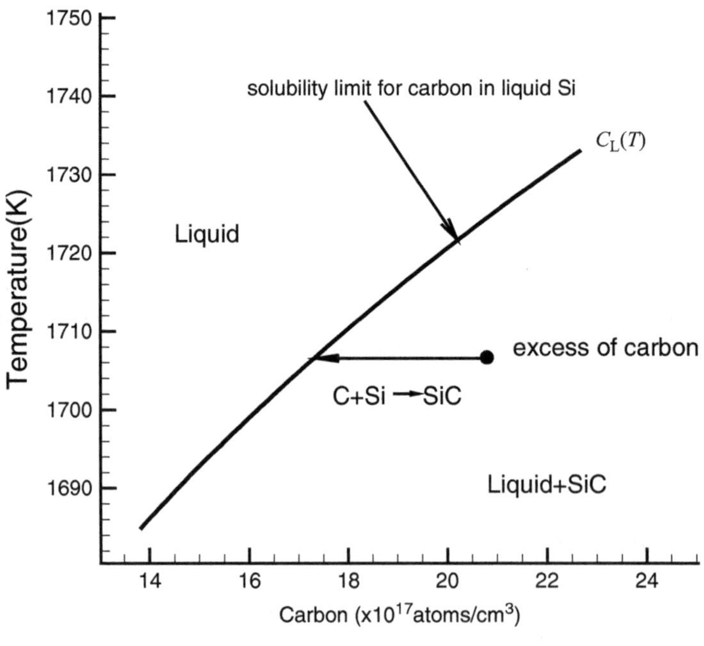

Fig. 4.5. Schematic of SiC particle precipitation with the Si–C phase diagram in C-rich domain

The units for carbon concentration and temperature in (4.2) are 10^{17} atoms cm^{-3} and K, respectively. With the molten silicon being solidified and the solidification interface moving upward in the crucible, during the solidification process, the carbon concentration in the melt increases due to the small segregation coefficient of carbon in silicon. If the carbon concentration exceeds the local solubility limit, excessive carbon precipitates and the following chemical reaction occurs:

$$\mathrm{Si} + \mathrm{C} \rightarrow \mathrm{SiC} \tag{4.3}$$

Thus, the substitutional carbon is reduced and the same amount of SiC particles is generated in the melt. The formation rate of SiC particles and the destruction rate of substitutional carbon are equal and proportional to the super-saturation degree of substitutional carbon and the speed of the chemical reaction (4.3):

$$G_{\mathrm{SiC}} = -G_{\mathrm{C}} = \alpha(C_{\mathrm{C}} - C_{\mathrm{L}}(T)) \quad \text{when} \quad C_{\mathrm{C}} > C_{\mathrm{L}}(T), \tag{4.4}$$

$$G_{\mathrm{SiC}} = -G_{\mathrm{C}} = 0 \quad \text{when} \quad C_{\mathrm{C}} \leq C_{\mathrm{L}}(T), \tag{4.5}$$

where G_{SiC} and G_{C} are the formation rates of SiC particles and substitutional carbon, respectively. Minus values denote the destruction rate. The coefficient α is the factor correlating the particle formation rate and chemical reaction rate in (4.3).

With these assumptions, the governing equations for the concentrations of substitutional carbon and SiC particles in the melt can be written as follows:

$$\frac{\partial C_{\mathrm{C}}}{\partial t} + \mathbf{V} \cdot \nabla C_{\mathrm{C}} = \nabla \cdot (D_{\mathrm{C}} \nabla C_{\mathrm{C}}) + G_{\mathrm{C}} \quad \text{for substitutional carbon,} \tag{4.6}$$

$$\frac{\partial C_{\mathrm{SiC}}}{\partial t} + \mathbf{V} \cdot \nabla C_{\mathrm{SiC}} = G_{\mathrm{SiC}} \quad \text{for SiC particles.} \tag{4.7}$$

The initial conditions for both impurities in the melt are defined as follows:

$$C_{\mathrm{C}} = C_0 \quad \text{for substitutional carbon,} \tag{4.8}$$

$$C_{\mathrm{SiC}} = 0 \quad \text{for SiC particles,} \tag{4.9}$$

where C_0 is the carbon concentration in the feedstock of silicon. The zero mass flux condition is applied at crucible walls and the top of the melt for both impurities. At the solidified ingot interface, the segregation effect is taken into account for substitutional carbon and the continuity condition is applied for SiC particles. The segregation coefficient of carbon in silicon was set to 0.07.

Figure 4.6a, b shows the substitutional carbon and SiC particle distributions, respectively in a cross section of the solidified ingot. It can be seen that the particle precipitation begins at the center, when the fraction solidified reaches about 30%. The SiC particles are clustered at the center-top region of the ingot, where the concentration of substitutional carbon is almost constant. This distribution pattern is due to the m-c interface shape, which is concave to the solid side throughout the solidification process.

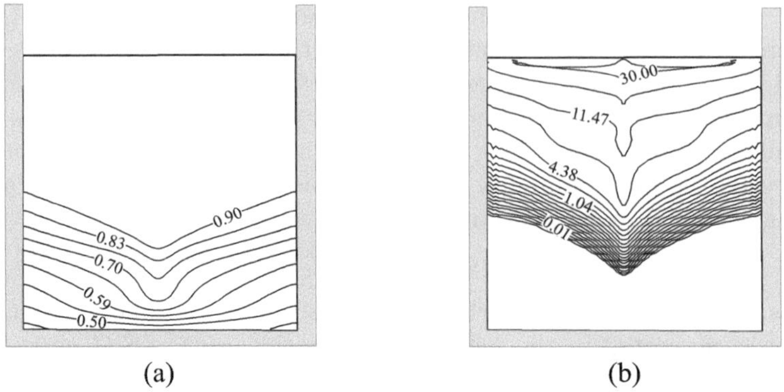

Fig. 4.6. Substitutional carbon and SiC particle distributions in a cross-plane of the ingot solidified in a fast-cooling process. $C_0 = 6.3 \times 10^{17}$ atoms cm^{-3}. *Contour lines* are plotted in exponential distribution. Unit of concentration is 10^{17} atoms cm^{-3} (**a**) substitutional carbon; (**b**) SiC particles

The impurity level in feedstock of silicon has a significant impact on conversion efficiency of an mc-Si solar cell. The author carried out a series of computations for the same fast-cooling solidification process but with different carbon concentrations in silicon feedstock (C_0). The obtained concentration distributions of substitutional carbon and SiC particles along the center axis of the solidified ingot are compared in Fig. 4.7 [18] with the carbon concentration in silicon feedstock ranging from 1.26×10^{16} atoms cm^{-3} (equivalent to 0.1 ppmw) to 6.30×10^{17} atoms cm^{-3} (equivalent to 5.0 ppmw). When C_0 is 1.26×10^{16} atoms cm^{-3}, no SiC particles are precipitated in the ingot and there is only a very thin layer at the top, rich in substitutional carbon. When C_0 increases above 1.26×10^{17} atoms cm^{-3}, the content of SiC particles increases significantly in magnitude as well as in space in the solidified ingot. Thus, it is necessary to control the carbon concentration in silicon feedstock to less than 1.26×10^{17} atoms cm^{-3}, which is equivalent to 1.0 ppmw.

Figure 4.8 [19] shows the distributions of substitutional carbon and SiC particles along the center axis of the ingots solidified in two solidification processes: fast-cooling process and slow-cooling process. The carbon concentration in silicon feedstock is 1.26×10^{17} atoms cm^{-3} for both processes. It is noticeable that both the substitutional carbon concentration and the SiC particle concentration are lower along the center axis of the ingot solidified in the slow-cooling process than that in the fast-cooling process. The region with high concentrations of both carbon and SiC particles at the upper portion is also much smaller in the ingot solidified in the slow-cooling process. The following two mechanisms may control this phenomenon. First, during the slow-cooling process of solidification, the solidification rate is small. The uniformity of impurities in the melt can, therefore, be improved due to the

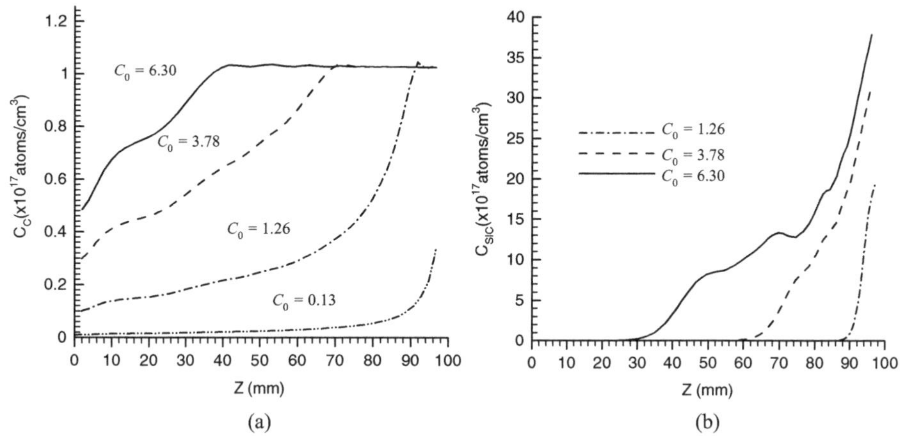

Fig. 4.7. Substitutional carbon and SiC particle distributions along the center axis of the ingots solidified in a fast-cooling process with different carbon concentrations in silicon feedstock. Unit of C_O is 10^{17} atoms cm^{-3}. (**a**) substitutional carbon; (**b**) SiC particles

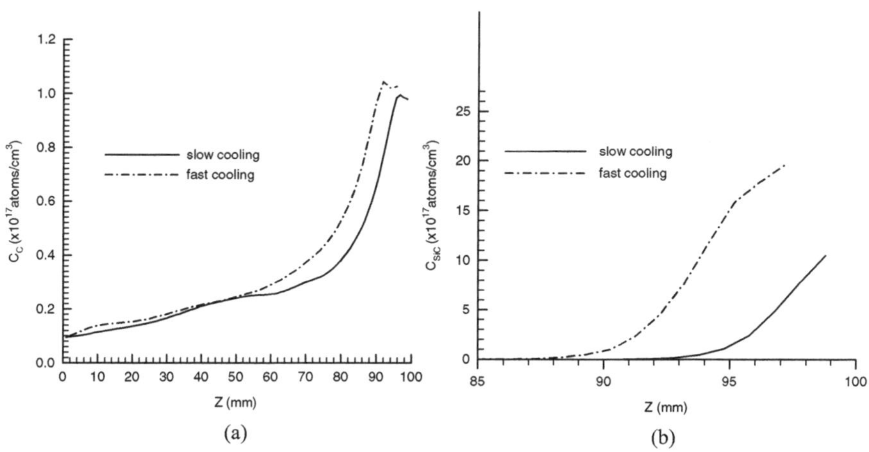

Fig. 4.8. Comparison of substitutional carbon and SiC particle distributions along the center axis of ingot between two solidification processes with different cooling rates. C_O is 1.26×10^{17} atoms cm^{-3}. (**a**) substitutional carbon; (**b**) SiC particles

longer period of diffusive and convective mixing. This is preferable for delaying occurrence of SiC particle precipitation in the melt.

Second, in the fast-cooling solidification process, the m-c interface is concave to the crystal and the melt flows from the periphery to the center along the solidification front surface. Thus, the impurities in the melt and the SiC particles generated at the interface are transported to and accumulated in the center area. However, in the slow-cooling solidification process, the interface changes to be convex after the fraction solidified reaches 70%.

The melt flows along the solidification front surface from the center to the periphery of crucible. Thus, the impurities in the melt and SiC particles generated at the interface are transported to and accumulated in the periphery region. This is preferable to homogenize the radial distribution of impurities and delaying occurrence of SiC particle precipitation in the melt.

As a result, the substitutional carbon and SiC particle distributions are obtained in a cross plane of the solidified ingot as shown in Fig. 4.8. Both substitutional carbon and SiC particles are clustered in a smaller periphery-top region of the ingot. This indicates that we can control the distributions of impurities in the solidified ingot by optimizing the process conditions or furnace configurations.

Iron in silicon is another important impurity causing reduction in conversion efficiency of solar cells. Figure 4.9a, b shows the distribution of iron concentration in a solidified silicon ingot that had been cooled for 1 h during the cooling process and a line profile of iron concentration along the center of silicon crystalline in the z-direction. Areas with high iron concentration were formed at the top of the melt. This is due to the segregation phenomenon, by which iron is segregated from the melt to the solid; therefore, such areas with high concentration of iron were formed at the end of solidification. Moreover, areas with a high concentration of iron were formed close to the crucible walls. Such areas were formed by diffusion, which occurred during and after solidification. This is based on the small activation energy of iron diffusion in the solid of silicon.

The profile shows that the concentration decreases rapidly near the bottom of a crucible to inside the crystal and then gradually increases up to a position of 80 mm, due to the segregation effect of iron during the solidification process.

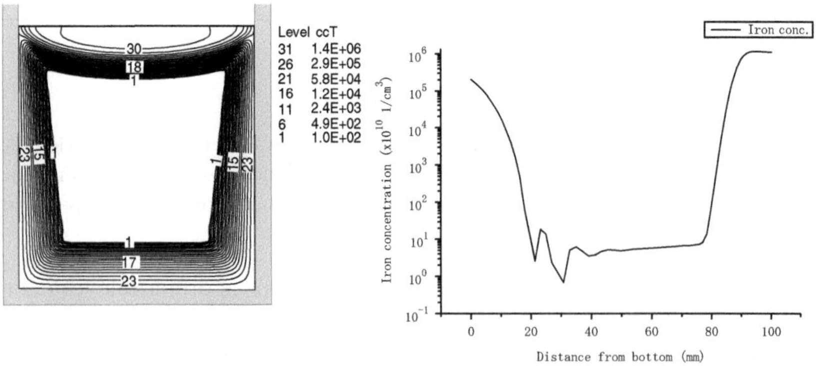

Fig. 4.9. (**a**) the distribution of iron concentration in a solidified silicon ingot, (**b**) a line profile of iron concentration along the center of silicon crystalline in the z-direction

Subsequently, the concentration rapidly increases near the top of the crystal. The key point for producing highly efficient solar cells is how to reduce areas with high concentrations of iron in a crystal, since iron acts as a lifetime killer of minority carriers in the crystal.

4.4 Three-Dimensional Effects of Solidification

There have been many papers concerning the computational studies of unidirectional solidification for solar cells, in which the growth system was imposed to be axisymmetric. However, the actual crystal shape is square, calculation of square-shaped crystals is necessary. When square crucibles are used, the configuration of the furnace becomes asymmetric, and heat transfer in the furnace consequently becomes three-dimensional. Three-dimensional (3D) global modeling is, therefore, necessary for the investigation of m-c interface shape with square crucibles [20].

There have been no works using global analysis to investigate the effect of crucible shape on m-c interface shape. The author developed a steady code with 2D and 3D global models for the unidirectional solidification process used for cylindrical and square crucibles, respectively.

The author carried out calculations to investigate the m-c interface shapes with cylindrical and square crucibles and to investigate the influence of crucible shape on m-c interface shape using 2D and 3D global analyses.

Three-dimensional global analysis with square crucibles requires large computational resources because of the huge number of 3D structured grids. To overcome this difficulty, a 2D/3D mixed discretization scheme is employed to reduce the requirement of computational resources [11, 12]. The domains in the central area of the furnace, in which the configuration and heat transfer are nonaxisymmetric, are discretized in a 3D way.

The local 3D computational grid in the domains in the central area of the furnace is established as shown in Fig. 4.10 [21]. The other block regions that are away from the central area of the furnace, in which the configuration and heat transfer are axisymmetric, are discretized in a 2D way. The 2D computational grid in other block regions of the furnace is the same as that used in the 2D global model as shown in Fig. 4.1.

The assumptions used in the 3D global analysis are, (2), (3) and (4), stated above. The computation method used for 3D global analysis is the same that used for 2D global analysis.

Figure 4.11a [22] shows the m-c interface shape and temperature distribution of the melt and crystal obtained by 3D global analysis. The m-c interface shape in the case of using square crucibles also becomes concave to the melt. Figure 4.11b, c shows the m-c interface shape and temperature distribution of the melt and crystal in the X-o-Z plane and D-o-Z plane, respectively. The magnitude of deformation of the m-c interface, ΔZ, is defined as the difference between heights at the center and edge of the

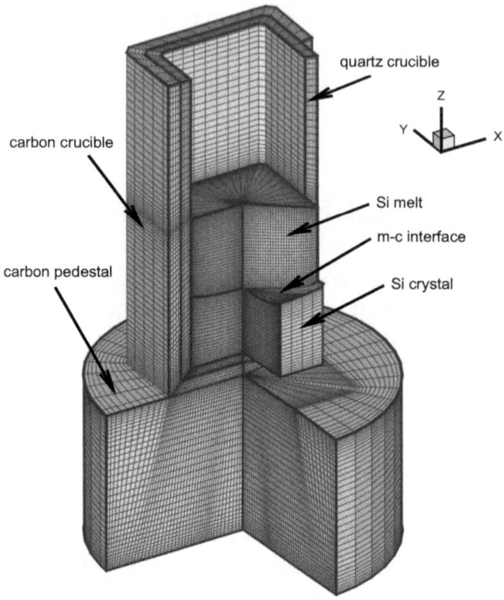

Fig. 4.10. The local 3D computational grid in the domains in the central area of the furnace

m-c interface. Values of ΔZ are 3.5 and 6.9 mm in the X-o-Z plane and D-o-Z plane, respectively. Maximum deformation of the m-c interface is observed near the corner of the crucible.

Figure 4.12 [22] shows the heat flux distribution on the vertical wall of the melt and crystal in the case of using square crucibles. We found that outgoing heat flux near the corner of the crucible becomes larger than that in other areas because cooling of the melt around the corner is enhanced due to two adjacent crucible walls near the corner. Furthermore, tilt angle of heat flux at the m-c interface increases with increase in the amount of outgoing heat flux through the vertical walls. Consequently, deformation of the m-c interface at the corner becomes large. Therefore, the m-c interface deforms three-dimensionally because heat flux has three-dimensionality.

4.5 Summary

The unidirectional solidification method has both merits and demerits. Controlling the process is a key issue for achieving solidification process that can produces solar cells with high conversion efficiency. Numerical modeling is important for realizing such a process with optimization. A 3D global model and code were developed for analyzing 3D features of a unidirectional solidification process with square crucibles. A 2D/3D mixed discretization scheme

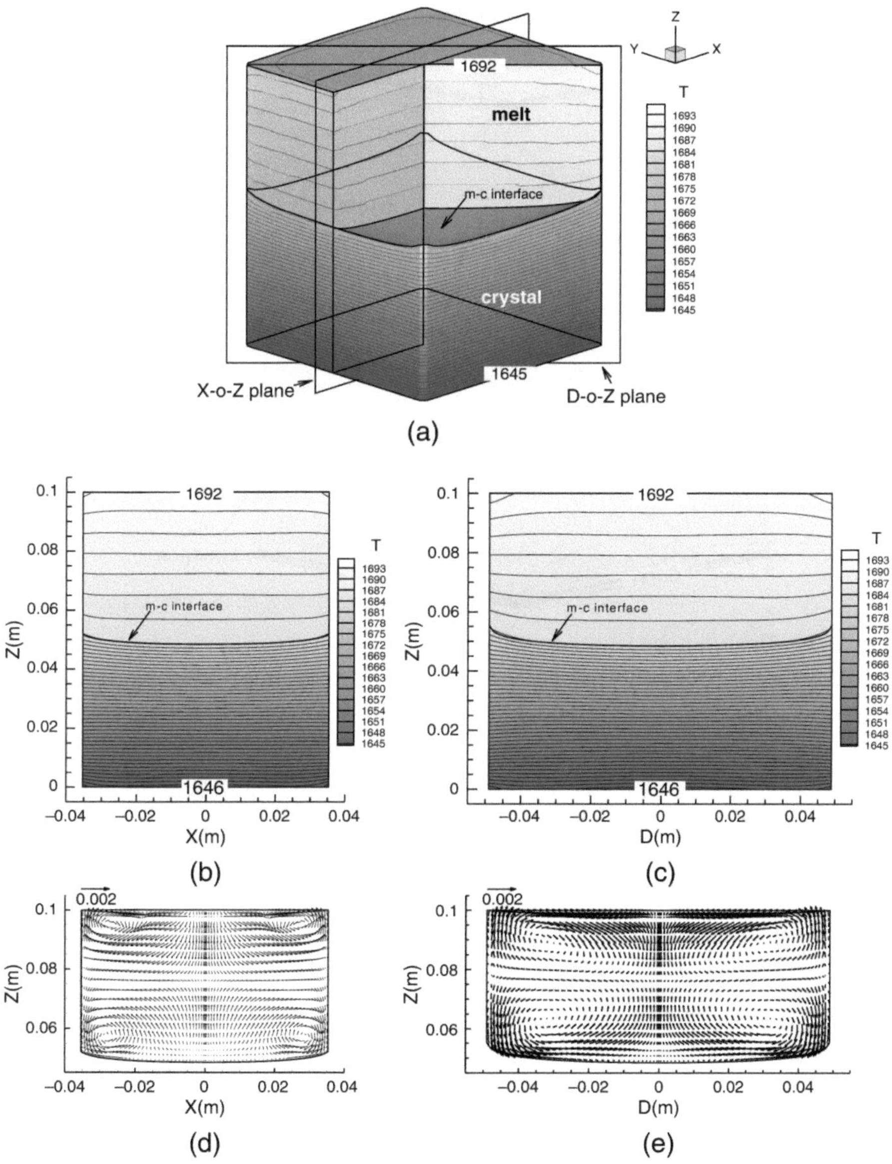

Fig. 4.11. (**a**) Temperature distribution of melt and crystal obtained by 3D global analysis. (**b**) Temperature distribution of the *X-o-Z* plane. (**c**) Temperature distribution of the *D-o-Z* plane [unit: K]. (**d**) Velocity fields of the melt flow of the *X-o-Z* plane [unit: $m\,s^{-1}$]. (**e**) Velocity fields of the melt flow of the *D-o-Z* plane [unit: $m\,s^{-1}$]

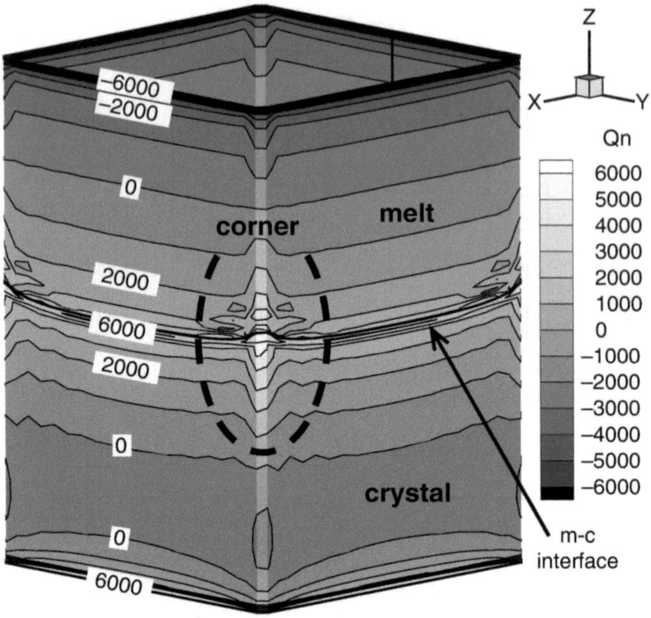

Fig. 4.12. Heat flux distribution on the vertical wall of the melt and crystal obtained by 3D global analysis [unit: $\mathrm{W\,m^{-2}}$]

was employed. An effective and efficient algorithm was developed for calculating view factors in radiation modeling. Thus, 3D global modeling can be carried out with moderate requirement of computation resource by a common PC. Some results of 3D features of a case study were presented.

Acknowledgment

This work was supported by a NEDO project, a Grant-in-Aid for Scientific Research (B) 14350010 and a Grant-in-Aid for the creation of innovation through business-academy-public sector cooperation from the Japanese Ministry of Education, Science, Sports and Culture.

References

1. D. Franke, T. Rettelbach, C. Habler, W. Koch, A. Müller, Sol. Energy Mater. Sol. Cells **72**, 83 (2002)
2. K. Fujiwara, W. Pan, N. Usami, K. Sawada, M. Tokairin, Y. Nose, A. Nomura, T. Shishido, K. Nakajima, Acta Mater. **54**, 3191 (2006)
3. N. Stoddard, B. Wu, L. Maisano, R. Russel, J. Creager, R. Clark, J. Manuel Ferrandez, in 18th Workshop on Crystalline Silicon Solar Cells and Modules: Materials and Processes, pp. 7–14, 2008, ed. by B.L. Sopiri

4. M. Ghosh, J. Bahr, A. Müller, in *Proceedings of the 19th European Photovoltaic Solar Energy Conference*, Paris, 2004, p. 560
5. D. Vizman, S. Eichler, J. Friedrich, G. Müller, J. Cryst. Growth **266**, 396 (2004)
6. A. Krauze, A. Muiznieks, A. Muhlbauer, Th. Wetzel, W.v. Ammon, J. Cryst. Growth **262**, 157 (2004)
7. L.J. Liu, K. Kakimoto, Cryst. Res. Technol. **40**, 347 (2005)
8. K. Kakimoto, L.J. Liu, Cryst. Res. Technol. **38**, 716 (2003)
9. J.J. Derby, R.A. Brown, J. Cryst. Growth **74**, 605 (1986)
10. F. Dupret, P. Nicodeme, Y. Ryckmans, P. Wouters, M.J. Crochet, Int. J. Heat Mass Transfer **33**, 1849 (1990)
11. M. Li, Y. Li, N. Imaishi, T. Tsukada, J. Cryst. Growth **234**, 32 (2002)
12. V.V. Kalaev, I.Yu. Evstratov, Yu.N. Makarov, J. Cryst. Growth **249**, 87 (2003)
13. L.J. Liu, K. Kakimoto, Int. J. Heat Mass Transfer **48**, 4492 (2005)
14. L.J. Liu, S. Nakano, K. Kakimoto, J. Cryst. Growth **282**, 49 (2005)
15. L.J. Liu, K. Kakimoto, Int. J. Heat Mass Transfer **48**, 4481 (2005)
16. L.J. Liu, S. Nakano, K. Kakimoto, J. Cryst. Growth **292**, 515 (2006)
17. V. Voronkov, J. Cryst. Growth **56**, 625 (1982)
18. K. Arafune, Y. Ohshita, M. Yamaguchi, Physica B **134**, 236 (2006)
19. L.J. Liu, S. Nakano, K. Kakimoto, J. Cryst. Growth **310**, 2192 (2008)
20. H. Laux, Y. Ladam, K. Tang, E.A. Meese, in *Proceedings of the 21st EUPVSEC*, Dresden, Germany, 2006
21. H. Miyazawa, L.J. Liu, S. Hisamatsu, K. Kakimoto, J. Cryst. Growth **310**, 1034 (2008)
22. H. Miyazawa, L.J. Liu, S. Hisamatsu, K. Kakimoto, J. Cryst. Growth **310**, 1142 (2008)

5

Mechanism of Dendrite Crystal Growth

Kozo Fujiwara* and Kazuo Nakajima

Abstract. Fundamental understanding of crystal growth behaviors from Si melt is significant for the researchers, who are involved in the development of crystal growth technologies. In the light of Si crystals for solar cells, it is imperative to improve the crystal-quality of Si-multicrystal ingot grown by casting, because it is widely used for solar cell substrates in the present and future. Faceted dendrite has unique structural features and has a potential to be used for controlling the crystal structure in Si-multicrystal ingots. In addition, basically, its growth behavior is fascinating. In this chapter, the growth mechanism of Si faceted dendrite will be described with recent experimental results.

5.1 Introduction

No one doubts that Si-multicrystal is one of the most important materials, along with Si single crystal, for the substrate of solar cells in the future, although other variety of materials will be developed. Currently, a casting method based on unidirectional solidification is commercially operated for producing Si-multicrystal ingots. To prevent metal impurity incorporations, purity of raw Si materials, crucible coating and furnace design have been improved. To reduce dislocation density and thermal stress in an ingot, thermal history during/after crystallization is controlled. Despite such efforts, the solar cell performance of Si-multicrystals is inferior to that of Si single crystals. This fact suggests that there is still room for improvement of crystal-quality of Si-multicrystal ingots. Therefore, the researchers do have to keep trying to improve the crystal-quality of Si-multicrystal ingots and to develop a technology for obtaining such high-quality Si-multicrystal ingots.

Fine control of macro- and microstructures in Si-multicrystal ingots may be needed for the future. The crystal structure of Si-multicrystal is largely different from Si single crystal, as illustrated by formation of grain boundaries and the distribution of crystallographic orientations on a wafer surface.

Generally, some grain boundaries act as recombination centers of photocarriers [1, 2]. It has also reported that dislocation clusters exist around grain boundaries [3] and that subgrain boundaries are generated from grain boundaries, which will be described in detail in Chap. 6 in this book. To reduce the grain boundary density in Si-multicrystal ingots, size of crystal grains should be controlled to be larger during crystallization process. Random distribution of crystallographic orientations on a wafer surface is also not convenient for solar cells. A surface textural structure for light trapping is easily formed by anisotropic chemical etching in Si single crystal solar cells [4]. However, it is difficult to form a uniform surface textural structure in a Si-multicrystal wafer due to the differences in etching rate and morphology between grains. Therefore, another special technique, at higher cost, should be used for obtaining such a good textural structure [5]. If Si-multicrystal ingots with large size grains oriented in one direction were grown by casting method, low-cost and high-performance solar cells will be realized.

Recently, an idea was proposed to obtain structure-controlled Si-multicrystal ingots by casting method [6]. The concept named "dendritic casting method" is schematically shown in Fig. 5.1. Faceted dendrite growth is promoted along the bottom of the crucible in the earlier stage of casting. The growth rate of faceted dendrite is much faster than that of normal crystals, and then large-size grains are formed at the bottom of an ingot. Further, crystallographic orientation vertical to the upper surface of a faceted dendrite is limited to <112> or <110>, due to the unique structural features of faceted dendrites. Therefore, if all grains were formed by faceted dendrites,

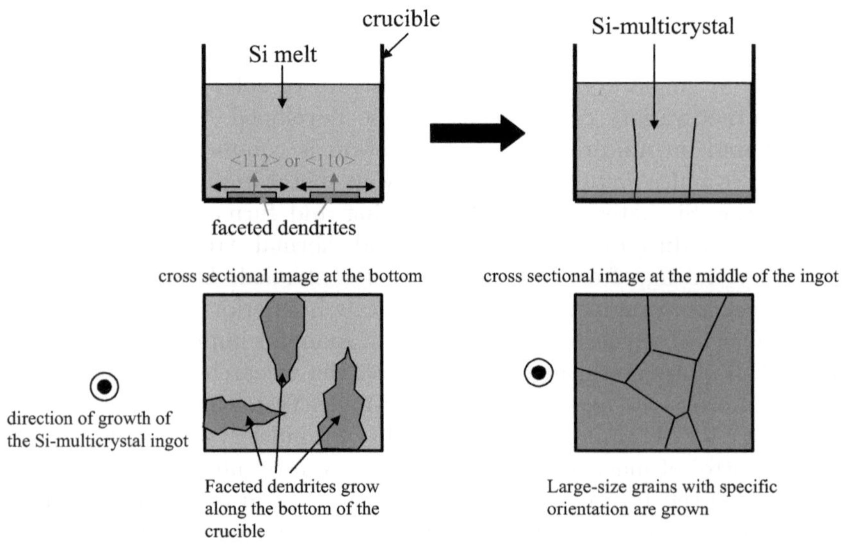

Fig. 5.1. Growth concept for obtaining a Si-multicrystal ingot with large-size grains oriented to one direction

a structure with large-size grains oriented specific direction is formed at the bottom of an ingot. After that, crystallization should be promoted to upper direction from the surface of the dendrites-structure at the bottom. In this way, a Si-multicrystal ingot with large-size grains oriented specific orientation will be obtained. Actually, such a structure-controlled Si-multicrystal ingot was obtained for a small-size ingot by the dendritic casting method [6], and its higher solar cell performance was demonstrated [6]. This technology has a potential to be applied to the growth of industrial-size ingots because the growth concept is very simple: only the control of crystal growth mechanism is required in the earlier stage of casting. In this chapter, the growth mechanism of faceted dendrite will be fundamentally described.

5.2 Twin-Related Dendrite Growth in Semiconductor Materials

Dendrite crystals grow during crystallization from liquid or vapor phase in almost all materials containing metals, semiconductors, oxides, and organic materials. In particular, twin-related dendrites of Si or Ge are so-called "faceted dendrites." The study of faceted dendrite has had a long history since the first report by Billig [7]. Long Ge-ribbons propagated in the <211> directions and bounded by well-developed {111} habit faces were grown by pulling from a seeded undercooled melt [8–10]. The most unique feature of faceted dendrite is existence of {111} parallel twins at the center of faceted dendrites [8–12]. Si faceted dendrites have exactly the same features with Ge faceted dendrite [13–16]. Figure 5.2 shows a typical surface morphology of Si faceted dendrite grown from undercooled melt, which was observed by SEM and analyzed by electron back scattering diffraction pattern (EBSP) method [16]. Two parallel twins shown by white lines in the image exist at the center of the faceted dendrite. Therefore, it is defined that the growth of faceted dendrites is related to parallel twins. The parallel twin formation during melt growth will be described in next section.

The growth model of faceted dendrites was proposed in 1960 [17, 18]. Recently, more real growth model was proposed through the in situ observation experiments of growth processes of Si faceted dendrite [19], although the previous model has been widely accepted. In Sect. 5.4, it will be described how faceted dendrite grows, and the role of parallel twins will be explained.

Si faceted dendrite is applied for growing ribbon sheet crystals [20] and Si-multicrystal ingots for solar cells [6,21]. In the dendritic web growth, after a seed crystal is dipped in the undercooled melt, a faceted dendrite propagates on the melt surface and then, a ribbon sheet crystal is pulled up. However, in the dendritic casting method, faceted dendrites are grown along the bottom of the crucible in the earlier stage of casting. Grain size will be large due to the faster growth of faceted dendrite as compared with equiaxed grains. In addition, grain orientation will be almost the same because the upper

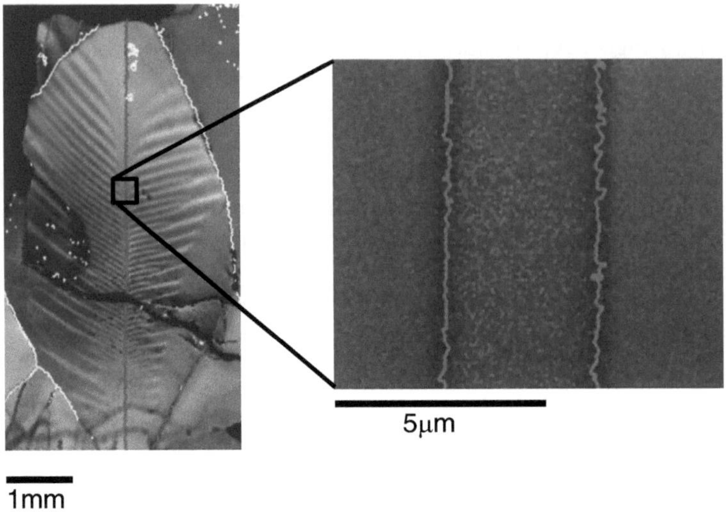

Fig. 5.2. SEM image of Si faceted dendrite grown from undercooled melt. EBSP analysis was performed to find parallel twins (*shown by white lines*). Parallel twins exist at the center of the faceted dendrite [16]

direction of faceted dendrites is limited to <110> or <112>. Subsequently, crystallization will be promoted to upper direction started on the faceted dendrites. In this way, structure-controlled Si-multicrystal ingots are obtained. In both technologies, effective promotion of faceted dendrites in the earlier stage of crystal growth is essential for obtaining high-quality crystals.

5.3 Formation Mechanism of Parallel Twins During Melt Growth Processes

Two prerequisites for growing faceted dendrite are well known:

1. More than two {111} parallel twins in a growing crystal
2. Sufficient undercooling in the melt

Researchers of materials science can easily imagine deformation twins, which are generated in a crystal when subjected to external stress. However, the parallel twins formation related to faceted dendrite growth during melt growth processes of Si has not been well understood, although a few ideas were reported [15, 22]. Fujiwara et al. directly observed the growth behavior of Si crystal and showed how faceted dendrite grew [16]. Figure 5.3 shows images of the solid–liquid growth interface during the melt growth of Si [16]. The most stable {111} faces are appeared on the growth surface, and then the zigzag shaped faceted interface is formed. The shape of the interface changed from Fig. 5.3a–d with decreasing melt temperature. The faceted interface was first

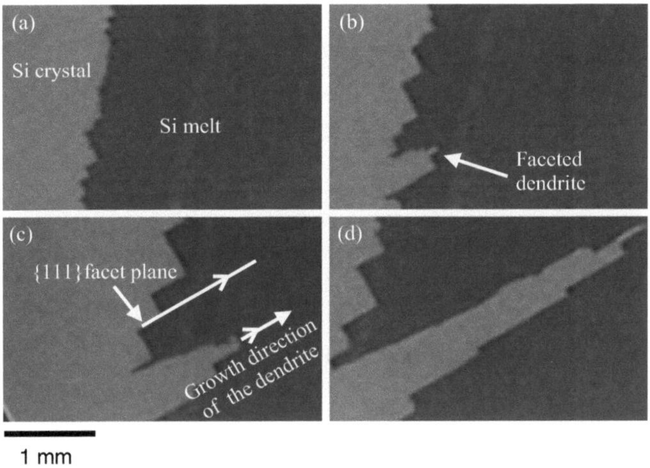

Fig. 5.3. Growth process of Si crystal from undercooled melt [16]. (**a**) Crystal is growing with faceted interface. (**b**)–(**d**) Faceted dendrite grows from a part of faceted interface. The direction of the growth of faceted dendrite is parallel to the {111} facet face on the interface [16]

observed, as shown in Fig. 5.3a. At this time, a faceted dendrite did not grow because the melt was not sufficiently cool, although parallel twins might have already been formed in the crystal. When the melt temperature had decreased sufficiently, a faceted dendrite grew in a constant direction from part of the faceted interface, as shown in Fig. 5.3b–d. Note that the direction of growth of the faceted dendrite was parallel to the facet plane, as shown in Fig. 5.3c. This means that the parallel twins at the center of the faceted dendrite are formed parallel to the facet plane. This fact is significant to consider the parallel twins formation.

Figure 5.4 schematically shows a model for generating parallel twins during the melt growth of Si [16]. Figure 5.4a shows a solid/liquid growth interface of Si. It has been explained theoretically that the crystal growth interface of Si is faceted and that {111} planes appear on the surface [23]. Such a faceted interface of Si was directly observed during in situ experiments as shown in Fig. 5.3 [16, 24, 25]. The shape of the faceted interface is dependent on the crystal growth orientation; an example of this is shown in Fig. 5.4. Since the growth interface is faceted, crystal growth is always promoted on the {111} facet planes. When an atom attaches on a facet plane with a twin relationship, a layer that maintains the twin relationship is formed on the facet plane after lateral growth, and then one twin boundary is generated on the layer, as shown in Fig. 5.4b. One can expect that the atoms are often deposited on the growth interface with twin relationship because the grain boundary energy of Si {111} twin boundary is close to zero ($\sim$30 mJ m^{-2}) [26]. On one hand, such the twin configuration of adatoms may be rearranged so that

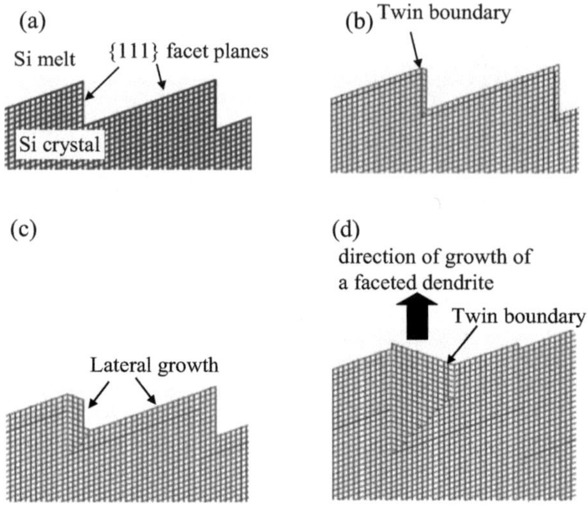

Fig. 5.4. Model of parallel-twin formation during crystal growth from Si melt. The growth interface is faceted as shown in (**a**). when a twin boundary is accidentally formed on a {111} facet face, another twin boundary is formed parallel to the first twin after lateral growth is promoted (from (**b**) to (**d**)). The direction of growth of the faceted dendrite should be parallel to the {111} facet plane in this model [16]

adatoms obey the epitaxial configuration at the crystal surface to minimize the Gibbs free energy of the system when the driving force for crystallization is very low at near equilibrium condition. On the other hand, when the driving force is large, high undercooling, the growth of the layer with twin relationship should be promoted because of the faster growth kinetics and the larger energy gain due to the crystallization. If the crystal growth continues in the lateral-growth mode (Fig. 5.4c), we note that another twin boundary may form parallel to the previous twin, as shown in Fig. 5.4d. In this model, the formation of two parallel twins is certain when one twin is formed on a facet plane at the growth surface. When the undercooling of the melt is sufficient, faceted dendrite growth starts from these two parallel twins. In this mechanism, the growth direction of the faceted dendrite is parallel to the facet planes on the growth interface, as shown in Fig. 5.4d. Thus, this model explains how and why parallel twins are generated in the growing crystal. This agrees with the explanation of the experimental result shown in Fig. 5.3.

The required undercooling for growing faceted dendrite was also investigated [6, 15, 27]. Some experimental data showed that the range of undercooling was $10\,\mathrm{K} < \Delta T < 100\,\mathrm{K}$. When the degree of undercooling is larger than 100 K, twin free dendrite is appeared with changing the growth mode from lateral growth to continuous growth [15, 28].

5.4 Growth Mechanism of Si Faceted Dendrite

The growth model of faceted dendrites was proposed in 1960 [17, 18]. It was well considered about the role of parallel twins on the growth of faceted dendrite. However, no one has obtained the direct evidences to prove the growth model. Recently, Fujiwara et al. succeeded in observing the growth process of faceted dendrite in detail, and a growth model was presented on the basis of experimental evidence [19]. Here, the experimental results of direct observations are shown and both the models are explained.

Figure 5.5a shows a typical growth behavior of a Si faceted dendrite [19]. It is shown that the faceted dendrite grew faster than the rest of the crystal. The faceted dendrite continuously propagated both in the direction of rapid growth and the direction perpendicular to the rapid growth. Crystallographic orientation analysis of a small area at the center of the dendrite (indicated by the small box in the right image of Fig. 5.5a showed that the two parallel twins exist (Fig. 5.5b)), and rapid-growth direction of the faceted dendrite was <112> and that the direction of the side growth was <111>, as schematically shown in Fig. 5.5. Importantly, the faceted dendrite grew not only in the rapid-growth direction of <112> but also in the <111> direction perpendicular to the rapid-growth direction. In Fig. 5.5a, it is in fact shown that the growth occurs also in the direction parallel to the {111} planes. This is an observation from the direction parallel to {111} twins in the faceted dendrite.

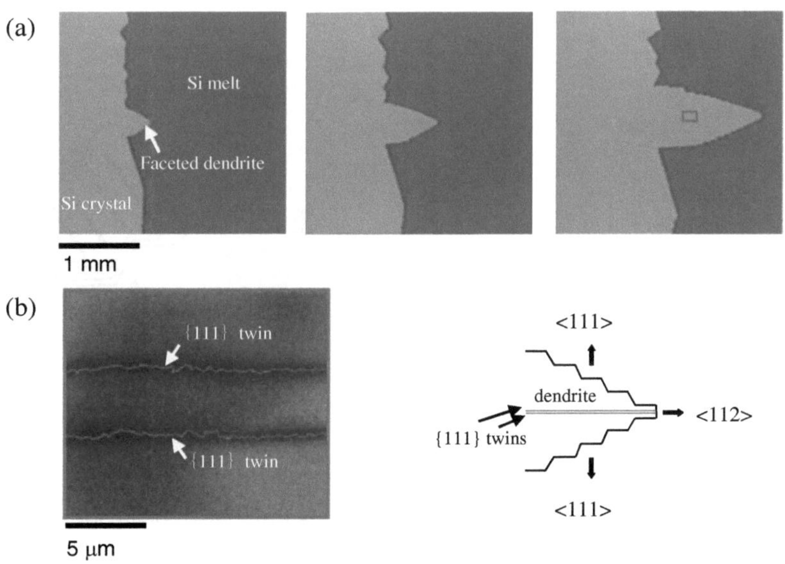

Fig. 5.5. (**a**) A typical growth behavior of a Si faceted dendrite. (**b**) Result of EBSP analysis in the center of the faceted dendrite. Two parallel twins exist. Growth orientation of the faceted dendrite is schematically shown [19]

To obtain more information on the growth behavior of the faceted dendrite, another observation was performed from perpendicular to the {111} twins. Figure 5.6a shows the experimental procedure [19]. A piece of Si {111} wafer was set in a crucible, and the sample was carefully heated to melt in such a way that an unmelted part remained. Then, the sample was cooled rapidly to promote dendrite growth. Crystal growth started from the unmelted "seed" wafer. In this experiment, when the dendrite appears during crystal growth, the {111} parallel twins are generated parallel to the {111} crystal surface. The direction of observation of the growing dendrite is schematically summarized in Fig. 5.6a. Figure 5.6b shows the growth process of a faceted dendrite. It revealed the unique growth behavior of a faceted dendrite [19]. Note that triangular corners with an angle of 60° were formed at the tip of the dendrite and also that the direction of the corners alternately changed from outward to forward to the direction of growth, as highlighted in Fig. 5.6b. Such a triangular corner is not formed in the previous model of Hamilton and Seidensticker [17]. From Figs. 5.5a and 5.6b, one thus obtained three important facts for the growth behavior of faceted dendrites; (1) a faceted dendrite can propagate in the <111> direction, which is perpendicular to the rapid-growth direction of <112>, (2) triangular corners with an angle of 60° are formed at the tip of the faceted dendrite, and (3) the direction of the 60° corners changes during growth.

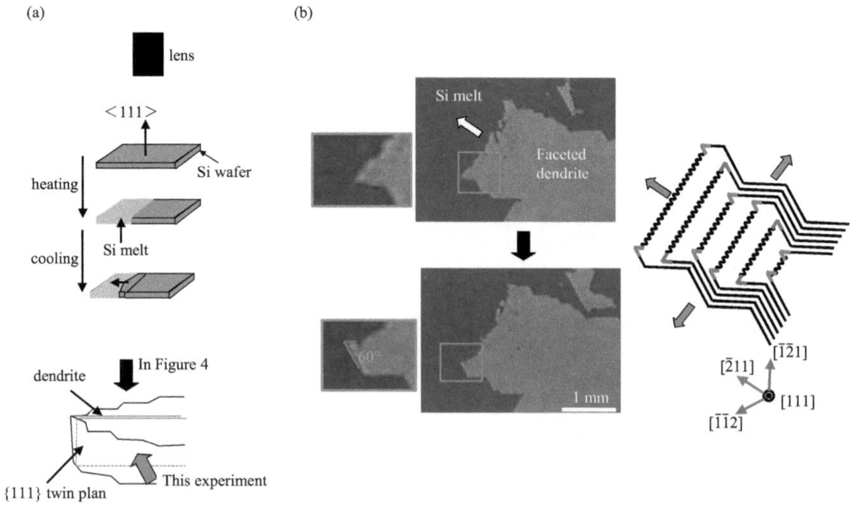

Fig. 5.6. (**a**) Experimental procedure for observing the growth behavior of a faceted dendrite perpendicular to the {111} twins in the faceted dendrite. The direction of observation of the growing dendrite is schematically shown. (**b**) Growth behavior of Si faceted dendrite observed perpendicular to the {111} twins. Note that a triangular corner of angle 60° is formed at the growth tip and that the direction of the corner changes with growth. The growth processes are schematically shown [19]

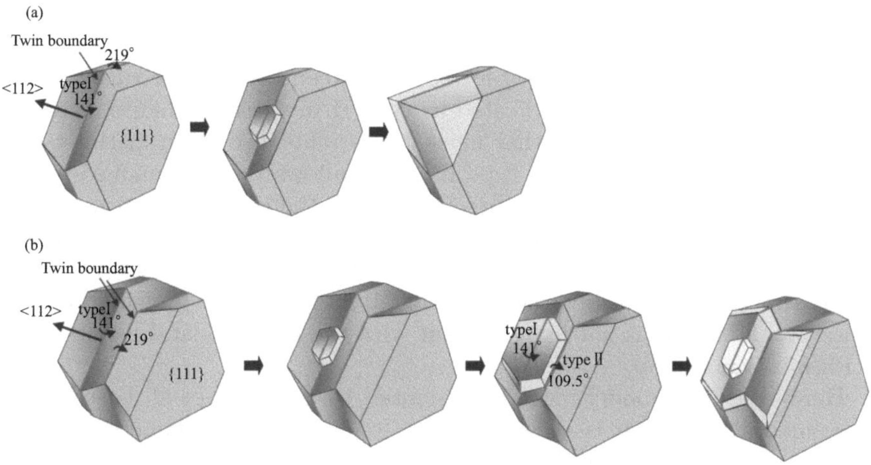

Fig. 5.7. (**a**) Schematic images of growth of a crystal with one twin. The crystal is bounded by {111} habit planes. It is considered that the crystal is growing in one direction for the sake of simplicity. A reentrant corner of angle 141° (type I) appears at the growth surface. Rapid growth occurs at the corner until a triangular corner is formed. (**b**) Schematic images of the growth of a crystal with two twins, which was proposed by Hamilton and Seidensticker [19]. They assumed that nucleation events easily occurred at the reentrant type I corner and that crystal growth on the {111} flat surface was difficult. Nucleation at the type I corner leads to the formation of a new reentrant corner of angle 109.5°, marked type II (shown in the third image). They considered that nucleation also occurred at the type II corner, which permitted continuous propagation of the crystal in the lateral direction before the type I corner disappeared (*right*). Importantly, no triangular corner is formed during dendrite growth in their model [19]

Here, the growth of the Si crystal with one and two twins is reviewed according to the explanation of Hamilton and Seidensticker [17]. It is well known that {111} habit planes appear on the crystal surface during the crystallization of Si [17,23]. Figure 5.7a shows the equilibrium form of the crystal with one twin, which is bounded by {111} planes. Now, we consider the situation that the crystal is growing in one direction for the sake of simplicity. One reentrant corner with an external angle of 141° (type I) appears at the growth surface. Nucleation readily occurs at the reentrant corner compared with that at {111} flat surfaces [17,18]. Therefore, the crystal rapidly grows at the reentrant corner, and a triangular crystal with a 60° corner is finally formed. Rapid growth has ceased at this time due to the disappearance of the reentrant corner. This is why the faceted dendrite does not appear when the crystal has only one twin. Next, it is explained that the growth model of a crystal with two parallel twins presented by Hamilton and Seidensticker [17], which corresponds to the growth model of a faceted dendrite presented in 1960. Figure 5.4b shows a two-twin crystal bounded by {111} habit planes.

They assumed that the crystal growth on {111} flat surface hardly occurred. Rapid growth occurs at the reentrant type I corner, similar to that in the crystal with one twin. The new layer forms a new reentrant corner with an angle of 109.5° at the next twin, indicated as type II in the third figure. Hamilton and Seidensticker considered that nucleation events also occur at this type II corner. Thus, the creation of a type II corner allows the continuous propagation of the crystal in the lateral direction before the type I corner disappears, in contrast to the crystal with only one twin. Most important in their model, the reentrant type I corner does not disappear during dendrite growth, which means that no triangular corners appear at the tip of the growing dendrite. However, the model is not in agreement with the experimental results shown in Figs. 5.5a and 5.6b.

Here, a growth model recently presented by Fujiwara et al. based on the experimental evidences is shown in Fig. 5.8 [19]. In the explanation, the two twins are distinguished by labeling them $twin_1$ and $twin_2$ (Fig. 5.8a). Rapid growth at the type I corner leads to the formation of a triangular corner with an angle of 60° at the growth tip of the faceted dendrite (Fig. 5.8b, c), similar to that in the crystal with one twin, which was observed in experiment. Crystal growth can continue on the {111} flat surface, although the rapid growth is inhibited due to the disappearance of the reentrant corner. In the previous model, crystal growth on the {111} flat surface seemed hardly to

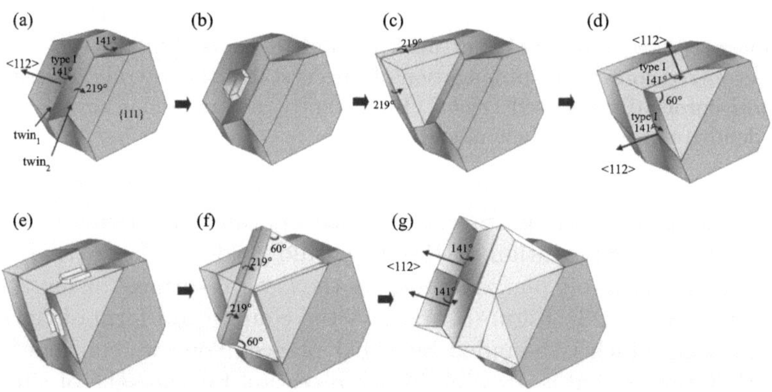

Fig. 5.8. (**a**) Equilibrium form of a crystal with two twins, which is bounded by {111} habit planes. (**b**), (**c**) A triangular corner is formed due to the rapid growth at the type I corner at $twin_1$. Crystal growth can continue on the {111} flat surface, although the rapid growth is inhibited because of the disappearance of the type I corner. (**d**) When the triangular crystal propagates across $twin_2$, two type I corners are newly formed at $twin_2$. (**e**), (**f**) Rapid growth occurs at the two type I corners again, and triangular corner is formed. (**g**) After propagation of the crystal, a type I corner is formed at $twin_1$. The faceted dendrite continues to grow by repeating the process from (**a**) to (**g**) [19]

occur. However, it was shown that significant crystal growth occurred on the {111} surface in the undercooled melt (shown in Fig. 5.6a). After propagation of the crystal, note that two type I corners are newly formed on the growth surface at $twin_2$ (Fig. 5.8d). Thus, rapid growth occurs there again, and triangular corners with an angle of 60° are formed in the same manner as before (Fig. 5.8e, f). The direction of the 60° corner has been changed as shown in Fig. 5.5d–f, which is in agreement with the experimental results. Crystal growth is promoted on the {111} flat surface again, leading to the formation of a new reentrant type I corner at $twin_1$ (Fig. 5.8g). The faceted dendrite continues to grow by repeating the same processes and forming the 60° corner at the growth tip. In the previously presented growth models [17,18], they assumed that the crystal propagated laterally owing to the formation of the type II corner at $twin_2$ and that crystal growth on the {111} flat surface was negligible. In such processes, the faceted dendrite propagates only in the rapid-growth direction. However, the faceted dendrite grew not only in the direction of rapid growth but also perpendicular to that direction, which means that crystal growth on the {111} flat surface readily occurs in undercooled melt. The significance of the existence of two twins is not only the formation of type II corners, but also the alternate formation of type I corners at each twin. The model shown in Fig. 5.8 fully explains the experimental evidence of the growth behavior of the faceted dendrite. Furthermore, the twin-related growth model can be applied not only to Si but also for other faceted materials.

References

1. A. Fedotov, B. Evtodyi, L. Fionova, Yu. Ilyashuk, E. Katz, L. Polyak, Phys. Stat. Sol. (a) **119**, 523 (1990)
2. Z. Wang, S. Tsurekawa, K. Ikeda, T. Sekiguchi, T. Watanabe, Interf. Sci. **7**, 197 (1990)
3. B. Ryningen, K.S. Sultana, E. Stubhaug, O. Lohne, P.C. Hjemas, in *Proceedings of the 22th European Photovoltaic Solar Energy Conference*, Milan, 2007, p. 1086
4. F. Restrepo, C.E. Backus, IEEE Trans. Electron Devices **ED-23**, 1195 (1976)
5. H.F.W. Dekkers, F. Duerinckx, J. Szlufcik, J. Nijs, in *Proceedings of the 16th European Photovoltaic Solar Energy Conference*, Glasgow, 2000, p. 1532
6. K. Fujiwara, W. Pan, N. Usami, K. Sawada, M. Tokairin, Y. Nose, A. Nomura, T. Shishido, K. Nakajima, Acta Mater. **54**, 3191 (2006)
7. E. Billig, Proc. R. Soc. **A229,** 346 (1955)
8. E. Billig, P.J. Holmes, Acta Metall. **5**, 53 (1957)
9. E. Billig, Acta Metall. **5**, 54 (1957)
10. A.I. Bennett, R.L. Longini, Phys. Rev. **116**, 53 (1959)
11. N. Albon, A.E. Owen, J. Phys. Chem. Solids **24**, 899 (1962)
12. D.R. Hamilton, R.G. Seidensticker, J. Appl. Phys. **34**, 1450 (1963)
13. S. Ohara, G.H. Schwuttk, J. Appl. Phys. **36**, 2475 (1965)
14. T.N. Tucker, G.H. Schwuttk, J. Electrochem. Soc. **113**, C317 (1966)
15. K. Nagashio, K. Kuribayashi, Acta Mater. **53**, 3021 (2005)

16. K. Fujiwara, K. Maeda, N. Usami, G. Sazaki, Y. Nose, K. Nakajima, Scripta Mater. **57**, 81 (2007)
17. D.R. Hamilton, R.G. Seidensticker, J. Appl. Phys. **31**, 1165 (1960)
18. R.S. Wagner, Acta Metall. **8,** 57 (1960)
19. K. Fujiwara, K. Maeda, N. Usami, K. Nakajima, Phys. Rev. Lett. **101**, 055503 (2008)
20. D.L. Barrett, E.H. Myers, D.R. Hamilton, A.I. Bennett, J. Electrochem. Soc. **118**, 952 (1971)
21. K. Fujiwara, W. Pan, K. Sawada, M. Tokairin, N. Usami, Y. Nose, A. Nomura, T. Shishido, K. Nakajima, J. Cryst. Growth **292**, 282 (2006)
22. R-Y. Wang, W-H. Lu, L.M. Hogan, Metall. Mater. Trans. **28A**, 1233 (1997)
23. K.A. Jackson, *Growth and perfection of crystals* Wiley, New York, 1958), p. 319
24. K. Fujiwara, K. Nakajima, T. Ujihara, N. Usami, G. Sazaki, H. Hasegawa, S. Mizuguchi, K. Nakajima, J. Cryst. Growth **243**, 275 (2002)
25. K. Fujiwara, Y. Obinata, T. Ujihara, N. Usami, G. Sazaki, K. Nakajima, J. Cryst. Growth **266**, 441 (2004)
26. M. Kohyama, R. Yamamoto, M. Doyama, Phys. Stat. Sol. **B138**, 387 (1986)
27. K. Fujiwara, K. Maeda, N. Usami, G. Sazaki, Y. Nose, A. Nomura, T. Shishido, K. Nakajima, Acta Mater. **56,** 2663 (2007)
28. G. Devaud, D. Turnbull, Acta Metall. **35**, 765 (1987)

6

Fundamental Understanding of Subgrain Boundaries

Kentaro Kutsukake*, Noritaka Usami, Kozo Fujiwara, and Kazuo Nakajima

Abstract. Generally, Si multicrystals, which are grown by a casting method using a crucible, contain many grain boundaries (GBs) and crystal grains with various orientations. Since the grain size has increased as a result of improving in the growth technique, instead of GBs, subgrain boundaries (sub-GBs) have become major defects acting as recombination centers for photogenerated carriers. In this chapter, the study of sub-GBs in Si multicrystals is comprehensively reviewed with the authors' current results.

6.1 Introduction

Subgrain boundaries (sub-GBs) are major and basic defects in metallic materials, particularly in constructional materials, and have been well investigated, because they strongly affect the mechanical properties of the materials. In Si multicrystals, sub-GBs have also recently become major defects, since the grain size has increased as a result from improving the growth technique. In this section, we present the results of investigations of sub-GBs in metallic materials, which can be applied to Si multicrystals used in solar cells.

Sub-GBs, which are sometimes called small-angle boundaries, are boundaries between subgrains with small crystallographic misorientations in the crystal grains. Sub-GBs are two-dimensional defects similar to GBs, however, they are distinguished from GBs by their mechanical, electrical, and magnetic properties. From the viewpoint of the microstructure, sub-GBs consist of dislocations with array-like configurations that minimize their elastic energy as illustrated in Fig. 6.1 [1]. In other words, clustered dislocations are rearranged into sub-GBs, as a result of their motion. For the simplest dislocation model of sub-GBs, the distance, D, between adjacent dislocations is expressed as

$$D = \frac{b}{2\sin\left(\frac{\Delta\theta}{2}\right)}, \tag{6.1}$$

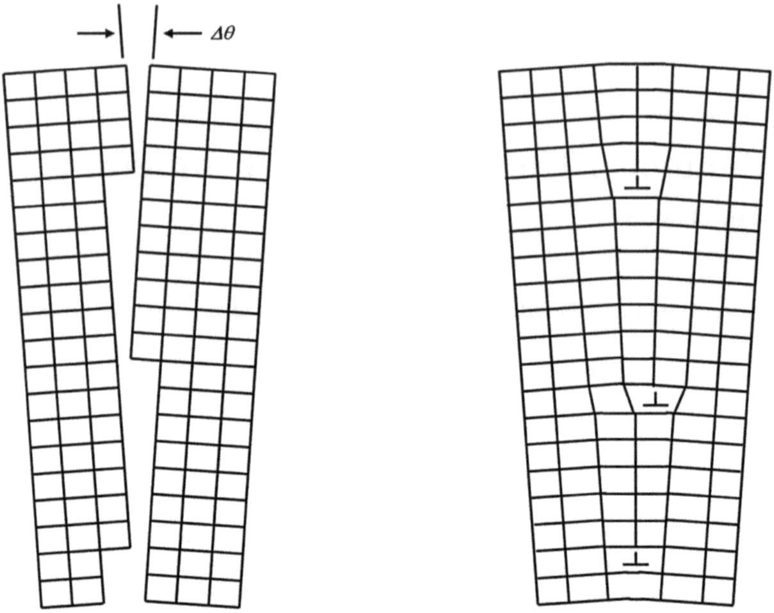

Fig. 6.1. Schematic illustration of dislocation model of sub-GB

where $\Delta\theta$ is the misorientation between the two subgrains sandwiching a sub-GB, and b is the length of the Burgers vector of the dislocations. This expression shows that the dislocation density on a sub-GB increases with increasing misorientation, which results in the variation of the properties of sub-GBs, as discussed in the following sections. The type of dislocations comprising a sub-GB systematically changes with the type of sub-GB. In the case of a tilt boundary, i.e., when the rotation axis of the misorientation is parallel to the boundary plane, a sub-GB consists of an array of edge dislocations. However, in the case of a twist boundary, i.e., when the rotation axis of the misorientation is perpendicular to the boundary plane, the sub-GB consists of a network of screw dislocation arrays. With increasing misorientation, the array-like dislocation structure vanishes and the microstructure of the boundary becomes similar to the structure of a random (general) GB, which results in the distinguishing properties of the sub-GB vanishing and the properties becoming similar to those of random (general) GBs. The physically meaningful maximum misorientation, $\Delta\theta_{\mathrm{max}}$, i.e., the boundary between sub-GBs and GBs, has been estimated to be 15° using the Brandon criterion [2], which is expressed as

$$\Delta\theta_{\mathrm{max}} = \theta_0 \Sigma^{-1/2}, \tag{6.2}$$

where θ_0 is an experimentally determined constant (generally 15° is used for metallic materials) and Σ is the sigma value in coincidence site lattice (CSL) theory ($\Sigma = 1$ for sub-GBs). For more detailed descriptions of the microstructures of sub-GBs and dislocations, the books of Hirth and Lothe [1] and Sutton and Ballufi [3] can be referred.

6.2 Structural Analysis of Subgrain Boundaries

As mentioned in the previous section, a sub-GB is a boundary between subgrains with small crystallographic misorientation. Therefore, to investigate sub-GBs crystallographically, the detection of small misorientations is necessary. Table 6.1 summarizes the methods of analyzing crystallographic orientation from the viewpoints of angular resolution and spatial resolution.

Scanning electron microscopy (SEM) equipped with electron backscattering pattern (EBSP) analysis is widely used to investigate multicrystalline structures. In particular, it is a powerful tool for visualizing the grain orientation and grain shape, and for characterizing the GB character. One of the applications of Si multicrystals is to study CSL boundaries, which are the GBs with a special grain configuration and a low GB energy. By combining EBSP analysis with electron beam induced current (EBIC) measurement, Tsurekawa and coworkers revealed that CSL boundaries in Si multicrystals, in particular $\Sigma 3$ boundaries, exhibit less electrical activity than random GBs [4, 5]. Chen et al. also combined EBSP analysis with EBIC measurement and reported that the low electrical activity of CSL boundaries is preserved even after contamination with iron [6]. As remarked here, SEM-EBSP analysis has helped us to understand the properties of CSL boundaries in Si multicrystals. It can be applied to detect sub-GBs, and electrical investigations of the sub-GBs characterized by SEM-EBSP analysis have been reported [7, 8]. However, the angular resolution of SEM-EBSP analysis is about 1° under low magnification. SEM-EBSP analysis, therefore, cannot detect sub-GBs with small misorientation; however, sub-GBs affect the electrical performance of solar cells even when the misorientation is less than 1°.

Transmission electron microscopy (TEM) is an attractive tool for microscopically observing the individual dislocations comprising sub-GBs. The

Table 6.1. Angular resolution and spatial resolution for different methods for analyzing crystal orientation

	SEM-EBSP	TEM	XRD
Spatial resolution (μm)	<1	<0.01	>100 (Lab. scale) <1 (Synchrotron)
Angular resolution (degree)	~1 (low magnification)	~0.05 (Kikuchi line)	<0.01 (Rocking curve)
Comments	Good visualization	Individual dislocations can be observed	Very high angular resolution

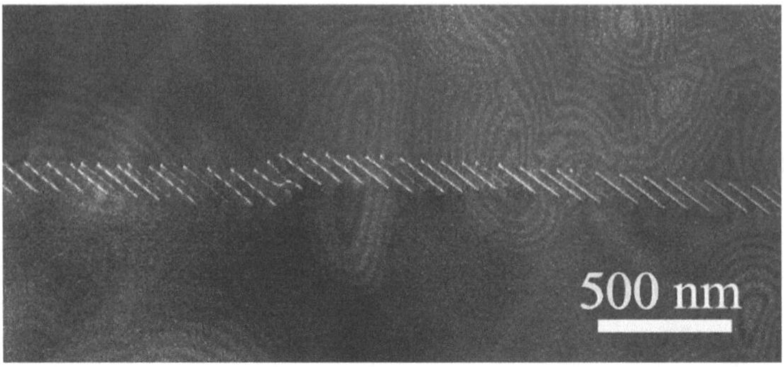

Fig. 6.2. TEM image of sub-GB

array-like configuration of dislocations on a sub-GB, which is schematically illustrated in Fig. 6.1, can be observed by TEM, as shown in Fig. 6.2 [9]. Moreover, the crystal orientation can be analyzed by combining a TEM image with a Kikuchi line pattern or a diffracted electron-beam pattern. Such analysis has confirmed the validity of (6.1), which expresses the relationship between misorientation and the distance between adjacent dislocations. However, while the angular resolution of TEM is sufficiently high for sub-GB investigation, its measurable area is too small relative to the surface area of a solar cell wafer.

Recently, crystal orientation analysis through spatially resolved X-ray rocking curves has been performed to detect structural modifications in local structures in multicrystals with angular resolution better than 0.01° [10]. This method of analysis can be applied to investigate sub-GBs in Si multicrystals, if combined with an appropriate sample. The sample should contain crystal grains with diameter comparable to or larger than the incident X-ray diameter, which is typically less than 1 mm, for alignment of the optical configuration. In addition, ideally, the crystal grains are controlled to have the same orientation to allow the measurement of many crystal grains without changing the optical configuration.

Kutsukake et al. [11] reported the distribution of sub-GBs in Si multicrystals in a wafer-scale area using Si multicrystals grown by the dendritic casting method [12,13] to contain crystal grains aligned in the [110] direction. Figure 6.3 illustrates the optical configuration used in the X-ray rocking curve analysis. The rocking curve measurements were carried out by scanning the incident angle, ω, with a fixed diffraction angle, 2θ, so that the diffraction from each subgrain can be distinguished. Figure 6.4 shows a typical rocking curve profile containing multiple peaks. Since each diffraction peak originates from one subgrain, the number of sub-GBs within the incident X-ray spot is estimated to be $(n-1)$, where n is the number of peaks. In addition, the maximum angular difference, $\Delta\omega$, is regarded as the sum of the angular differences of sub-GBs in the spot. By performing the above measurements on the sample surface, the sub-GB density can be imaged. Figure 6.5 shows (a)

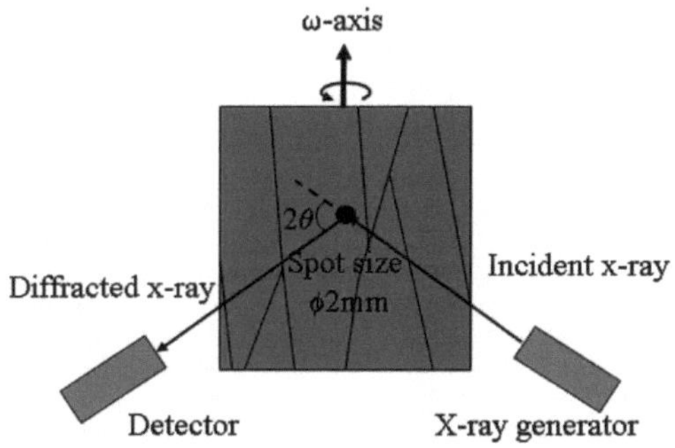

Fig. 6.3. Schematic illustration of optical configuration used in X-ray rocking curve analysis

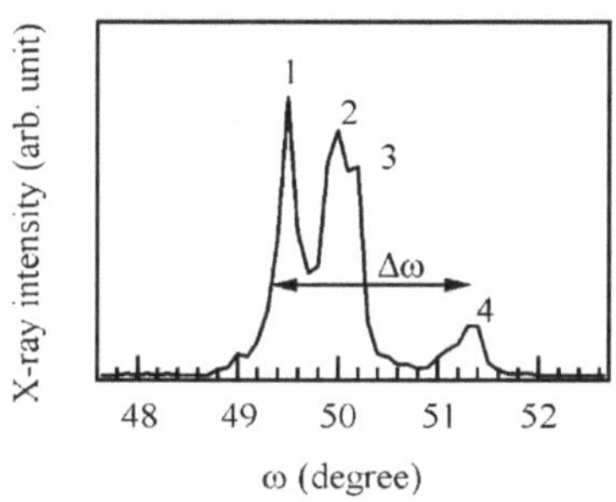

Fig. 6.4. Typical rocking curve profile containing multiple peaks

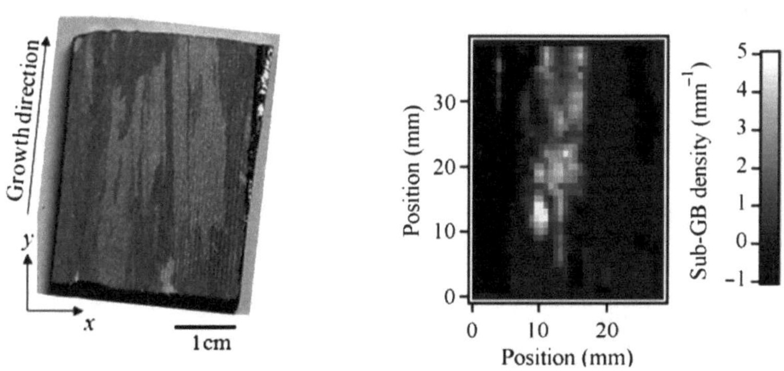

Fig. 6.5. (**a**) Photographic image of a sample and (**b**) distribution of sub-GBs

a photographic image of a sample and (b) the spatial distribution of sub-GB density, in which the dark region with the sub-GB density of −1 corresponds to positions where the diffraction condition was not satisfied and the gray region with the sub-GB density of 0 corresponds to single-crystals without

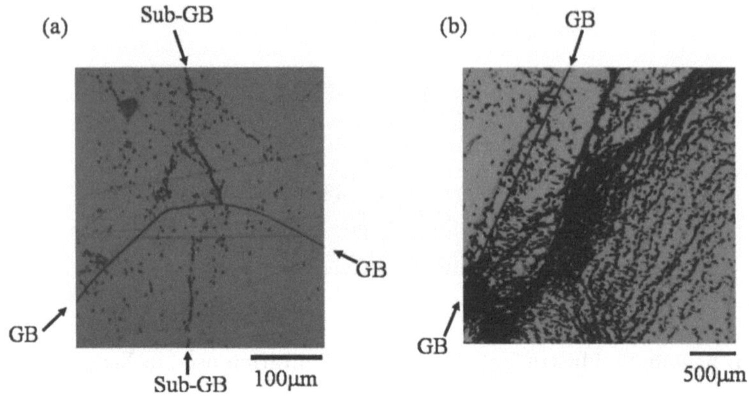

Fig. 6.6. Typical images of (**a**) linelike and (**b**) clustered etch pits

any subgrains. It was found that sub-GBs are not homogeneously distributed but are densely located within a narrow area that extends in the growth direction over several GBs. This tendency appears to be due to the same mechanism by which dislocations tend to lengthen in the growth direction during crystal growth to decrease its length. As demonstrated above, X-ray rocking curve measurement is useful for the structural investigation of sub-GBs; however, its spatial resolution is much larger than that of SEM and TEM. Therefore, it is important to choose and combine methods of crystal orientation analyses depending on the subject and sample.

We briefly comment on etch pit observation, even though it is not a method of crystal orientation analysis. Numerous etch pit studies have been reported for Si wafers, in which dislocations and their clusters were observed using the etching solutions of Sirtl [14], Dash [15], Sopori [16], and others. In Si multicrystals, etch pits, i.e., dislocations, are frequently observed in the vicinity of GBs and inclusions such as Si_3N_4, SiC and so forth [17]. This suggests that GBs and inclusions act as a source of dislocations. The mechanism of dislocation generation is discussed in detail in the following section. Figure 6.6 shows typical images of line-like and clustered etch pits etched by Sopori solution. As mentioned above, line-like etch pits correspond directly to sub-GBs. Moreover, in the area of clustered dislocations, multiple peaks are frequently observed in the X-ray rocking curve profile. This suggests that clustered dislocations are reconstructed into sub-GB structures in the microscopic scale.

6.3 Electrical Properties of Subgrain Boundaries

Investigation of the electrical properties of sub-GBs is an important subject for improving the performance of Si multicrystal solar cells. In particular, the carrier recombination activity of sub-GBs has been well investigated through EBIC measurements and as well as by photoluminescence (PL) and electroluminescence (EL) spectroscopy and imaging.

Ikeda et al. investigated the effect of the misorientation on carrier recombination activity using a unique method; they formed artificial sub-GBs with a controlled misorientation using a wafer-bonding technique and measured the carrier recombination activity through EBIC analysis [18]. They revealed that sub-GBs show active carrier recombination even if the misorientation is only 0.1°, which increases with increasing the misorientation and saturates when the misorientation is approximately 2°. This suggests that the electrical properties of sub-GBs vary with their structure, i.e., their misorientation. Among other EBIC studies, the effects of contamination reported by Chen et al. should be noted [7]. They measured the change in the EBIC of sub-GBs with a misorientation of about 1°, CSL boundaries and random GBs before and after contamination with iron. The electrical activity of sub-GBs increased after contamination and, more importantly, sub-GBs exhibited the largest electrical activity both before and after contamination compared with CSL boundaries and random GBs. This means that sub-GBs have a greater effect than GBs influencing solar cell performance.

Sugimoto et al. investigated intragrain defects in Si multicrystals through PL mapping tomography [8]. They measured the spatial distribution of PL intensity filtered with a transmission band of 1,050–1,230 nm to extract the band edge emission. By repeating this measurement and sample thinning by chemical etching, they obtained three-dimensional images of the PL intensity. PL dark patterns, which indicate a plane-like structure extending in the crystal growth direction, were observed in regions where the minority carriers have a short diffusion length, and they confirmed that the pattern originated from the intragrain defects. By performing low-temperature PL spectroscopy, SEM-EBSP analysis and etch pit observation, they concluded that the defects are metal contaminated dislocation clusters that originate from sub-GBs. It should be noted that the GBs did not exhibit strong PL dark patterns, in contrast to the pattern from sub-GBs. This means that sub-GBs are more critical defects than GBs in terms of carrier recombination sites in solar cells.

Recently, PL and EL imaging techniques using a high-spatial-resolution CCD camera have been developed to rapidly obtain luminescence images [19, 20]. PL imaging can be performed on wafer samples without a solar cell structure, but for quantitative analysis the inhomogeneities of excitation light, surface light reflectance and surface carrier recombination should be considered. For EL imaging, a solar cell structure is necessary. In other words, solar cell parameters can be estimated. Würfel et al. proposed a technique for detecting shunting sites using the ratio of EL images taken with different pass filters. They showed that areas with a small ratio of shorter wavelength EL can be assigned as shunting sites [21].

To clarify the impact of sub-GBs on solar cell performance, in particular, the shunting effect, Kutsukake et al. performed EL imaging on a small solar cell sample, in which the distribution of sub-GBs was specified by using X-ray

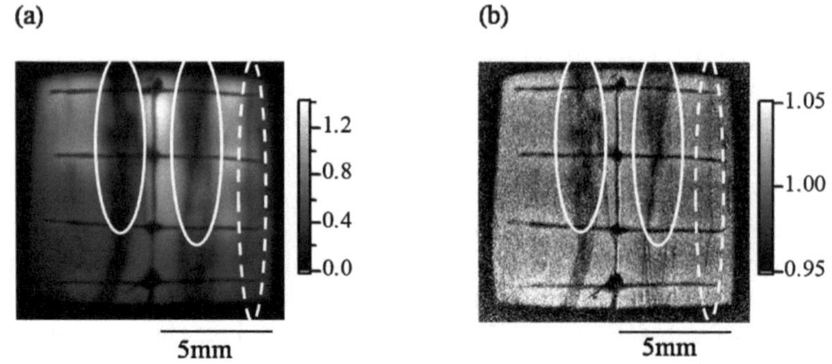

Fig. 6.7. (**a**) EL image taken with a 960-nm-long pass filter and (**b**) ratio of EL intensities taken with 800-and 960-nm-long pass filters. The areas in *solid ellipses* correspond to areas containing sub-GBs, and the areas in *broken ellipses* correspond to area with random GBs

rocking curve measurements [11]. Figure 6.7 shows (a) an EL image taken with a 960-nm-long pass filter and (b) the ratio of EL intensities taken with 800- and 960-nm-long pass filters. In the image taken with a 960-nm-long pass filter, areas with dense sub-GBs are shown as dark area, similar to the case of PL imaging. Further, in the image showing the EL ratio, the area with dense sub-GBs was found to be dark, i.e., the ratio of the shorter wavelength EL was small. On the basis of the report of Würfel et al., sub-GBs are expected to act as shunts in solar cells. Importantly, the area with dense sub-GBs was found to be darker than that with random GBs. This shows that the shunting effect of sub-GBs is stronger than that of random GBs.

As reviewed above, sub-GBs are concluded to be most serious defects preventing the improvement of overall performance of solar cells based on Si multicrystals. We, therefore, suppress the generation of sub-GBs on the basis of fundamental knowledge of its mechanism. In the next section, the generation mechanism of sub-GBs is discussed.

6.4 Origin of Generation of Subgrain Boundaries: Model Crystal Growth

Sub-GBs are expected to be formed during a high-temperature growth process by the stabilization of dislocations, to form array-like configuration that minimizes the self-energy. To form dislocations, the existence of a source with a discontinuous crystal lattice is required, because extremely large stress, typically in the order of the modulus of transverse elasticity, is necessary to form a dislocation from a perfect crystal lattice. In the case of Si multicrystal growth, possible sources are the inner wall of the crucible, inclusions and GBs.

In particular, regarding GBs, numerous observations and theoretical models of dislocation generation have been reported for metals and their compounds. For example, in situ direct observations of a thin sample with unidirectional tensile or compressive stress through TEM [22] and X-ray topography [23] have revealed that a GB ledge structure submitted to concentrated stress can generate dislocations. In analytical treatments of the generation of dislocations from GBs, Varin et al. calculated the critical shear stress required to generate dislocations by consideration of the strain field caused by extrinsic GB dislocations [24], and Bata and Pereloma calculated the critical shear stress by treating the GBs as an array of edge dislocations [25].

On the basis of former studies of metals and their compounds, in the growth process of bulk Si multicrystals, GBs are expected to be the source of lattice dislocations, i.e., the source of sub-GBs. However, there is a large difference between the conditions of a thin sample in a chamber and a bulk sample in a furnace; thus, there is no guarantee that the phenomena observed in TEM and X-ray topography take place during bulk crystal growth. Therefore, defects and structure characterization tracking in the growth direction for a grown bulk Si multicrystal is a reliable method of investigating the mechanism of sub-GB generation. However, systematic investigation is difficult for a Si multicrystal grown by a practical casting method owing to the random and complicated shape and crystal orientation of crystal grains.

To control the shape and crystal orientation of crystal grains, model crystal growth was performed using purposely designed seed crystals, which consist of several Si single-crystals with a controlled configuration between adjacent pairs of crystals. An example of an arrangement of seed crystals is illustrated in Fig. 6.8. Two Si (110) single-crystal columns with a diameter of 32 mm were cut every 5 mm, and pieces of crystal from the different Si columns were alternately placed in the bottom of the crucible. As illustrated by this example,

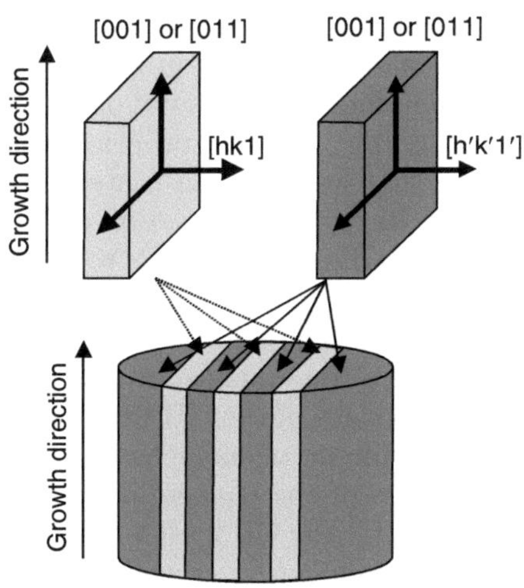

Fig. 6.8. Schematic illustration of the configuration of seed crystals

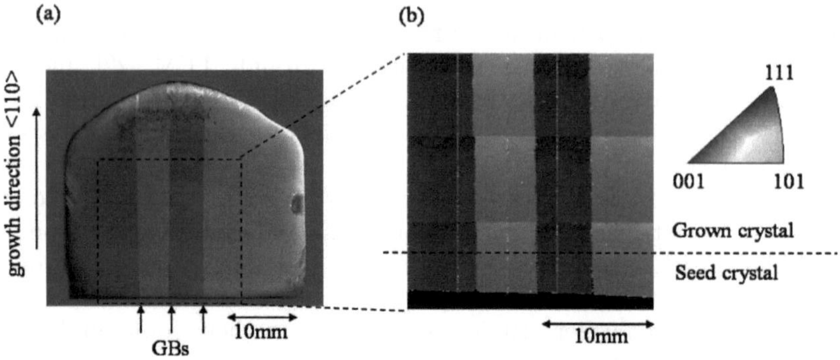

Fig. 6.9. (**a**) photographic image and (**b**) EBSP image of the cross section of a sample containing three GBs with a relative crystal orientation of 103°

this growth method enables the formation of arbitrary GB structures, i.e., GB planes and relative crystal orientations, which can be controlled by varying the arrangement of the cut planes. The seed crystals and poly-Si sources were arranged in a SiO_2 crucible coated with Si_3N_4 powder. The crucible was raised to a position in the temperature gradient in the furnace used for crystal growth, where all the sources and half of the seed crystals melted then lowered to induce epitaxial growth on the seed crystals.

Figure 6.9 shows (a) a photographic image and (b) an EBSP image of the cross section of a sample containing three GBs with a relative crystal orientation of 103°. As the EBSP image shows, the crystal grew epitaxially on the seed crystals, and GBs were formed in a line almost parallel to the growth direction. However, no change in the GB structure or the generation of sub-GBs was detected at the angle resolution of SEM-EBSP analysis, which is generally in the order of 1°.

Figure 6.10 shows the distribution of the crystal orientation measured by XRD at positions around one of the GBs in the sample shown in Fig. 6.9. The peak positions corresponding to the crystal orientations were normalized by the value at the grain center. The crystal orientations in the vicinity of the GB were shifted from that at the grain center with increasing distance from the seed. This change in crystal orientation could have been produced only by the generation of sub-GBs, i.e., sub-GBs were generated from the GB as the crystal growth proceeded. Similar phenomena were also observed in the vicinity of the other GBs, which had almost the same structure, in the same grown crystals. Although the amplitudes and directions of the change in crystal orientation on both sides of the GB were almost same in the grown crystal shown in Fig. 6.9, different phenomena were observed in the other grown crystals. The directions of the change in crystal orientation on both sides of the GB were opposite for the GB with a relative crystal orientation of 138°, and the amplitude of the change in crystal orientation was significantly small around the GB with relative crystal orientation of 90°. This means that

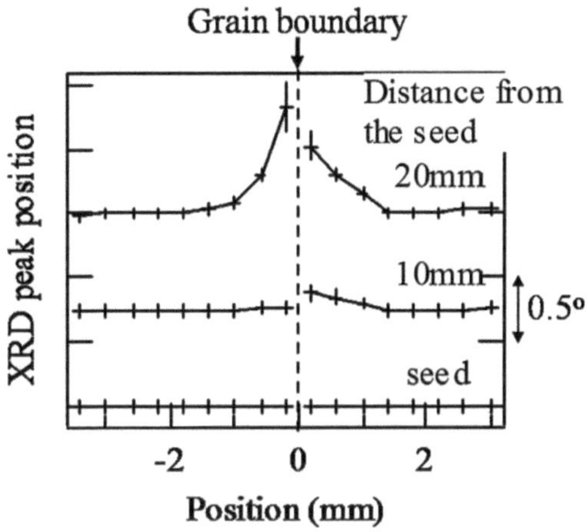

Fig. 6.10. Distribution of peak positions in ω scan profiles of XRD measurements at positions around a GB in the same sample shown in Fig. 6.9

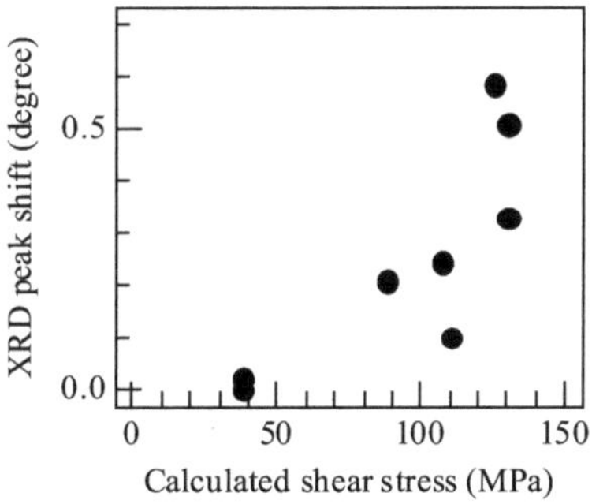

Fig. 6.11. Relationship between calculated amplitude of local shear stress on the slip plane near a GB and the amplitude of XRD peak shift

the amount and direction of the Burgers vector of dislocations comprising the sub-GBs depend on the GB structure.

As reported in the previous studies on metallic multicrystals, local shear stress might be a driving force of sub-GB generation from GBs during crystal growth. To confirm this and to clarify the origin of the local shear stress, finite-element stress analysis was used to model the multicrystalline structure grown in this study. Figure 6.11 shows the relationship between the calculated amplitude of local shear stress on the slip plane near the GB and the amplitude of the XRD peak shift. In this calculation, isotropic displacement was applied to the crystal edges as a boundary condition. A strong correlation can be seen; a multicrystalline structure with a large XRD peak shift, i.e., a large number of generated dislocations are observed, exhibits large local shear stress. This suggests that local shear stress is an important factor in sub-GB generation and that it originates from the isotropic deformation of the crystal. No correlation was found between the calculated local shear stress and the experimental results when anisotropic displacement was employed as a boundary condition.

Another factor related to the sub-GB source should be commented. Sub-GBs were also observed in the vicinity of participations, such as Si_3N_4, SiC, SiO_2 and so forth. This suggests that participations also act as a source of sub-GBs. Furthermore, local modifications of the GB structure should be considered. GB changes its structure as crystal growth proceeds so that its energy is decreased, which results in the generation of sub-GBs [26].

6.5 Summary

Studies of sub-GBs in Si multicrystals are comprehensively reviewed from the viewpoints of structure analysis, electrical properties and the generation mechanism. Sub-GBs consist of dislocations with an array-like configuration, are distributed locally and extend in the growth direction of the crystal. In terms of electrical properties, sub-GBs are concluded to be the most serious defects preventing the improvement of overall performance of solar cells based on Si multicrystals. Sub-GBs are generated during the crystal growth process. The sources of sub-GBs are concluded to be GBs and inclusions. Model crystal growth and finite-element stress analysis revealed that local shear stress caused by isotropic deformation is a key factor in the generation of sub-GBs.

References

1. J.P. Hirth, J. Lothe, *Theory of Dislocations*, 2nd edn. (Wiley, New York, 1982)
2. D.G. Brandon, Acta. Metall. **14**, 1478 (1966)
3. A.P. Sutton, R.W. Balluffi, *Interfaces in Crystalline Materials* (Oxford Science, Oxford, 1995)
4. Z.J. Wang, S. Tsurekawa, K. Ikeda, T. Sekigushi, T. Watanabe, Interf. Sci. **7**, 197 (1999)
5. S. Tsurekawa, T. Watanabe, Solid State Phenomena **93**, 333 (2003)
6. J. Chen, T. Sekiguchi, D. Yang, F. Yin, K. Kido, S. Tsurekawa, J. Appl. Phys. **96**, 5490 (2004)
7. J. Chen, T. Sekiguchi, R. Xie, P. Ahmet, T. Chikyo, D. Yang, S. Ito, F. Yin, Scripta Mater. **52**, 1211 (2005)
8. H. Sugimoto, K. Araki, M. Tajima, T. Eguchi, I. Yamaga, M. Dhamrin, K. Kamisako, T. Saitoh, J. Appl. Phys. **102**, 054506 (2007)
9. H.Y. Wang, N. Usami, K. Fujiwara, K. Kutsukake, K. Nakajima, Acta Materialia **57**, 3268 (2009)
10. N. Usami, K. Kutsukake, K. Fujiwara, K. Nakajima, J. Appl. Phys. **102**, 103504 (2007)
11. K. Kutsukake, N. Usami, K. Fujiwara, K. Nakajima, J. Appl. Phys. **105**, 044909 (2009)

12. K. Fujiwara, Y. Obinata, T. Ujihara, N. Usami, G. Sazaki, K. Nakajima, J. Cryst. Growth **262**, 124 (2004)
13. K. Fujiwara, W. Pan, N. Usami, K. Sawada, M. Tokairin, Y. Nose, A. Nomura, T. Shishido, K. Nakajima, Acta Mater. **54**, 3190 (2006)
14. E. Sirtl, A. Adler, Z. Metallkunde **52**, 529 (1961)
15. W.C. Dash, J. Appl. Phys. **27**, 1193 (1956)
16. B.L. Sopori, J. Electrochem. Soc. **131**, 667 (1984)
17. B. Ryningen, K.S. Sultana, E. Stubhaug, O. Lohne, P.C. Hjemås, in *Proceedings of the 22nd EU PVSEC*, Milan, 2007, p. 1086
18. K. Ikeda, T. Sekiguchi, S. Ito, M. Takebe, M. Suezawa, J. Cryst. Growth **210**, 90 (2000)
19. H. Sugimot, M. Tajima, Jpn. J. Appl. Phys. **46**, L339 (2007)
20. T. Fuyuki, H. Kondo, T. Yamazaki, Y. Takahashi, Appl. Phys. Lett. **86**, 262108 (2005)
21. P. Würfel, T. Trupke, T. Puzzer, E. Schäffer, W. Warta, S.W. Glunz, J. Appl. Phys. **101**, 123110 (2007)
22. L.E. Murr, Mater. Sci. Eng. **51**, 71 (1981)
23. J. Gastaldi, C. Jourdan, G. Grange, Philos. Mag. **57**, 971 (1987)
24. R.A. Varin, K.J. Kurzydlowski, K. Tangri, Mater. Sci. Eng. **85**, 115 (1987)
25. V. Bata, E.V. Pereloma, Acta Metall. **52**, 657 (2004)
26. K. Kutsukake, N. Usami, K. Fujiwara, Y. Nose, T. Sugawara. T. Shishido, K. Nakajima, Mater. Trans. **48**, 143 (2007)

7

New Crystalline Si Ribbon Materials for Photovoltaics

Giso Hahn*, Axel Schönecker, and Astrid Gutjahr

Abstract. The objective of this chapter is to review, for photovoltaic application, the current status of crystalline silicon ribbon technologies as an alternative to wafers originating from ingots. Increased wafer demand, the current silicon feedstock shortage and the need of a substantial module cost reduction are the main issues that must be faced in the booming photovoltaic market. Ribbon technologies make excellent use of the silicon, as wafers are crystallised directly from the melt in the desired thickness and no kerf losses occur. Therefore, they offer a high potential to significantly reduce photovoltaic electricity costs when compared to wafers cut from ingots. Nevertheless, the defect structure present in the ribbon silicon wafers can limit material quality and cell efficiency.

7.1 Ribbon Growth

To provide an answer to the rapidly growing demand for cheap and easily available solar electricity, the photovoltaic industry is looking for more cost efficient and faster manufacturing processes for all steps of the value chain from silicon feedstock to crystallisation and wafering, cell processing and finally module fabrication. Silicon ribbons, as an alternative to silicon wafers that have been crystallised in the form of ingots and then cut into wafers, have, due to their higher silicon usage and potential for high production speed, significant advantages with respect to cost and throughput. However, a more complex technology and different wafer properties that often result in slightly lower conversion efficiencies are obstacles to be overcome.

In principle, all ribbon processes have the characteristics that almost all silicon supplied into the process is converted into wafers. There is also no wafer cutting from a block, although in dependence upon the production method separation of larger sheets into wafers may be needed. Depending upon the type of crystallisation, silicon ribbon growth processes are either relatively slow processes, which can be run on rather low-cost equipment or high-speed processes operated with rather complex machinery.

Driven by the motivation of increasing the Si yield in wafer manufacturing and avoiding the time and energy consuming and, therefore, costly steps of ingot growing and wafer cutting, research and development of methods to crystallise silicon directly in the planar form of a wafer have been going on for four decades [1–21] (see [22] for an early overview). It was only recently that some of the ribbon technologies reached maturity, and manufacture on a megawatt scale, such as in the case of the edge-defined film-fed growth (EFG [23]) and the string ribbon (SR [24]) technologies could emerge. Other technologies such as the silicon film [25], the ribbon growth on substrate (RGS [17]), crystallisation on dipped substrate (CDS [26]) and the ribbon on a sacrificial template (RST [5]) are under development in pilot demonstration phases.

In the following, the different principles of silicon ribbon production methods will be outlined and more detailed information will be given on typical technologies in each category that are either in production or under development or demonstration.

Apart from wafer characteristics as geometry, size, surface roughness, the most important electronic characteristics of a silicon wafer for use in a solar cell are the resistivity in the dark and the minority charge carrier lifetime. However, in opposite to cast silicon wafers or CZ material, the as-grown characteristics of ribbon silicon material are slightly different. The resistivity of the wafer, which is determined by the type and amount of doping material that is added during crystal growth, is often lower than in other wafers for solar applications. If boron doping is used, it turned out that the highest average efficiency on ribbon silicon material is reached in 2–3 Ω cm material, while cast silicon material is typically doped with boron in the 0.5–2 Ω cm range.

The same is true for minority carrier lifetime. Typical as-grown lifetimes in ribbon silicon are lower than in cast or CZ doped wafers. Ribbon silicon wafers therefore depend upon solar cell processing steps to improve their electrical characteristics during the solar cell process (gettering and passivation). Thus, monitoring the minority carrier lifetime and the respective diffusion length during the solar cell process has proven to be a valuable instrument in the development of the ribbon materials.

7.2 Description of Ribbon Growth Techniques

To understand the potential of the different photovoltaic silicon ribbon technologies, a closer look at the wafer growth technology, wafer characteristics and behaviour in the cell process are all necessary.

In the past, the different ribbon Si technologies were classified in several ways:

- By the shape of the meniscus built up at the liquid–solid interface (see Fig. 7.1)

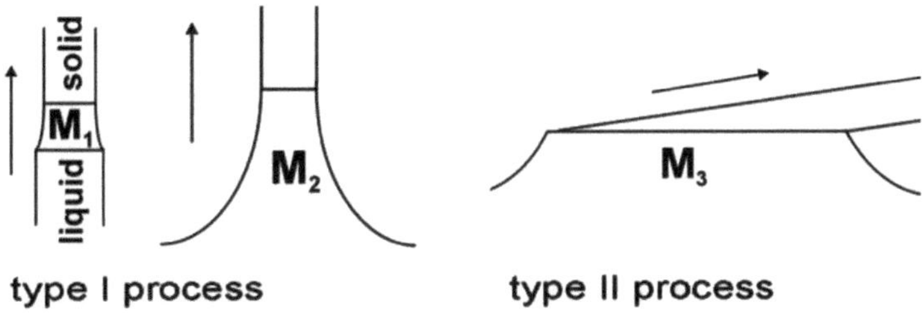

Fig. 7.1. Classification of silicon ribbon technologies according to the shape of the meniscus at the liquid–solid interface [22]. For M1, the lower part of the meniscus is formed, e.g. by a shaping die, whereas M2 has a broad base at the free surface of the liquid. Both M1 and M2 represent ribbon techniques, where the crystallisation front moves in the direction of ribbon transport (type I). M3 is characterised by a large liquid–solid interface and represents the techniques with wafer transport almost perpendicular to the crystal growth direction (type II)

- By the transport direction of the solidified ribbon with respect to the movement of the liquid–solid interface during crystallisation. *Type I*: liquid–solid interface moves in line with ribbon transport direction (e.g. EFG, SR). *Type II*: liquid–solid interface moves almost perpendicular to the ribbon transport direction (e.g. RGS, CDS)
- By the seeding of the silicon crystals; either continuously seeding in the case of ribbons in contact with a cold substrate (type II), or by only initial seeding after which the crystal growth continues (type I)
- By the way the crystallisation heat is removed. Solidification heat is mainly removed by contact with a "cold" material in type II, or by conduction through the solidified silicon wafer, which is radiating in a colder environment in type I

In an ideal silicon ribbon growth technology, the wafer characteristics are completely determined by the way the crystallisation heat (latent heat of fusion) is extracted.

Obviously, the circumstances during crystal growth have a major impact on the crystal structure and on the chemical, electrical and mechanical properties of the silicon ribbon. Silicon-melt preparation and especially the dissolution of impurities from the crucible material and the wafer cooling procedure are also crucial.

This will be demonstrated for three typical examples representing the most important silicon ribbon growth processes: the edge-defined film-fed growth and the string ribbon process where the crystallisation interface moves in line with the ribbon transport direction as typical representatives of type I, and the ribbon growth on substrate technique where the liquid–solid interface moves almost perpendicular to the ribbon transport direction as type II process.

7.2.1 Type I

For type I technologies, the crystallisation heat is transported by a temperature gradient from the liquid–solid interface through the solidified wafer to a colder area on the wafer. From there the heat is removed to the surroundings via radiation or other cooling mechanisms. In this type of process, the crystal growth speed is constant and is controlled by the heat flux through the wafer. In this case, the maximum pulling velocity, v_{p}, (growth rate) can be calculated as

$$v_{\mathrm{p}} = \frac{1}{L\rho_{\mathrm{m}}}\left(\frac{\sigma\varepsilon\,(W+t)\,K_{\mathrm{m}}T_{\mathrm{m}}^{5}}{Wd}\right)^{1/2} \tag{7.1}$$

with L being the latent heat of fusion, ρ_{m} the density of the crystal at melting temperature, σ the Stefan–Boltzmann constant, ε the emissivity of the crystal, K_{m} the thermal conductivity of the solid crystal at the melting temperature, T_{m}, W the ribbon width, and d the ribbon thickness [27]. For 300 μm thick ribbons, (7.1) predicts a maximum growth rate of ~8 cm min^{-1}.

Technical growth rates are much lower due to the maximum tolerable thermal stress limiting the maximum tolerable temperature gradient in the ribbon. The temperature gradient, reached in the silicon ribbon is around 1,000 K cm^{-1} close to the liquid–solid interface (for EFG [28]). This gradient causes the stress, which increases with d^2T/dy^2 (y is the growth direction). The resulting dislocation formation and buckling are critical, limiting the realised growth speeds to ~2 cm min^{-1} for plane ribbons of ~300 μm thickness, and to somewhat lower speeds for thinner ribbons (compared to the theoretical value of ~8 cm min^{-1}). The thermal environment around the newly formed ribbon above the crucible is crucial for the stress. Stacks of thermal shields and afterheaters are used to control and reduce stress [29]. The best approach is to confine stress as much as possible to the liquid–solid interface, where plastic flow can occur.

As the crystal growth is based upon the crystal structure of the already solidified silicon, the silicon ribbon exhibits long crystals in the ribbon growth direction with horizontal dimensions from the mm up to cm range. Rapidly growing crystal orientations are preferred. Depending on the growth velocity and initial seeding, even mono-crystalline material is possible [1].

7.2.1.1 Edge-Defined Film-Fed Growth (Type I Technology)

In the EFG process, commercialised at Schott Solar (under development in the past by Mobil Tyco, ASE and RWE), the silicon ribbon is pulled to heights of up to 7 m from the top of a graphite die. The molten silicon is fed through the die by capillary action (shape of meniscus: M1). Extensive temperature control by radiation shields, cold shoes and afterheating realises maximum temperature gradient where plastic flow is possible, to allow for a maximum growth rate as described above [30, 31] (Fig. 7.2).

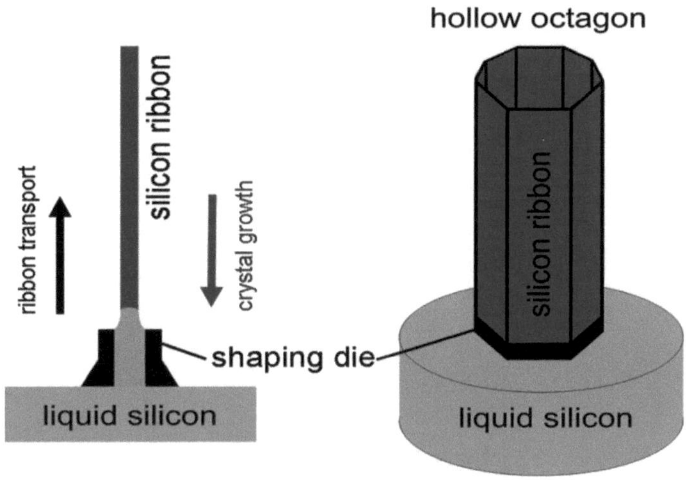

Fig. 7.2. Schematic drawing of the edge-defined film-fed growth (EFG) technology as an example of a type I process. Liquid silicon is lifted by capillary forces through the die, where the silicon ribbon is pulled (*left*). Closed forms with 8 or 12 facets (side width 12.5 cm) are used for the shape of the foil in order to avoid edge effects

The silicon is contained in a graphite crucible, which contains only about 1 kg of silicon. It is continuously replenished, with up to 200 kg Si supplied during a growth run. The solid silicon ribbon is rich with carbon but contains very little oxygen. Crystal grain dimensions are typically in the order of a few mm in width but can reach great lengths in crystal growth direction. The as-grown diffusion length is related to the purity of the graphite parts. The tubes (8 or 12 facets) are cut into wafers ($12.5 \times 12.5\,\mathrm{cm}^2$) by a laser. Currently, tests for EFG wafers with reduced thickness well below 300 μm are carried out [32].

7.2.1.2 String Ribbon (Type I Technology)

The string ribbon technology was invented at the national renewable energy laboratory and at Arthur D. Little, commercialised by Evergreen Solar Inc. It uses high-temperature resistant strings, which are drawn at a distance of 8 cm through a crucible with liquid silicon. They pull up a Si meniscus (M2) of about 7 mm height, which crystallises to become the ribbon. In contrast to the EFG technique (where temperature near the liquid–solid interface must be controlled to ± 1 K), temperature control near the liquid–solid interface is less critical (± 10 K is tolerated) and allows the use of more cost-effective furnace designs [24] (Fig. 7.3).

The result is a silicon ribbon with a typical dislocation density of less than $10^5\,\mathrm{cm}^{-2}$. The main defects in the central area of the ribbon are twins. High-angle grain boundaries occur at the edges due to heterogeneous nucleation.

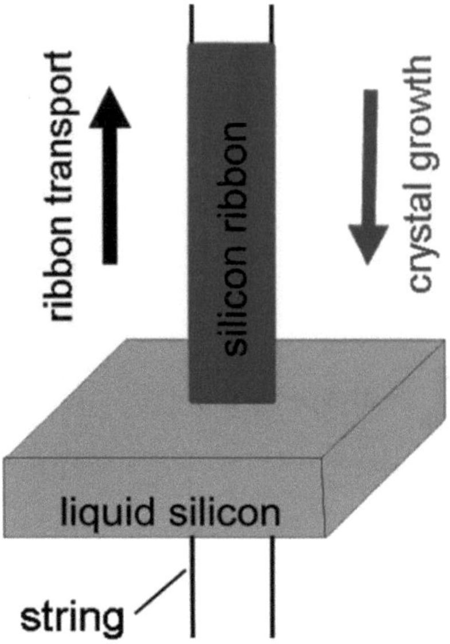

Fig. 7.3. Schematic drawing of the string ribbon (SR) process. Two strings are pulled through the silicon melt. They define the edges of the silicon film. Currently two strings are pulled in parallel ("Gemini" technique), with a new geometry of four strings in line coming into production (so called "Quad" process)

Typical grain size for 200 μm thick ribbons is in the cm range. As with EFG, oxygen concentration is low, but carbon concentration is reduced.

Similar to EFG, an afterheater construction around the crystallisation area is used to reduce thermal stress [29]. As for EFG material, the string ribbon technology uses a small silicon melt crucible in combination with continuous melt replenishment. To overcome the limited throughput of only one ribbon per furnace, compared to 8 or 12 ribbons for EFG, two ribbons [33] and even four with new crucible design [34] can be grown simultaneously in one furnace. The ribbon is cut into $8 \times 15\,\text{cm}^2$ wafers.

7.2.2 Type II

In the case of type II silicon ribbons, the heat is removed from the liquid–solid interface through the solidified wafer into the cold substrate. In contrast to the type I technologies, heat removal through the thin wafer with a large cross section is more effective, resulting in a much higher growth rate.

In this case, ribbon growth speed can be expressed as

$$v_\text{p} = \frac{4\alpha K_\text{m} s}{(2K_\text{m} - \alpha t)\,\text{d}L\rho_\text{m}} \Delta T, \tag{7.2}$$

where α is the effective coefficient of heat transfer, s the length of the liquid–solid interface (in the pulling direction), and ΔT the temperature gradient between melt and substrate [17]. For $\Delta T = 160°C$, (7.2) predicts maximum growth rates in the order of 600 cm min^{-1}. This indicates that techniques with a large liquid–solid interface have the potential of very high pulling rates and, therefore, a higher throughput as compared to techniques of type I.

Due to the silicon crystal growth being in contact with the substrate, type II ribbons have completely different characteristics when compared to type I ribbons. As crystal seeding takes place on the substrate, the wafers typically have small, columnar grains with random orientation. The crystal growth velocity is time dependant. The position of the liquid–solid interface follows a square-root dependence with faster initial crystal growth velocity, when the liquid silicon is in direct contact with the substrate, and slower growth occurring with increasing thickness of the wafer due to the additional heat transport through the solidified silicon. The principle of the type II crystal growth can be described by the "classical Stefan problem" [35]. This assumes that a liquid at uniform temperature T_l, which is higher than the melting temperature, T_m, is confined to a half space $x > 0$. At time $t = 0$, the boundary surface ($x = 0$) is lowered to a temperature T_0 below the melting temperature (i.e. contact with the cold substrate) and maintained at this temperature. As a result, solidification starts at the surface $x = 0$ and a liquid–solid interface $s(t)$ moves into positive x-direction. Under these assumptions, the heat conduction equations can be solved and the position $s(t)$ of the liquid–solid interface in time is described by

$$s(t) = \lambda\sqrt{\alpha_s t}, \tag{7.3}$$

where α_s is the thermal diffusivity of the solid phase and λ is the solution to the equation

$$\frac{\exp\left(\lambda^2/4\right)}{\operatorname{erf}\left(\lambda/2\right)} + \frac{b}{\sqrt{a}}\frac{T_m - T_l}{T_m - T_0}\frac{\exp\left(-\lambda^2/4a\right)}{\operatorname{erfc}\left(\lambda/2\sqrt{a}\right)} - \frac{\lambda\sqrt{\pi}h_{sf}}{2c_{ps}(T_m - T_0)} = 0 \tag{7.4}$$

with the parameters:

b: ratio of liquid to solid heat conductivity
a: ratio of liquid to solid heat diffusivity
L: latent heat of fusion
c_{ps}: specific heat capacity of the solid phase

In general, crystal growth is more complex than the system outlined above due to the behaviour of the T_0 (temperature at the bottom of the solidified wafer), which in general is not constant, the temperature dependence of the material characteristics, and the often turbulent flow in the liquid silicon melt. The variable growth speed results in thickness dependent material characteristics due to processes like velocity dependent effective segregation of metallic impurities.

In contrast to type I crystal growth, where a relatively large temperature gradient in the solidified silicon is the driving force for crystallisation, the temperature gradient through the solidified silicon in type II processes can

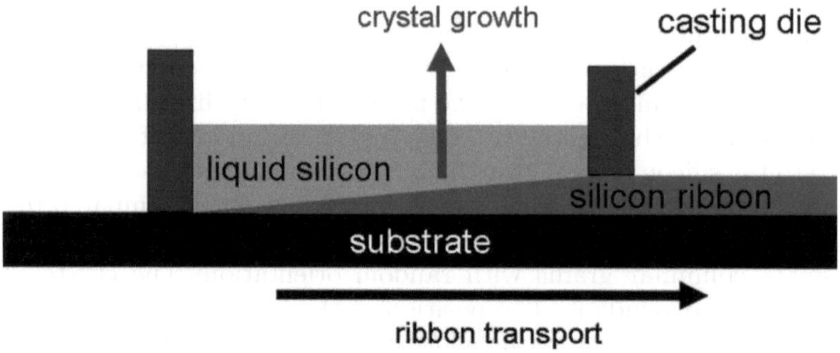

Fig. 7.4. Schematic drawing of the ribbon growth on substrate (RGS) process, typical of type II technology. Preheated substrates are transported at ribbon pulling speed underneath a casting frame filled with liquid silicon. The crystallisation heat is removed into the colder substrate and a 300 μm thick silicon film is grown. After the silicon ribbon is removed, the substrate is re-used in the process

be very small. Therefore, in principle it is possible to grow wafers with lower thermal stress, provided that other process parameters such as wafer cooling or other mechanical stresses are controlled. Examples for this type of silicon ribbon technologies are the RAFT process, which was under development by Wacker in the 1980s, CDS, which is in the prototype phase by Sharp Solar and the RGS process, which is presented as a typical example in the following.

7.2.2.1 Ribbon Growth on Substrate (type II)

RGS was developed by Bayer AG in the 1990s. It is now under development by RGS Development B.V. in co-operation with Deutsche Solar AG, Sunergy Investco B.V. and ECN (Energy Research Centre of the Netherlands). The principle of the RGS process is that a series of graphite based substrates move at high velocity (typically $10\,\mathrm{cm\,s^{-1}}$ or $6\,\mathrm{m\,min^{-1}}$) under a casting frame, which contains liquid silicon. Substrate and casting frame define the size of the wafers and the solidification front (Fig. 7.4).

The crystal growth speed can be controlled by the heat extraction capacity of the substrate material. During cooling, the difference in thermal expansion coefficient between substrate material and Si ribbon causes a separation of the silicon ribbon from the substrate and allows the re-use of the substrate material.

Similar to the other ribbon technologies, RGS material is rich in carbon concentration due to the refractory materials used and, nowadays, low in oxygen concentration.

7.2.3 Comparison of Growth Techniques

Data related to crystal growth of SR and EFG (type I) and RGS (type II) are given in Table 7.1.

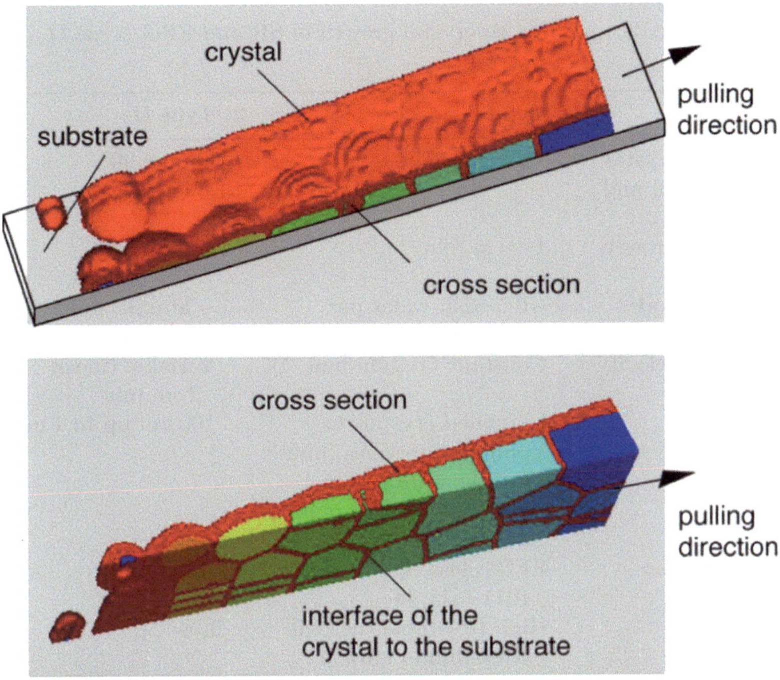

Fig. 7.5. Simulation results of the seeding behaviour and the columnar crystal growth of RGS wafers [36]

7.3 Material Properties and Solar Cell Processing

As nearly all the developed ribbon silicon growth techniques result in multicrystalline material, crystal defects play a major role in the solar cell efficiencies obtained on ribbon silicon wafers. There is a general trend for higher defect concentrations with faster ribbon growth, but even within wafers of one growth technique, there is an inhomogeneous defect distribution. In this section, we will introduce the known relevant defects for the three materials under closer examination (EFG, SR and RGS) and their impact for solar cell processing. Especially, interaction between different types of defects must be taken into account to understand the behaviour of the different ribbon silicon materials within the cell process.

7.3.1 Refractory Materials

Apart from crystallisation conditions, the materials that are used in ribbon growth equipment and the atmospheric environment are important factors that influence the wafer characteristics. In most of the ribbon technologies, the solidification area is in close proximity to refractory materials, such as the shaping die in EFG or the casting frame and substrate in RGS. In contrast to

Table 7.1. Some data related to crystal growth of SR and EFG (type I) and RGS (type II)

	Type I	Type II
Angle between crystal growth direction and ribbon transport	180°	almost 90°
Typical ribbon growth velocities	1–2 cm min^{-1}	600 cm min^{-1}
Typical annual wafer output	~0.5 mill. wafer per machine	~20 mill. wafer per machine
Crystallisation velocity	Constant (1–2 cm min^{-1})	Variable (mean 2 cm min^{-1})
Crystal size	Extended crystals in pulling directions, mm up to cm dimensions perpendicular to pulling direction	100 μm up to 1 mm range
Crystal orientation	EFG: close to {011} [43,44]	Random
Thermal stress	High (tolerable thermal stress limits growth velocity)	Low

ingot casting, it is not possible to discard the silicon that was crystallised in contact with the refractory material. In addition, other technologies, such as SiO evaporation in the Cz process are not possible due to the close proximity of the crystallisation process to refractory material. Therefore, much research has done into the development of silicon resistant refractory materials with the aim of minimising the resulting contamination [37]. The result was that most ceramic materials produced in combination with metal-oxide binders cannot be used due to the very low contamination tolerance of silicon wafers to metallic impurities [38,39]. The same is true for ceramics including doping elements, such as B in BN or Al in SiAlON. Today, only quartz [40] and graphite based crucibles [41] with or without additional coatings based on silicon compounds (SiN, SiC) are in common use as refractory materials.

The behaviour of quartz crucibles and the interaction with liquid silicon was examined thoroughly to control the oxygen content in Cz wafers [42]. Important factors are the dissolution of quartz in liquid silicon and the evaporation of SiO from the melt. This results in a transport process of SiO_2 via the silicon melt into the gas phase and respectively to crucible dissolution and silicon melt contamination.

The important interaction of liquid silicon with graphite crucibles and the formation of a SiC interface layer is a topic of ongoing research that is important for the further development of silicon ribbons. It is generally assumed that the initial contact of pure liquid silicon with graphite leads to the dissolution of graphite in the silicon [43]. From reactive wetting experiments, there are indications that this dissolution of graphite is a very rapid process, as is the diffusion of carbon in liquid silicon [44, 45]. Normally, this behaviour should result in a carbon saturated silicon melt, but scanning electron microscopy analyses of the silicon–graphite interface show the existence of a SiC layer at the interface [46]. The growth of the SiC layer takes place in two different growth regimes: the initial one with the linear kinetics of an interface-reaction limited process, followed by a slower process with approximately parabolic kinetics, which can be explained by a growth process that is limited by carbon diffusion through the SiC layer. Therefore, after initial growth of the SiC interface layer, the further dissolution of carbon from the graphite crucible is kinetically hindered by diffusion through a SiC interface layer.

The carbon concentration in the wafers in relation to the growth conditions is of high technological interest. Experiments [47] show that the solubility of carbon in liquid silicon equilibrated with SiC can be described by

$$\log\left([C]/\text{mass\%}\right) = 3.63 - \frac{9{,}660}{T} \quad \text{for } T\text{:1,723–1,873 K.} \tag{7.5}$$

At the silicon melting point, this should result in carbon concentration of $9.1 \times 10^{18}\,\text{cm}^{-3}$. As carbon solubility in solid silicon is $3.5 \times 10^{17}\,\text{cm}^{-3}$ at the melting point of Si, the high carbon solubility in the liquid silicon should result in carbon supersaturated solidified silicon or the formation of SiC, either residually in the melt or incorporated in the silicon crystal.

Despite the high dissolution of carbon in liquid silicon, it is technically possible to produce silicon ribbons with substitutional carbon concentrations lower than $1 \times 10^{18}\,\text{cm}^{-3}$ or even in the $5 \times 10^{17}\,\text{cm}^{-3}$ range in a graphite environment [41]. As carbon contamination is important for the electrical and mechanical properties of the silicon wafer, the growth of silicon ribbons with carbon content well below the liquid solubility is very much desired.

7.3.2 Ribbon Material Properties

The specific growth conditions of the silicon ribbons result in the material properties listed in Table 7.2. The higher optimum in resistivity than that used for standard ingot cast multi-crystalline wafers ($\sim 1\,\Omega\,\text{cm}$) may be related to the high carbon concentration. A possible recombination centre involving both boron and carbon might be responsible for the observed material degradation detected for lower resistivity material [48], although the underlying defect has not yet been identified.

EFG and RGS material both share very high carbon concentrations due to their contact to graphite containing materials near the liquid–solid interface (die or substrate). Oxygen concentration is very low for EFG and SR, and slightly higher for RGS. Grain size is smaller for RGS (0.1–0.5 mm) as there are

Table 7.2. Material properties of the three ribbon techniques under closer consideration

Material	Grain size	Dislocation density (cm^{-2})	Thickness (μm)	Resistivity (Ωcm)	[C] (cm^{-3})	[O] (cm^{-3})	As-grown L_{diff} (μm)
EFG	cm	10^4–10^5	300	2–4	10^{18}	$<5 \times 10^{16}$	10–300
SR	cm	10^4–10^5	200	3	5×10^{17}	$<5 \times 10^{16}$	10–300
RGS	<mm	10^5–10^7	300	3	10^{18}	4×10^{17}	~10

more nucleation sites on the substrate. Dislocation density in RGS generally tends to be higher than that of EFG and SR, which means that the possibility of stress free wafer growth is not yet fully realised.

Additionally, transition metals are present in all materials, although mostly in concentrations not limiting material quality. Nevertheless, some are effective recombination centres as point defects or in the form of precipitates and affect the as-grown material quality.

The effects of an isolated defect on material quality (e.g. recombination activity of a clean, undecorated dislocation, capture cross sections of point defects) are well known for many defects present in crystalline silicon material, but the interactions of the impurities or structural defects form a major challenge in getting an improved understanding of the complex situation in the solidified silicon ribbon. Currently, it is impossible to list a complete overview of the known interactions, but more information can be found in [49].

7.3.2.1 EFG and SR

Stress and Dislocations

Ribbon technologies with the plane of the liquid–solid interface perpendicular to the growth direction all suffer from built-in stress due to the varying thermal gradient in the solidified ribbon [50]. This stress can lead to the formation of areas with high dislocation density. In these areas, carrier lifetimes are reduced as shown by photoluminescence spectroscopy (PL) [51, 52] and transmission electron microscopy (TEM). High stresses can be detected also in areas with a low dislocation density [53]. Areas containing only twins without increased dislocation densities do not show reduced lifetimes. They are, however, highly stressed and there is evidence that this might be due to the incorporation of carbon into the twin boundaries [54].

It is known that clean dislocations without decoration reveal almost no recombination activity [55], but increasing decoration with impurities leads to recombination centres deep in the band gap, which significantly reduce carrier lifetime [56]. It can, therefore, be concluded that one of the most detrimental defects in EFG and SR apart from recombination active large angle grain boundaries are decorated dislocations.

Transition Metals

Transition metals are known to be recombination centres in crystalline silicon. Studies in deliberately contaminated EFG wafers have shown the detrimental impact of different metals such as Cr, Mo, V, Ti and Fe [57, 58]. As Ti, V and Mo are slowly diffusing in silicon, as they cannot be effectively gettered in the solar cell process. It could also be shown that Fe and Cr pair with the boron acceptor. The formation of the Cr–B pair always results in a decrease in lifetime and pairing can be revoked by a 200°C anneal [59, 60]. However, the harmfulness of Fe–B pairing as compared to interstitial Fe (Fe_i) depends on injection level. Under low injection, Fe_i leads to lower lifetimes than Fe–B pairs, whereas in high injection, Fe–B pairs show a higher recombination activity than Fe_i [61]. The detrimental effect of both impurities Fe and Cr is strongly dependant on B dopant concentration [59, 61] and might explain the need for the use of higher resistivity material for silicon ribbons in comparison with standard cast multi-crystalline wafers. As Fe and Cr in isolated form are fast diffusors, these elements can be gettered more easily, which is essential for improving material quality during solar cell processing.

7.3.2.2 RGS

Oxygen

Apart from a larger amount of grain boundaries due to the smaller grain size in RGS, there are other defects affecting as-grown material quality of this high-speed ribbon production technique. Older material (before 2003) was characterised by high carbon concentrations combined with a high oxygen content. The high oxygen concentration, in the form of interstitials, was responsible for the formation of lifetime reducing defects: apart from the formation of thermal donors at temperatures <600°C, SiO_x-containing new donors are formed in the temperature range of 600–900°C, when high interstitial oxygen levels are present [62]. This formation is enhanced by a high carbon concentration [63] and can drastically reduce carrier lifetimes in RGS [64].

Improvements in the RGS process resulted in lower oxygen concentrations, which are nowadays comparable with multi-crystalline wafers from cast ingots. Annealing steps for a deliberate oxygen precipitation to avoid new donor formation are, therefore, no longer necessary [65].

7.3.3 Ribbon Silicon Solar Cells

Progress in state-of-the-art solar cell processing has allowed the use of highly defected crystalline silicon wafers in industrial type solar cell production without a major reduction in cell efficiency. One prerequisite for obtaining

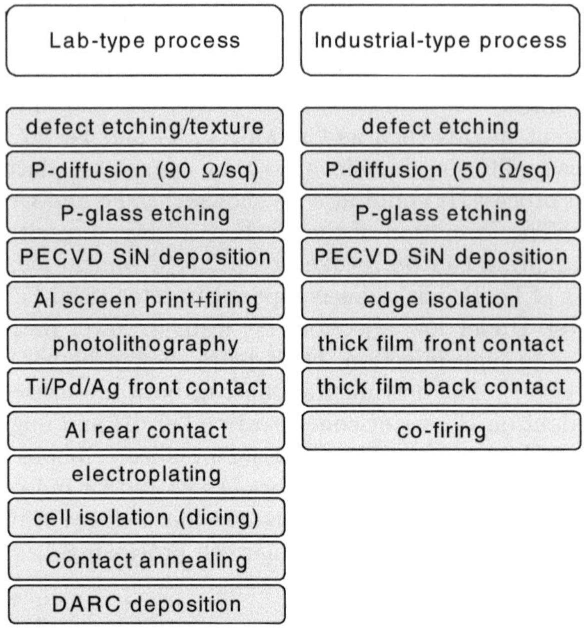

Fig. 7.6. Typical solar cell fabrication sequences used for lab-type processing (*left*) and industrial-type processing (*right*)

acceptable efficiencies is that material quality is improved during cell processing. This can be achieved by gettering and hydrogenation steps (see [49] for a detailed review of gettering and hydrogenation studies in ribbon silicon materials). A typical state-of-the-art industrial-type cell process for multi-crystalline silicon material includes the steps shown in Fig. 7.6: formation of the emitter by indiffusion of P at temperatures between 800 and 900°C for about ∼20 min, deposition of a SiN_x antireflection coating (ARC), and thick film metallisation at the front (Ag) and rear side (Al), followed by a firing step (700–850°C, <1 min) for BSF (back surface field) and contact formation. The challenge in processing defected, as-grown silicon ribbon wafers into highly efficient solar cells is to use lifetime improving steps within the cell process to achieve a significant increase in material quality.

It is well known that hydrogen incorporated in crystalline silicon can reduce the recombination activity of defects and increase minority carrier lifetimes. Therefore, hydrogenation techniques play a major role in improving the material quality of defected areas within all multi-crystalline wafers during solar cell processing. The importance of hydrogenation, especially for highly defected ribbon silicon materials for reaching sufficiently high-efficiencies is, therefore, evident.

7.3.3.1 Hydrogenation in Ribbon Silicon

In industrial solar cell processing, hydrogen is incorporated via a hydrogen-rich SiN_x layer deposited, e.g. by plasma-enhanced chemical vapour deposition (PECVD [66]). Depending on the NH_3/SiH_4 ratio used for deposition, which also influences the refractive index and absorption coefficient [67], up to ~30 at.% H can be incorporated into the SiN_x [68]. The deposited SiN_x layer is, therefore, not of stoichiometric composition. An annealing step following the SiN_x deposition can release atomic hydrogen into the bulk of the wafer [69–71]. Besides being a reservoir of hydrogen, the deposited PECVD SiN_x layer can simultaneously act as an antireflection coating and can provide excellent surface passivation [72]. Due to this threefold benefit for solar cell processing [73], PECVD SiN_x layers are now state-of-the-art in modern industrial fabrication of multi-crystalline solar cells.

Although gettering techniques improve mainly good quality areas of the as-grown wafer, hydrogen treatment can improve areas of all quality significantly. Nevertheless, the achievable final lifetime after hydrogenation is not a function of lifetime in the as-grown state alone [74–77]. This clearly demonstrates the complex situation of inhomogeneous defect distribution in ribbon silicon. The efficiency of hydrogenation is strongly dependant on the underlying defects. Figure 7.7 shows lifetimes of a SR wafer in the as-grown state, after Al-gettering and after hydrogenation. Indicated are areas of comparable starting lifetimes, resulting in areas of significantly different lifetimes after the gettering and hydrogenation step. As dislocations contaminated with precipitates seem to play the major role in recombination activity, an explanation for the different behaviour are changes in chemical composition of these

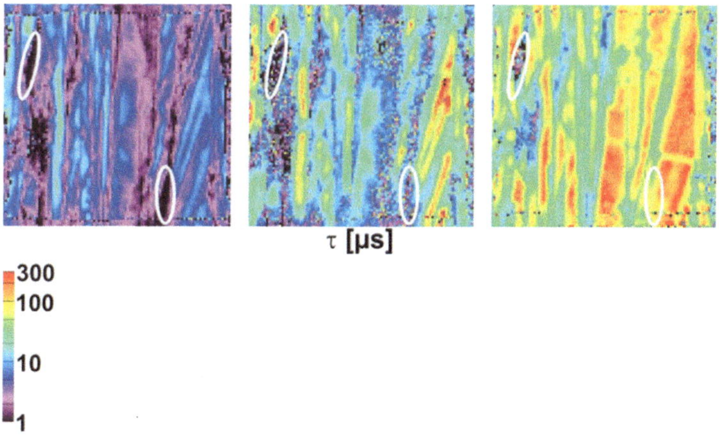

Fig. 7.7. Bulk minority carrier lifetimes of a $5 \times 5\,cm^2$ SR wafer as-grown (*left*), after P-gettering (*middle*) and after hydrogenation (*right*) with identical scaling. *Indicated* are areas of comparable as-grown quality, which significantly differ after hydrogenation [75]

precipitates. Dependant on the nature of the precipitates [55,56], hydrogenation is more or less effective in reducing recombination strength.

Retention of hydrogen at the defects is important to maintain the passivation properties. For temperatures >400°C, reactivation of recombination activity can be observed, due to thermal activation. Therefore, cooling rates after H-passivation are important. Hydrogenation from PECVD SiN_x layers with annealing temperatures in the range of 700–850°C can be significantly affected by cooling-down ramps. Experiments using rapid thermal processing (RTP) in combination with PECVD SiN_x layers revealed that higher lifetimes can be achieved by using fast cool-down ramps [78,79].

Hydrogen passivation of defects in RGS is also important, because as-grown minority charge carrier lifetimes are even lower than for EFG and SR. D-profiles obtained for RGS material differing only in interstitial oxygen concentration showed a reduced diffusion for higher oxygen contents [80]. This effect was also observed for other materials [81] and can be explained by oxygen trapping atomic hydrogen. Whereas, for low-oxygen materials like EFG, hydrogen diffuses through the whole wafer within ~30 min at 350°C this takes several hours in the case of oxygen-rich RGS [81]. More details concerning trapped hydrogen diffusion in defected silicon can be found in [49,82].

7.3.3.2 Solar Cell Processing

Fabrication of solar cells using ribbon silicon wafers has to be adapted to the material needs to reach satisfactory conversion efficiencies. As for all multicrystalline silicon wafers, material quality should be improved during cell processing to cope with the defect structure present in the as-grown material. The implementation of gettering and hydrogenation steps into the solar cell process is, therefore, crucial, as the efficiency obtainable for solar cells from a given material is important for cost-effectiveness.

When considering efficiencies, two types of cell processes have to be distinguished, namely lab-type (small area cells, determination of materials potential, disregarding fabrication costs) and industrial-type processing (large area cells, transferable into mass production).

EFG and SR Solar Cells

Both EFG and SR wafers are fabricated commercially and solar cells have been processed on a large scale, industrial base since 1994 (EFG) and 2001 (SR), respectively. Progress for SR and EFG record efficiencies have developed in parallel, again demonstrating their comparable material quality (Fig. 7.8).

As material quality is inhomogeneous even after gettering and hydrogenation (Fig. 7.7), solar cell results are affected by both good and bad areas. Cell performance in areas of low diffusion length is limited due to recombination in the bulk, whereas rear surface recombination, S_b, can limit carrier collection in good quality areas.

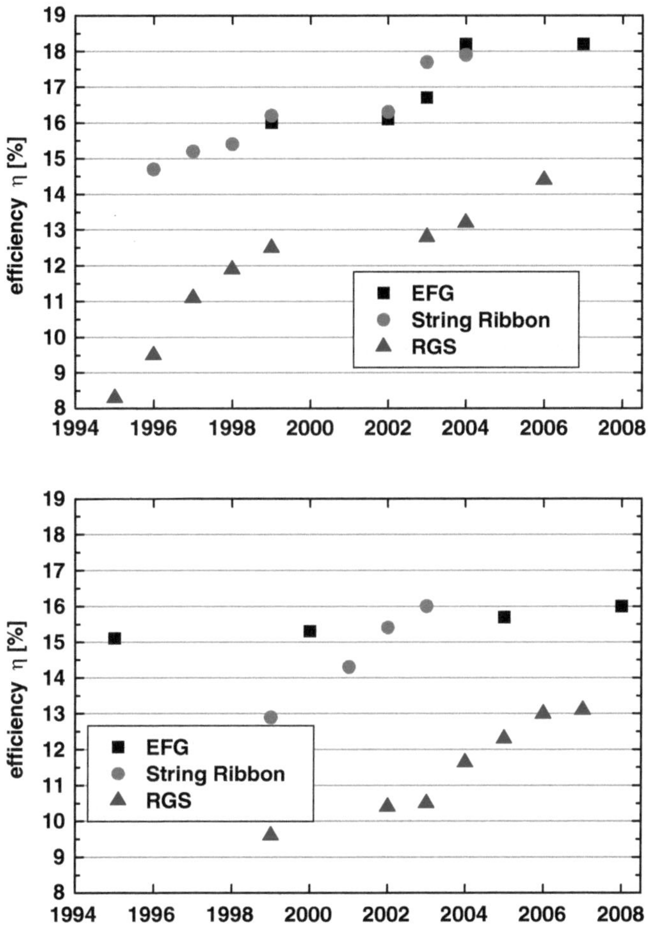

Fig. 7.8. Progress in record efficiencies of EF, SR and RGS solar cells using lab-type (*left*) and industrial-type processing (*right*). Some of the data are from [83]

Areas of lower material quality are of special interest, as these regions normally limit cell efficiency. Lower quality areas can be improved mainly by hydrogenation [74, 76], therefore, bulk defect passivation kinetics have been studied intensively. It was shown that retention of hydrogen at the defect sites can only be obtained for temperatures <400°C, if no capping layer is present [84, 85]. Similar results have been obtained for cells with a SiN_x layer on top of the emitter [86, 87].

Record efficiencies of 18.2% for EFG and 17.9% for SR were obtained using lab-type processing including double layer ARC (DARC) [85, 88]. Large area solar cells processed according to an industrial-type fabrication scheme show significantly lower efficiencies. Record values of 15.7% for EFG (10 × 10 cm^2

[89]) and 16.0% for SR ($8 \times 10\,\mathrm{cm}^2$ [90]) have been reported. One significant aspect to be considered in industrial-type ribbon silicon solar cell processing is the need of a surface texture for EFG and SR which is different to standard ingot cast multi-crystalline materials, due to the missing saw damage. As alkaline texturing can not be applied due to anisotropic preferential etching at grain boundaries, reflection after cell processing with a standard PECVD SiN_x as single layer antireflection coating is significantly higher than for other materials, if no surface texture is applied. Recently, studies have been carried out to check the potential of adapted acidic etch solutions [91,92] and plasma texturing [88,93]. With an industrial-type process for textured EFG, even top cell efficiencies up to 16% have been demonstrated recently [94].

RGS Solar Cells

Cell processing of RGS wafers must be adapted to the wafer surface morphology. There are two main differences compared to EFG and SR wafers. Firstly, the latter exhibit uneven surfaces on both sides, whereas RGS wafers have a flat back side due to the use of a substrate during crystallisation. Secondly, impurities can segregate with the liquid–solid interface and are frozen at the RGS wafer front side, which is uneven. Therefore, this side has to be treated prior to cell processing to remove this impurity-rich layer. Progress in processing can be seen in Fig. 7.8 and is caused by improvements in both material quality as well as cell processing. Efficiencies are significantly lower as compared to EFG and SR, mainly due to the lower values of L_{diff} caused by higher defect densities.

For practical reasons, RGS wafers from the current lab-type machine yielding ten wafers in a single casting process were cut down from $8.6 \times 13\,\mathrm{cm}^2$ to $5 \times 5\,\mathrm{cm}^2$ for industrial-type processing. Using this wafer format, efficiencies of 13.1% have been reached [95]. This process involved a mechanical planarisation step to flatten the front surface and to remove the defect-rich surface layer. Efficiencies of 14.4% [95] have been reached by applying lab-type processing techniques. The RGS technique offers the possibility to reduce as-grown wafer thickness down to the $100\,\mu\mathrm{m}$ regime. First solar cells processed from these thin wafers reached efficiencies in the 10–11% range. These solar cells reveal an extremely good yield of silicon needed for photovoltaic power production of below $3\,\mathrm{g_{Si}/W_p}$ [95].

7.4 Summary

Solar cells from ribbon silicon wafers are more cost-effective, when efficiencies are in the same range as for cells from costlier wafers originating from ingots. Record efficiencies for EFG and SR cells in the range of 18% are comparable to the best cells fabricated from multi-crystalline wafers from ingots, when lab-type processes of the same complexity are used. The same is true for industrial-type processing, with record efficiencies between 15 and 16% and mean values around 15% for EFG solar cells in production [96], slightly lower as for solar cells from ingot based wafers. From these data, it can be concluded that solar cells from EFG or SR wafers have a significant advantage concerning cost per W_p, provided a comparable yield is achieved.

RGS wafers have the advantage of a more cost-effective fabrication due to the high-production speed. The expectation is that, even if the efficiency is somewhat lower, the introduction of RGS would further reduce the costs per W_p of PV modules. Improvements in wafer quality and an increased understanding of the interaction between defects and solar cell processing are necessary to reach higher efficiency values.

In addition to the cost effectiveness of silicon ribbons, energy payback time (i.e. the time needed to produce the amount of energy that was consumed during the manufacturing of a solar system) is drastically reduced. In a recent life-cycle analysis of crystalline silicon wafer based PV systems, it was demonstrated that the energy pay-back times can be reduced by half (based on cut multi-crystalline wafers), by the use of RGS ribbons for systems in central Europe [97].

Due to the large economical and technological potential of silicon ribbons, their application in solar wafer production will be a major milestone in PV cost reduction. Thus, it is very likely that silicon wafer based PV module manufacturing will maintain the cost advantage over other upcoming technologies and, therefore, the role as the major PV technology.

Current tendencies for multi-crystalline wafer based solar cells are heading towards thinner and larger wafers. Ribbon silicon based wafer technologies have to deal with these developments in the future to maintain their cost effectiveness. As thin EFG and RGS wafers have already been produced on a laboratory scale with thicknesses of $<200\,\mu$m, their industrial application remains a topic for ongoing research.

References

1. S.N. Dermatis, J.W. Faust, IEEE Trans. Commun. Electron. **82**, 94 (1963)
2. J. Boatman, P. Goundry, Electrochem. Technol. **5**, 98 (1967)
3. T.F. Ciszek, Mater. Res. Bull. **7**, 731 (1972)
4. T. Koyanagi, in *Proceedings of the 12th IEEE PVSC*, Baton Rouge, 1976 (IEEE, Library of Congress, USA), p. 627
5. C. Belouet, J. Cryst. Growth **82**, 110 (1987)
6. I.A. Lesk, A. Baghadadi, R.W. Gurtler, R.J. Ellis, J.A. Wise, M.G. Coleman, in *Proceedings of the 12th IEEE PVSC*, Baton Rouge, 1976 (IEEE, Library of Congress, USA), p. 173
7. J.D. Heaps, R.B. Maciolek, J.D. Zook, M.W. Scott, in *Proceedings of the 12th IEEE PVSC*, Baton Rouge, 1976 (IEEE, Library of Congress, USA), p. 147
8. T.F. Ciszek, G.H. Schwuttke, J. Cryst. Growth **42**, 483 (1977)

9. T.F. Ciszek, J.L. Hurd, in *Proceedings of the 14th IEEE PVSC*, San Diego, 1980 (IEEE, Library of Congress, USA), p. 397
10. K.M. Kim, S. Berkman, M.T. Duffy, A.E. Bell, H.E. Temple, G.W. Cullen, *Silicon Sheet Growth by the inverted Stepanov Technique*, DOE/JPL-954465 (1977)
11. N. Tsuya, K.I. Arai, T. Takeuchi, K. Ohmor, T. Ojima, A. Kuroiwa, J. Electron. Mater. **9**, 111 (1980)
12. T.F. Ciszek, J.L. Hurd, M. Schietzelt, J. Electrochem. Soc. **129**, 2838 (1982)
13. H.E. Bates, D.M. Jewett, in *Proceedings of the 15th IEEE PVSC*, Kissimmee, 1981 (IEEE, Library of Congress, USA), p. 255
14. J.G. Grabmaier, R. Falckenberg, J. Cryst. Growth **104**, 191 (1990)
15. A. Beck, J. Geissler, D. Helmreich, J. Cryst. Growth **82**, 127 (1987)
16. A. Eyer, N. Schillinger, I. Reis, A. Räuber, J. Cryst. Growth **104**, 119 (1990)
17. H. Lange, I. Schwirtlich, J. Cryst. Growth **104**, 108 (1990)
18. M. Suzuki, I. Hide, T. Yokoyama, T. Matsuyama, Y. Hatanaka, Y. Maeda, J. Cryst. Growth **104**, 102 (1990)
19. Y. Maeda, T. Yokoyama, I. Hide, T. Matsuyama, K. Sawaya, J. Electrochem. Soc. Solid State Sci. Technol. **133**(2), 440 (1986)
20. Y. Komatsu, N. Koide, M.-J. Yang, T. Nakano, Y. Nagano, K. Igarashi, K. Yoshida, K. Yano, T. Hayakawa, H. Taniguchi, M. Shimizu, H. Takiguchi, Sol. Energy Mater. Sol. Cells **74**, 513 (2003)
21. M. Konagai, in *Proceedings of the 29th IEEE PVSC*, New Orleans, 2002 (IEEE, Library of Congress, USA), p. 38
22. T.F. Ciszek, J. Cryst. Growth **66**, 655 (1984)
23. F.W. Wald, in *Crystals: Growth, Properties, and Applications 5*, ed. by J. Grabmaier (Springer, Berlin, 1981), p. 147
24. W.M. Sachs, D. Ely, J. Serdy, J. Cryst. Growth **82**, 117 (1987)
25. J.S. Culik, F. Faller, I.S. Goncharovsky, J.A. Rand, A.M. Barnett, in *Proceedings of the 17*th *EC PVSEC*, Munich, 2001 (WIP, Munich), p. 1347
26. H. Mitsuyasu, S. Goma, R. Oishi, K. Yoshida, H. Taniguchi, in *Proceedings of the 23rd EU PVSEC*, Valencia, 2008, p. 1497
27. T.F. Ciszek, J. Appl. Phys. **47**(2), 440 (1976)
28. J.P. Kalejs, J. Cryst. Growth **128**, 298 (1993)
29. R.J. Wallace, J.I. Hanoka, S. Narasimha, S. Kamra, A. Rohatgi, in *Proceedings of the 26th IEEE PVSC*, Anaheim, 1997 (IEEE, Library of Congress, USA), p. 99
30. J.P. Kalejs, *Silicon Processing for Photovoltaics II*, ed. by C.P. Khattak, K.V. Ravi (Elsevier, Amsterdam, 1987), p. 185
31. C.K. Bhihe, P.A. Mataga, J.W. Hutchinson, S. Rajendran, J.P. Kalejs, J. Cryst. Growth **137**, 86 (1994)
32. J. Horzel, A. Seidl, W. Buss, I. Westram, F. Mosel, S. Guenther, J. Novak, J. Sticksel, G. Blendin, M. Jahn, M. Rinio, H. von Campe, W. Schmidt, in *Proceedings of the 22nd EU PVSEC*, Milan, 2007, p. 1394
33. L.W. Wallace, E. Sachs, J.I. Hanoka, in *Proceedings of the 3rd WCPEC*, Osaka, 2003 (WCPEC-3 Organising Committee, Arisumi Printing Inc., Japan), p. 1297
34. E. Sachs, D. Harvey, R. Janoch, A. Anselmo, D. Miller, J.I. Hanoka, in *Proceedings of the 19th EC PVSEC*, Paris, 2004, p. 552
35. S.H. Cho, J.E. Sunderland, J. Heat Transf. **91c**, 421 (1969)
36. M. Apel, D. Franke, I. Steinbach, Sol. Energy Mater. Sol. Cells **72**, 201 (2002)

37. A. Briglio, K. Dumas, M. Leipold, A. Morrison, *Flat Plate Solar Array Project: Volume III Silicon Sheets and Ribbons*, DOE/JPL 1012–125 (1986)
38. J. Fally, E. Fabre, B. Chabot, Rev. Phys. Appl. **22**, 529 (1987)
39. J.R. Davis, A. Rohatgi, R.H. Hopkins, P.D. Blais, P. Rai-Choudhury, J.R. McCormic, H.C. Mollenkopf, IEEE Trans. Electron Devices **ED-27**, 677 (1980)
40. R. Hull, *Properties of Crystalline Silicon* (Inspec, London, 1999)
41. R.E. Janoch, A.P. Anselmo, R.L. Wallace, J. Martz, B.E. Lord, J.I. Hanoka, in *Proceedings of the 28th IEEE PVSC*, Anchorage, 2000 (IEEE, Library of Congress, USA), p. 1403
42. H. Hirata, K. Hoshikawa, Jpn. J. Appl. Phys. **19**, 1573 (1980)
43. A. Schei, J. Tuset, H. Tveit, *Production of High Silicon Alloys* (Tapir Forlag, Trondheim, 1998)
44. J.-G. Li, H. Hausner, Scripta Metallurgica et Materialia **32**(3), 377(1995)
45. J.-G. Li, H. Hausner, J. Am. Ceram. Soc. **97**(4), 873 (1996)
46. R. Deike, K. Schwerdtfeger, J. Electrochem. Soc. **142**(2), 609 (1995)
47. K. Yanabe, M. Akasaka, M. Takeuchi, M. Watanabe, T. Narushima, Y. Iguchi, Mater. Trans. JIM **38**(11), 990 (1997)
48. J.P. Kalejs, Solid State Phenom **95–96**, 159 (2004)
49. G. Hahn, A. Schönecker, J. Phys. Condens. Matter **16**, R1615 (2004)
50. B. Chalmers, J. Cryst. Growth **70**, 3 (1984)
51. Y. Koshka, S. Ostapenko, I. Tarasov, S. McHugo, J.P. Kalejs, Appl. Phys. Lett. **74**, 1555 (1999)
52. S. Ostapenko, I. Tarasov, J.P. Kalejs, C. Hässler, E.-U. Reisner, Semicond. Sci. Technol. **15**, 840 (2000)
53. H.J. Möller, C. Funke, A. Lawerenz, S. Riedel, M. Werner, Sol. Energy Mater. Sol. Cells **72**, 403 (2002)
54. M. Werner, K. Scheerschmidt, E. Pippel, C. Funke, H.J. Möller, Phys. Conf. Ser. **180**, (2004)
55. K. Knobloch, M. Kittler, W. Seifert, J. Appl. Phys. **93**(2), 1069 (2003)
56. M. Kittler, W. Seifert, T. Arguirov, I. Tarasov, S. Ostapenko, Sol. Energy Mater. Sol. Cells **72**, 465 (2002)
57. J.P. Kalejs, L. Jastrzebski, L. Lagowski, W. Henley, D. Schielein, S.G. Balster, D. K. Schroder, in *Proceedings of the 12th EC PVSEC*, Amsterdam, 1994 (H.S. Stephens, Bedford), p. 52
58. J.P. Kalejs, B.R. Bathey, J.T. Borenstein, R.W. Stormont, in *Proceedings of the 23rd IEEE PVSC*, Louisville, 1993 (IEEE, Library of Congress, USA), p. 184
59. K. Mishra, Appl. Phys. Lett. **68**, 3281 (1996)
60. O. Klettke, D. Karg, G. Pensl, M. Schulz, G. Hahn, T. Lauinger, *Technical Digest 12th PVSEC*, Jeju, 2001 (International PVSEC-12, Seoul), p. 617
61. D. Macdonald, A. Cuevas, J. Wong-Leung, J. Appl. Phys. **89**, 7932 (2001)
62. V. Cazcarra, P. Zunino, J. Appl. Phys. **51**, 4206 (1980)
63. Y. Kamiura, F. Hashimoto, M. Yoneta, Phys. Stat. Sol. (a) **123**, 357(1991)
64. D. Karg, A. Voigt, J. Krinke, C. Hässler, H.-U. Hoefs, G. Pensl, M. Schulz, H.P. Strunk, Solid State Phenomena **67–68**, 33 (1999)
65. G. Hahn, S. Seren, D. Sontag, A. Schönecker, M. Goris, L. Laas, A. Gutjahr, in *Proceedings of the 3rd WCPEC*, Osaka, 2003 (WCPEC-3 Organising Committee, Arisumi Printing Inc., Japan), p. 1285
66. H.F. Sterling, R.C.G. Swann, Solid State Electron. **8**, 653 (1965)

67. H. Nagel, A.G. Aberle, R. Hezel, in *Proceedings of the 2nd WC PSEC*, Vienna, 1998 (European Commission, Ispra), p. 1422
68. R. Chow, W.A. Lanford, W. Ke-Ming, R.S. Rosler, J. Appl. Phys. **53**, 5630 (1982)
69. J. Szlufcik, K. De Clercq, P. De Schepper, J. Poortmans, A. Buczkowski, J. Nijs, R. Mertens, in *Proceedings of the 12th EC PVSEC*, Amsterdam, 1994 (H.S. Stephens, Bedford), p. 1018
70. L. Cai, A. Rohatgi, IEEE Trans. Electron Devices **ED-44**, 97 (1997)
71. C. Boehme, G. Lucovsky, J. Vac. Sci. Technol. **A 19**, 2622 (2001)
72. A. Aberle, *Crystalline Silicon Solar Cells, Advanced Surface Passivation and Analysis* (Centre for Photovoltaic Engineering, University of New South Wales, Sydney, 1999)
73. F. Duerinckx, J. Szlufcik, Sol. Energy Mater. Sol. Cells **72**, 231 (2002)
74. P. Geiger, G. Hahn, P. Fath, E. Bucher, in *Proceedings of the 17th EC PVSEC*, Munich, 2001 (WIP, Munich), p. 1715
75. P. Geiger, G. Kragler, G. Hahn, P. Fath, E. Bucher, in *Proceedings of the 29th IEEE PVSC*, New Orleans, 2002 (IEEE, Library of Congress, USA), p. 186
76. P. Geiger, G. Kragler, G. Hahn, P. Fath, E. Bucher, in *Proceedings of the 17th EC PVSEC*, Munich, 2001 (WIP, Munich), p. 1754
77. P. Geiger, G. Kragler, G. Hahn, P. Fath, Sol. Energy Mater. Sol. Cells **85**, 559 (2005)
78. V. Yelundur, A. Rohatgi, J.W. Jeong, J.I. Hanoka, IEEE Trans. Electron Devices **ED-49**(8), 1405 (2002)
79. J.-W. Jeong, Y.H. Cho, A. Rohatgi, M.D. Rosenblum, B.R. Bathey, J.P. Kalejs, in *Proceedings of the 29th IEEE PVSC*, New Orleans, 2002 (IEEE, Library of Congress, USA), p. 250
80. G. Hahn, P. Geiger, P. Fath, E. Bucher, in *Proceedings of the 28th IEEE PVSC*, Anchorage, 2000 (IEEE, Library of Congress, USA), p. 95
81. G. Hahn, W. Jooss, M. Spiegel, P. Fath, G. Willeke, E. Bucher, in *Proceedings of the 26th IEEE PVSC*, Anaheim, 1997 (IEEE, Library of Congress, USA), p. 75
82. S. Kleekajai, F. Jiang, M. Stavola, V. Yelundur, K. Nakayashiki, A. Rohatgi, G. Hahn, S. Seren, J. Kalejs, J. Appl. Phys. **100**, 093517 (2006)
83. A. Rohatgi, J.W. Jeong, Appl. Phys. Lett. **82**, 224 (2003)
84. T. Pernau, G. Hahn, M. Spiegel, G. Dietsche, in *Proceedings of the 17th EC PVSEC*, Munich, 2001 (WIP, Munich), p. 1764
85. A. Rohatgi, D.S. Kim, V. Yelundur, K. Nakayashiki, A. Upadhyaya, M. Hilali, V. Meemongkolkiat, *Technical Digest 14th PVSEC, Bangkok*, 2004 (Krissanapong Kirtikara, Bangkok), p. 635
86. J.-W. Jeong, A. Rohatgi, M.D. Rosenblum, J.P. Kalejs, in *Proceedings of the 28th IEEE PVSC*, Anchorage, 2000 (IEEE, Library of Congress, USA), p. 83
87. K. Nakayashiki, D.S. Kim, A. Rohatgi, B.R. Bathey, in *Technical Digest 14th PVSEC*, Bangkok, 2004 (Krissanapong Kirtikara, Bangkok), p. 643
88. M. Käs, G. Hahn, A. Metz, G. Agostinelli, Y. Ma, J. Junge, A. Zuschlag, D. Groetschel, in *Proceedings of the 22nd EU PVSEC*, Milan, 2007, p. 897
89. J. Horzel, G. Grupp, R. Preu, W. Schmidt, in *Proceedings of the 20th EC PVSEC*, Barcelona, 2005, p. 895
90. G. Hahn, A.M. Gabor, in *Proceedings of the 3rd WCPEC*, Osaka, 2003 (WCPEC-3 Organising Committee, Arisumi Printing Inc., Japan), p. 1289

91. G. Hahn, I. Melnyk, C. Dube, A.M. Gabor, in *Proceedings of the 20th EC PVSEC*, Barcelona, 2005, p. 1438
92. J. Horzel, H. Nagel, B. Schum, G. Wahl, A. Seidl, B. Lenkeit, S. Bagus, P. Roth, W. Schmidt, in *Proceedings of the 20th EC PVSEC*, Paris, 2004, p. 435
93. H.F.W. Dekkers, F. Duerinckx, L. Carnel, G. Agostinelli, G. Beaucarne, in *Proceedings of the 21st EU PVSEC*, Dresden, 2006, p. 754
94. G. Blendin, J. Horzel, A. Seidl, A. Teppe, K. Vaas, B. Schum, W. Schmidt, in *Proceedings of the 23rd EU PVSEC*, Valencia, 2008 (in press)
95. S. Seren, M. Kaes, G. Hahn, A. Gutjahr, A.R. Burgers, A. Schönecker, in *Proceedings of the 22nd EU PVSEC*, Milan, 2007, p. 854
96. W. Schmidt, B. Woesten, J.P. Kalejs, Prog. Photovolt. Res. Appl. **10**, 129 (2002)
97. E.A. Alsema, M.J. de Wild-Scholten, in *Proceedings of the 19th EC PVSEC*, Paris, 2004, p. 840

8

Crystal Growth of Spherical Si

Kosuke Nagashio* and Kazuhiko Kuribayashi

Abstract. The spherical Si single crystal with 1 mmϕ has intensively attracted technological interests, since the cutting loss required for Si wafer fabrication can be reduced by 20% in terms of the solar cell application. The basic understanding of crystal growth of Si single crystal ingots cannot be applied directly to spherical single crystals because the critical issue to be controlled is not growth, as for ingots, but nucleation from the undercooled melt for spheres. However, the nucleation is difficult to be controlled externally. In this chapter, our novel approach to grow spherical single crystals is presented after a short review of the historical background for spherical solar cells.

8.1 Historical Background

The recent marked increase in the demand for multicrystalline Si solar cells has caused a shortage in Si raw material, since the solar cells had been fabricated using irregular Si for IC/LSI and/or redundant Si raw material. This unlikely situation has brought much attention to spherical Si solar cells with diameters of ~1 mm, as shown in Fig. 8.1 [1], because the cutting loss required for Si wafer fabrication can be reduced by 20% [2].

The pioneering concept of spherical Si was published by Prince et al. at Hoffman Electronics Corp. in 1960 [3]. Detailed investigation of crystal growth of spherical Si was carried out by McKee at Texas Instruments, Inc. in 1982 [4]. Figure 8.2 shows a drop tube apparatus for producing spherical Si crystals by two-step process. In the first process (a), the Si ingot is melted in a crucible, and the spherical Si crystals are grown during free fall in the drop tube by ejecting the Si melt through an orifice of ~1 mm diameter at the bottom of the crucible. The surface morphology of the samples obtained by this first process is shown in Fig. 8.3, where (a) was nearly single crystal, while (b) was severe polycrystal. It was reported that almost all the samples were polycrystalline and that the difference in their microstructure was caused by the undercooling and/or cooling rate during solidification. To improve the crystallinity, the second process in Fig. 8.2b was introduced. In that method, the spherical

Fig. 8.1. Photograph of spherical Si solar cell [1]

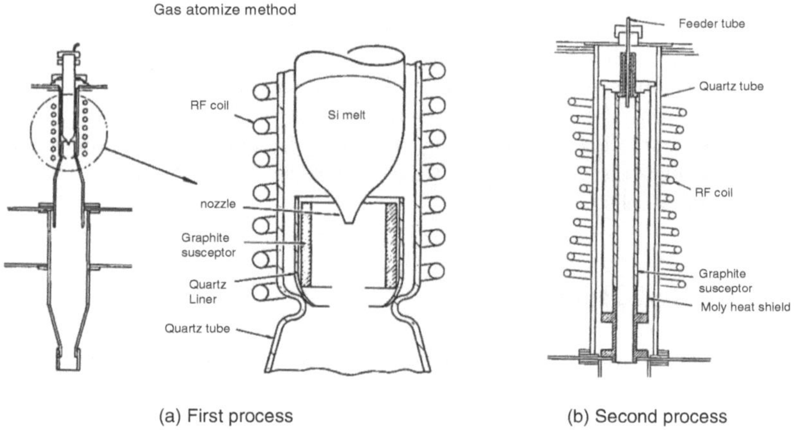

Fig. 8.2. Schematic drawings of McKee's two-step process [4]: (**a**) first process, and (**b**) second process

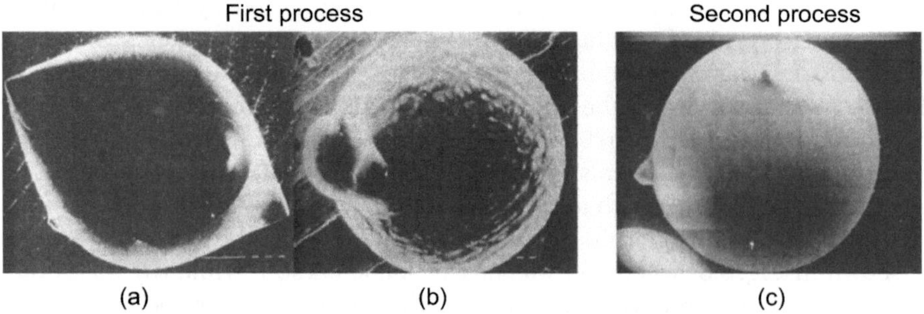

Fig. 8.3. Surface microstructure of spherical Si crystals obtained by (**a**, **b**) first process and (**c**) second process

Si single crystal was formed by remelting and regrowing the polycrystalline spheres in the second process. The shape of the sample after the second process was nearly spherical, and the existence of corns was characteristic of the remelted sample, as shown in Fig. 8.3c. The temperature of graphite susceptor was ~2,200°C, and its length was fairly large, since the sample was heated and melted by radiated heat from the graphite. It was reported that the second remelting process yielded greater than 95% single crystals. However, commercialization was not attained, might be due to the relatively complicated process control. In 1998, Ball Semiconductor Inc. tried to develop the spherical Si single crystal for solar cells using the similar two-step technique [5–7]. The disadvantage of high production costs caused by two-step processes was not solved.

Recently, many venture companies in Japan have been investigating the growth method to produce single crystal spheres directly by the first process [8–12]. When the droplet crystallizes during free fall, it experiences large undercooling prior to nucleation due to the elimination of the crucible wall, which is the main heterogeneous nucleation site. The instability of the growth interface due to large undercooling resulted in dendrite growth. This dendrite growth itself, however, is not the main reason for the severe polycrystallinity, since it is possible to grow a single crystal sphere by growing a single dendrite from a single nucleus throughout the droplet [13]. In spite of this fact, very fine grains have been reported experimentally, as shown in Fig. 8.3b. As mentioned later, this is caused by the fragmentation of dendrites, and an efficient technique to reduce the grain number has not been established. To solve this problem, the general technique to grow single crystal ingots by the Czochralski method cannot be simply applied for the growth of a spherical single crystal. A novel idea is required based on the detailed investigation of crystal growth of Si from the undercooled melt.

8.2 Crystal Growth from Undercooled Melt

The thermal history during crystallization of a droplet can be described on a dimensionless enthalpy – temperature diagram [14], as shown in Fig. 8.4. For simplicity, Newtonian cooling conditions are assumed, where the cooling process is controlled by the heat extraction from the droplet surface, i.e., with negligible temperature gradients inside the droplets. In this case, typically three kinds of thermal histories exist: (1) Isothermal growth: all the latent heat is removed by gas flow, and crystallization proceeds at the melting point, T_{M}, without undercooling. (2) General growth: droplets first experience the undercooling prior to nucleation, and the latent heat for crystallization is adiabatically released into the undercooled melt. Once the droplet temperature reaches T_{M}, the remaining melt crystallizes isothermally at T_{M}. (3) Isenthalpic growth (adiabatic growth): all the latent heat of crystallization is released into the undercooled melt. The term adiabatic growth means that the cooling by

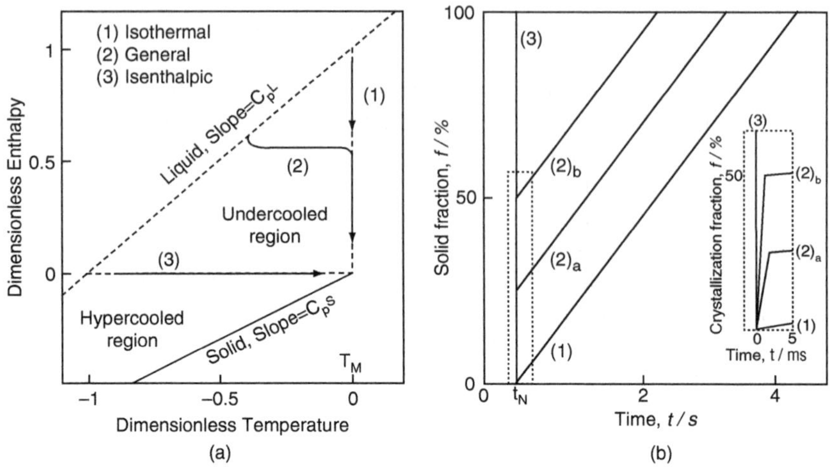

Fig. 8.4. (**a**) Dimensionless enthalpy – temperature diagram showing three crystallization paths. (**b**) Schematic diagram of solid fraction with time

gas flow can be neglected because the heat extraction rate into the undercooled melt is much larger. The undercooling for (3) is defined as the hypercooling limit (ΔT_{hyp}),

$$\Delta T_{\text{hyp}} = \Delta H_{\text{f}}/C_{\text{P}}^{\text{L}}. \tag{8.1}$$

At the hypercooling limit, the whole melt grows below T_{M} [15], unlike the general case where the remaining melt grows isothermally at T_{M}. ΔT_{hyp} for, e.g., Ni, a closely packed metal, is 446 K, while it is 1,977 K for Si because of the large enthalpy of fusion, ΔH_{f}. Therefore, it is not realistic that the Si melt undercools beyond ΔT_{hyp}.

Figure 8.4b shows the schematic diagram of the solid fraction with the lapse of time. In the isothermal case (1), if the nucleation occurs at t_{N}, the droplet completely crystallizes for several seconds at T_{M} by the cooling from the gas flow (growth rate is assumed to be constant, $\sim$1 mm s^{-1}). However, in the case of growth path (2) where the droplet is undercooled at t_{N}, the growth rate of $\sim$m s^{-1} is attained due to the large driving force for growth. Then, after the droplet temperature reaches T_{M}, the growth rate decreases drastically due to isothermal growth controlled by the cooling process of the gas flow. The growth rate in the undercooled melt depends on the degree of undercooling. In general, larger undercooling results in a higher growth rate ($(2)_{\text{b}} > (2)_{\text{a}}$), while the growth rates are the same after the sample temperature reaches T_{M}. Finally, for the isenthalpic case (3) where the melt is undercooled beyond ΔT_{hyp}, crystallization ends within several ms because the growth rate is controlled by the heat release rate of latent heat into the undercooled melt, which depends on the degree of undercooling.

In summary, compared with the single crystal ingot grown at a constant growth rate by the CZ method, the sample crystallized from the undercooled melt includes both nonequilibrium microstructure formed at a high growth rate from the undercooled melt and equilibrium microstructure formed at a low growth rate at T_M. The ratio of nonequilibrium to equilibrium microstructures depends on the degree of undercooling and is expressed as $f = \Delta T/\Delta T_{hyp}$. The mechanical strain is expected to be largely introduced in the nonequilibrium microstructure, which will be an important issue to be resolved in terms of solar cell properties.

8.3 Levitation Experiments: Polycrystallinity Due to Fragmentation of Dendrites

In the commercial application, the drop tube method, as mentioned above, is suitable for mass production. The detailed investigations, such as temperature measurement of each small droplet and in situ observation of microstructure formation are not easy to attain because each droplet is in free fall. Here, the levitation method, where an Si droplet with a diameter of ~8 mm can be levitated by electromagnetic force using an electro-magnetic levitator (EML), as shown in Fig. 8.5, is a powerful investigation technique because the controlled droplet position enables us to measure the surface temperature of the droplet by pyrometer and to observe the crystallization behavior in situ by a high-speed video camera (HSV) [16–18].

Figure 8.6 shows a typical cooling curve during recalescence for the sample crystallized at ΔT (=$T_M - T_N$) = 242 K, where T_N is the nucleation temperature. The release of the latent heats of nucleation and crystallization increased the sample temperature up to T_M, and the solid/liquid coexistence state followed at T_M. If this temperature profile is rotated clockwise by 90°, it can be seen that the shape of Fig. 8.6 is similar to a part of growth path (2).

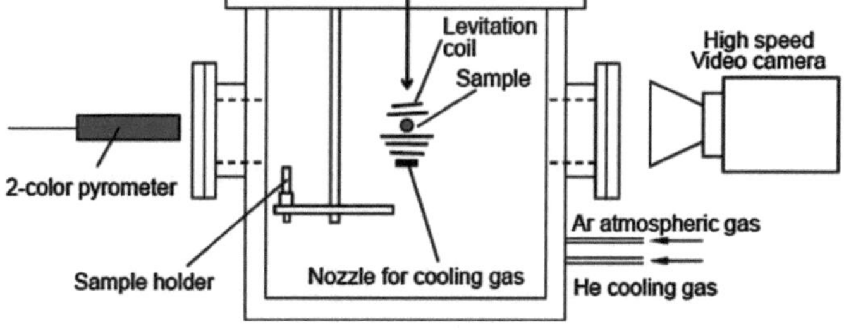

Fig. 8.5. Schematic of electromagnetic levitator with CO_2 laser heating system

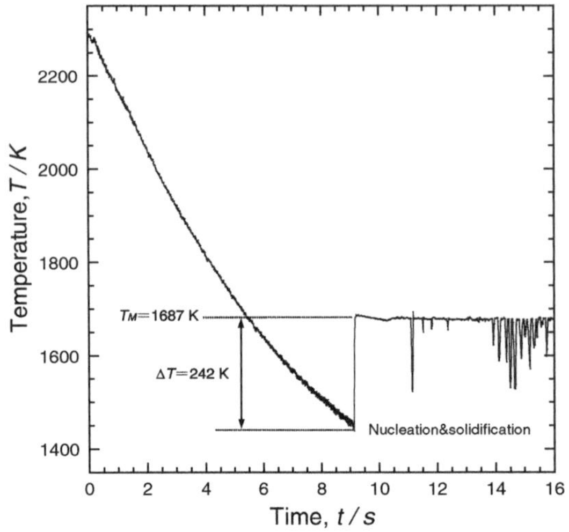

Fig. 8.6. Typical cooling curve during recalescence for the sample crystallized at $\Delta T = 242\,\mathrm{K}$

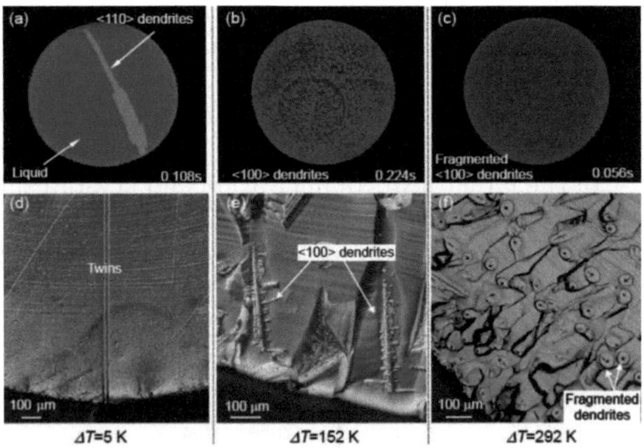

Fig. 8.7. Surface morphology observed during crystallization by HSV and cross-sectional microstructure by SEM for samples crystallized in EML at different undercoolings (**a**,**d**) 5 K, (**b**,**e**) 152 K, and (**c**,**f**) 292 K

Figure 8.7a–c shows the surface morphology observed during crystallization by HSV, and Fig. 8.7d–f shows cross-sectional microstructure by scanning electron microscopy (SEM) for Si samples crystallized at different undercoolings: (a,d), 5 K; (b,e), 152 K; and (c,f), 292 K. For sample (a), crystallized at low undercooling, <110> dendrites grew directly from the surface, while <100> dendrites covered the droplet surface for the sample crystallized at

$\Delta T > 100$ K. Here, the <uvw> dendrite indicates a dendrite growing in a <uvw> direction. Detailed analysis of the Si facet dendrites is described elsewhere [19–21]. Moreover, when the degree of undercooling exceeds 200 K, the connection of dendrites was not clear in the HSV image (c). As observed from the cross-sectional microstructures that corresponded to the HSV images for different undercoolings, the grain size decreased with increasing undercooling. Importantly, the large <100> dendrites were clearly observed in the sample at $\Delta T = 152$ K, while fragmentation of <100> dendrites occurred in the sample at undercoolings larger than 200 K. In this study, the sample was doped with B at 10^{20} cm^{-3}, which marked the shape of dendrites by the segregation of B. It was noted that the growth rate and microstructure at different undercoolings in B-doped Si sample were very similar to those in a pure Si sample [18].

In the case of crystallization from the undercooled melt, the driving force for the growth was initially much larger than that for isothermal crystallization because of the larger free energy difference between solid and liquid. The solid/liquid interface shape obtained a dendritic microstructure with a larger surface area, since the growth is controlled by the thermal diffusion of latent heat into the undercooled melt, and a larger surface area is effective to release the latent heat. However, once the sample temperature reached T_{M} and the driving force for the growth decreased, the larger surface area of dendrites produced the driving force for fragmentation by Gibbs–Thomson effect even during the isothermal solidification. The fragmented dendrites acted as nuclei for further crystallization, which resulted in a sample with very fine grains, as shown in Fig. 8.7f. In other words, the growth morphology of the dendrites determined the grain size in the final microstructure. Recently, the fragmentation process of Si dendrites was detected by time-resolved X-ray diffraction experiments using a synchrotron radiation source [22, 23]. The fragmentation of Si facet dendrites was complete within ~25 ms after crystallization for $\Delta T = 261$ K [23]. The fragmentation of dendrites has been observed widely in metallic systems and is considered to be one of the mechanisms for fine grained microstructure [24].

Figure 8.8 shows the grain size as a function of ΔT. The number of grains in a 1×1 mm^2 area was determined. The grain size decreased with increasing ΔT because of the fragmentation of dendrites. The severe polycrystallinity in Fig. 8.3b reported by Texas Instrument Inc. is indeed formed by the fragmentation of dendrites at $\Delta T > 200$ K. It is difficult to avoid the fragmentation, as it is completed within ~25 ms. However, if we focus on the solidification process of the <110> dendrites at low undercoolings, the <110> dendrites grow stably and slowly at the plateau period without fragmentation, as shown in Fig. 8.7a, d. Therefore, the strategy for single crystal sphere is to control the undercooling to below 100 K. This has been successfully demonstrated by enlarging the <110> dendrites grown forcedly using a trigger needle at low undercooling [25].

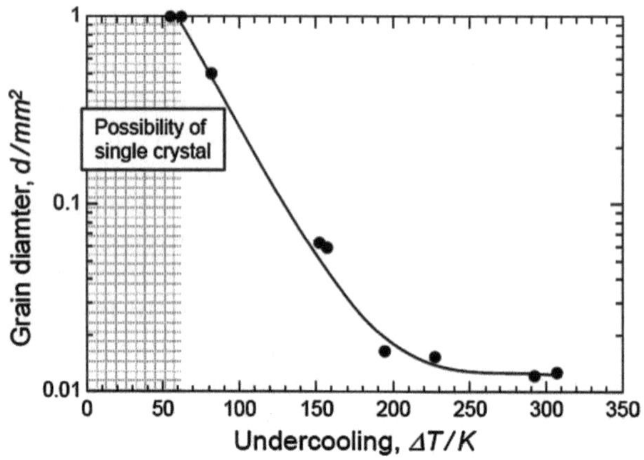

Fig. 8.8. Grain size as a function of ΔT

8.4 Spherical Si Crystal Fabricated by Drop Tube Method

Based on the fundamental study of crystallization behavior of Si droplets using EML, it is clearly shown that controlling the undercooling. Thus, the nucleation is the key to obtain single crystals directly using the first process of drop tube methodology. The following two methods are considered: (1) to introduce electrically inactive foreign particles as heterogeneous nucleation sites into the melt before ejection, and (2) to place a substrate tilted slightly from the normal direction just below the orifice for nucleation. As a primary experiment, both introducing foreign particles such as SiN, BN, Al_2O_3, and Y_2O_3 into the melt and placing a plate were attempted. However, they were not reproducible.

When we focus the work on spherical single crystals conducted by McKee [4], it is understood that the key concept for single crystallization of Si is hidden in the two-step process, as shown in Fig. 8.9a. The first step is the fabrication of a large number of spherical spheres with a uniform diameter using the drop tube method. The second step is single crystallization by remelting them during free fall through the cylindrical graphite susceptor. The idea underlying the second step, though not explicitly described in his paper, is speculated to be that the sample is only partially melted, leaving a small grain at the center as a nucleus for subsequent growth. In the second step, therefore, the droplet never experiences large undercooling because each droplet contains the nucleus within. The two-step process, however, is no longer a low-cost production process in terms of productivity. Here, if the two process steps are combined together into a single-step process, i.e., if the sample is ejected in the semisolid state at $\sim T_M$, where small Si solid particles exist in the melt, the droplets may include a single solid particle and crystallize at T_M without undercooling. Figure 8.9b shows the ejecting system with a quartz

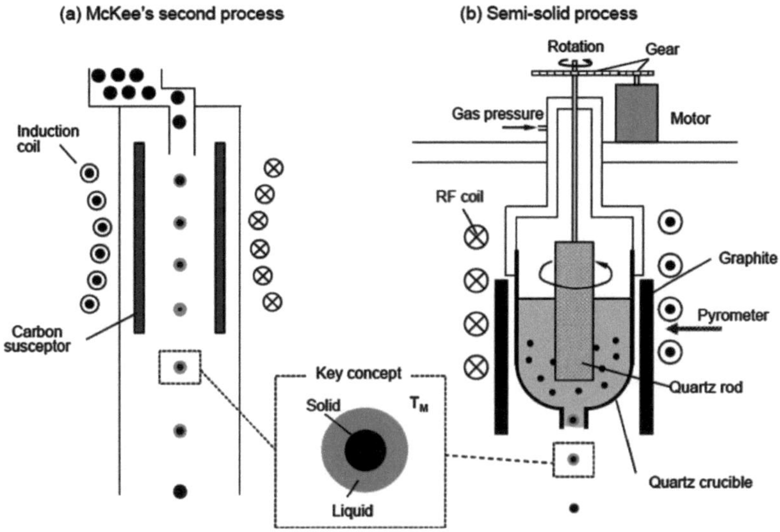

Fig. 8.9. Schematic drawings of (**a**) McKee's second process and (**b**) semisolid process

stirring apparatus for the semisolid process, which is placed on the top of a drop tube with the length of 26 m. The graphite susceptor is first heated by an RF induction coil, and pure Si lumps are heated and melted by its radiation. Moreover, the quartz rod can be rotated by external control to grind the crystallized large grains into small particles.

To evaluate the validity of this semisolid process, at first, a comparison with conventional method was carried out, where the melt superheated by 100 K was ejected through an orifice [26]. Figure 8.10 shows the temperature profile of the graphite susceptor before the ejection, which was monitored by the pyrometer. After complete melting, stirring began at (a) and the melt was cooled to $\sim T_M$. Subsequently, the Si melt was ejected in its semisolid state by Ar gas pressure after stirring at T_M for ~30 s. Alternatively, following the conventional method, when the temperature reached at the position of (a) (see Fig. 8.10), the Si melt was ejected while in a superheated condition.

The spherical samples collected in the bottom chamber were categorized into three groups using sieves according to their diameters, 355–600, 600–850, and 850–1,000 μm, with probabilities of roughly 75, 20, and 5%, respectively. Samples of three types of surface morphologies were observed for all the diameter ranges, even with the two different ejection methods applied in this study. Figure 8.11 shows the surface microstructures of three types of droplets in the range of 355–600 μm taken by SEM. In general, there exist heterogeneities with various catalytic potencies for nucleation in the melt. For the drop tube process, the melt was divided into a considerable number of small droplets, and the heterogeneities were statistically distributed

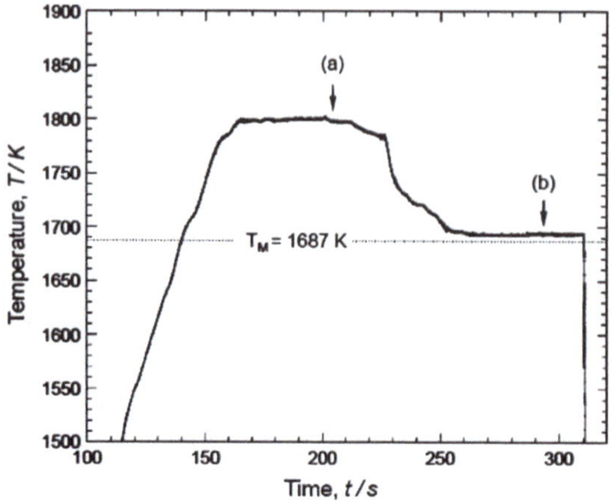

Fig. 8.10. Temperature profiles taken by pyrometer for two different ejection methods: (**a**) ejection above melting point, and (**b**) ejection in the semisolid state at T_{M}

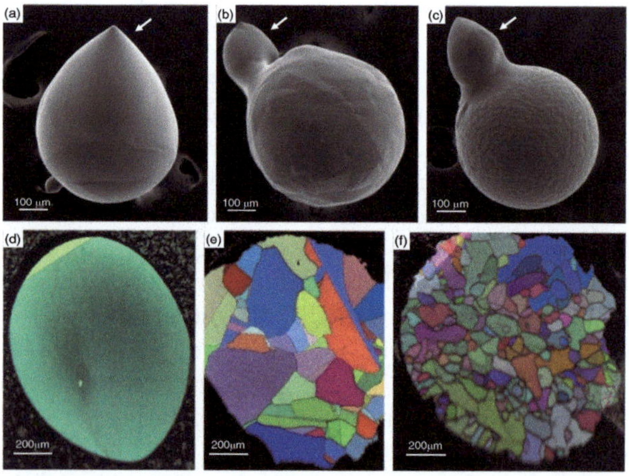

Fig. 8.11. (**a**)–(**c**) SEM images of three types of surface microstructure in range of 355–600 μm obtained by the drop tube method. (**d**)–(**f**) EBSP crystallographic orientation maps for cross-sections of typical samples, which corresponds with Fig. 8.3**a**–**c**, respectively

among some of them. Therefore, although almost all the droplets without heterogeneities experienced large undercoolings, some droplets that included the heterogeneities solidified at undercoolings determined by their catalytic potencies [27]. It is understood that the three types of samples solidified at different undercoolings. However, the shortcoming of the drop tube method is the difficulty of measuring the temperature of each droplet. To correlate the surface microstructure and undercooling, the surfaces of the droplets solidified at a predetermined undercooling in EML were compared with those in drop tube. This comparison suggests that the spherical samples (a), (b), and (c) solidified at low ($\Delta T < 100\,\mathrm{K}$), medium ($100 < \Delta T < 150\,\mathrm{K}$) and high ($\Delta T > 200\,\mathrm{K}$) undercoolings, respectively.

Extrusions were always observed for solidified samples, as shown by arrows in Fig. 8.11a–c. The formation of the extrusions is strongly related to the solidification process at the surface, since Si expands by a factor of 1.1 during solidification. At low undercooling (a), the expansion can be easily relaxed because the straight <110> dendrites that grow on the surface (see Fig. 8.7a) do not cover the droplet surface. However, at medium and high undercoolings (b) and (c), the droplet surface is completely surrounded by the <100> dendrites with 4-fold symmetry, as shown in Fig. 8.7b, c, after which a small meniscus of the liquid appears after breaking the thin solid layer. Therefore, it is possible to classify the approximate undercooling for each sample on the basis of surface morphology.

Figure 8.11d–f show the crystallographic orientation maps taken by an electron backscatter diffraction pattern (EBSP) apparatus for cross-sections of samples in Fig. 8.11a–c. It is clearly shown that the grain size decreased with increasing undercoolings from (a) to (c), since each grain is expressed by a corresponding color. Twins are mainly observed in the samples crystallized at low and medium undercoolings, as shown in Fig. 8.11d, e. The spherical grains observed in Fig. 8.11f indicate that the fragmentation of the <100> dendrites occurred, and that the undercooling was more than 200 K for this sample. In terms of solar cell performance, sample (a) is preferable because of the low number of grain boundaries, which act as carrier recombination sites.

Figure 8.12 shows the typical photoluminescence (PL) spectra of three types of samples with 1 kΩ cm at 4.2 K. "$\times$50" and "$\times$1" indicate the magnification of longitudinal axis. In the case of the sample crystallized at (a) low and (b) medium undercoolings, the boron bound transverse optical phonon (B_{TO}), which is generally observed in the high crystallinity sample, was observed at 1.09 eV. However, it was not observed in the sample crystallized at high undercooling (c), but D_1–D_4 lines related to dislocations were observed. Therefore, the dislocation density in (c) would be very large, compared with (a) and (b). A strong peak at $\sim$0.78 eV can be seen in (c), but it is not identified at this moment. Moreover, from the similarity of the spectra of (a) and (b), it can be said that a single crystal such as (a) is not necessary for solar cells, and a polycrystalline sample (b) is acceptable. This resulted from the fact that grain boundaries mainly consist of $\Sigma 3$ twin boundaries. Qualitatively, crystallinity was seen to deteriorate with increasing volume fraction of nonequilibrium microstructure.

Figure 8.13 shows the probabilities of the three types of spheres for the two ejection methods. It is noted that these probabilities were obtained for the samples with diameters of 355–600 μm as approximately 75% of the collected samples belong to this diameter range. As mentioned above, three types of spheres can be obtained by both ejection methods. When the melt was

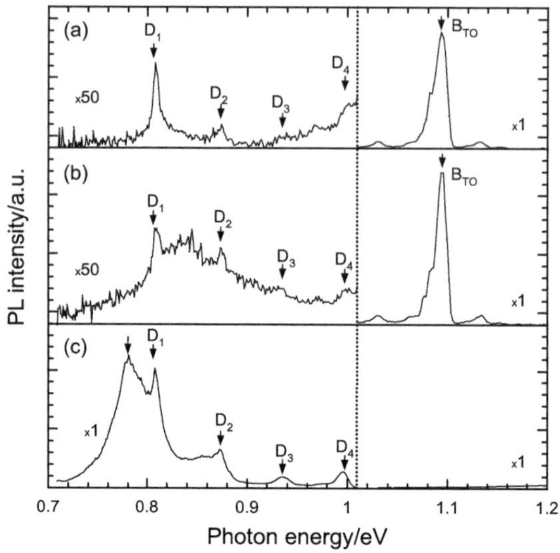

Fig. 8.12. Photoluminescence measurement at 4.2 K on samples crystallized at (**a**) low undercooling, (**b**) medium undercooling, and (**c**) large undercooling

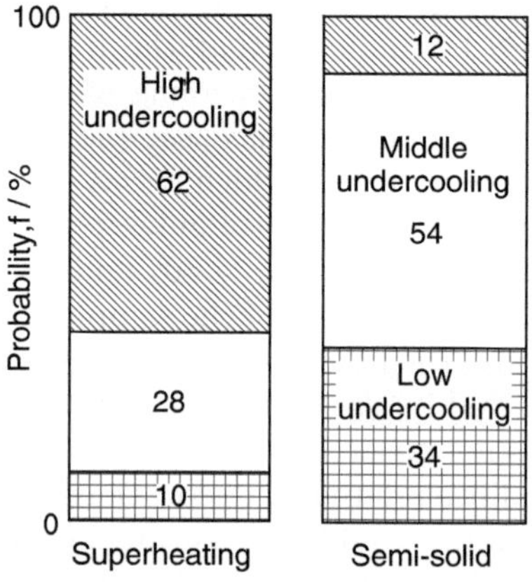

Fig. 8.13. Probabilities of three types of spheres with diameters of 355–600 μm for two ejection methods

ejected at ~100 K above T_M, the percentage of spheres crystallized at low undercoolings was ~10%. However, the percentage increased to 34% using the semisolid process. This indicated that the semisolid process is effective in fabricating spherical Si crystals with few grains. Recently, Liu et al. tried to introduce nucleation sites by blowing pure Si powder at falling droplets and reported the improvement of the probability for spheres crystallized at low undercoolings [28].

Finally, another advantage of the semisolid process is described, based on simple heat transfer calculation during free fall. In the case of the sample superheated above T_M, the falling distance is more than 10 m, since the sample is undercooled due to suppression of nucleation. However, for the semisolid process, where each droplet includes a tiny solid particle before ejection, the falling distance required for complete crystallization is considerably reduced to 3 mm for $f_S > 0.1$ because of the lack of undercooling. This information is useful in terms of capital investment.

8.5 Summary

The recent approach to obtain single crystal spheres directly from the first process, using a drop tube, was presented after a short review of the historical background for spherical solar cells. The basic understanding of crystal growth of Si single crystal ingots has been accumulated over a long time period. However, it cannot be applied directly to spherical single crystals because the critical issue to be controlled is not growth, as for ingots, but nucleation for spheres. However, the nucleation is difficult to control externally. The present semisolid process focuses on ejection at the state without undercooling, not on the external control of nucleation. The semisolid process has been developed as one of the casting techniques for light metals such as Al–Si alloys. In the case of alloys, the equilibrium solid fraction in the solid/liquid coexisting region can be retained by adjusting both the temperature and initial composition. However, for a pure substance, Si, the solid fraction is a function of time at T_M. The controlling technique of the solid fraction is a critical issue for further study.

Acknowledgements

The authors thank Prof. M. Tajima, ISAS/JAXA for PL measurements. This work was financially supported by a Grant-in-Aid for Scientific Research from the Ministry of Education, Culture, Sports, Science and Technology, Japan.

References

1. private communication, K. Taira, Kyosemi co
2. N. Tanaka, Nikkei Microdevices, 2006, March, 25–45 [in Japanese]
3. M.B. Prince, 14th Annual Proceedings of Power Sources Conference, 26 (1960)
4. W.R. McKee, IEEE Trans. Components Hybrids Manuf. Technol. **5**, 336 (1982)
5. Ball Semiconductor Technology conference, San Francisco, CA, 1998 (It is available at http://www.ball.co.jp/sale/index.html)
6. R. Toda, Ceramics **36**, 133 (2001) [in Japanese]
7. M. Yoshida, Nikkei Electron. **Nov**, 173 (2000) [in Japanese]
8. T. Minemoto, C. Okamoto, S. Omae, M. Murozono, H. Takakura, Y. Hamakawa, Jpn. J. Appl. Phys. **44**, 4820 (2005)

9. M. Murozono, Energy **37**, 46 (2004) [in Japanese]
10. K. Taira, N. Kogo, H. Kikuchi, N. Kumagai, N. Kuratani, I. Inagawa, S. Imoto, J. Nakata, M. Biancardo, Tech. Dig. **PVSEC-15**, 202 (2005)
11. S. Masuda, K. Takagi, Y.-S. Kang, A. Kawasaki, J. Jpn. Soc. Powder Metall. **51**, 646 (2004)
12. W. Dong, K. Takagi, S. Masuda, A. Kawasaki, J. Jpn. Soc. Powder Metall. **53**, 346 (2006)
13. Y.S. Sung, H. Takeya, K. Togano, Rev. Sci. Instrum. **72**, 4419 (2001)
14. C.G. Levi, R. Mehrabian, Metall. Trans. A **13A**, 221 (1982)
15. K. Nagashio, K. Kuribayashi, Acta Mater. **49**, 1947 (2001)
16. T. Aoyama, Y. Takamura, K. Kuribayashi, Metall. Mater. Trans. A **30A**, 1333 (1999)
17. Z. Jian, K. Nagashio, K. Kuribayashi, Metall. Mater. Trans. A **33A**, 2947 (2002)
18. K. Nagashio, H. Okamoto, K. Kuribayashi, I. Jimbo, Metall. Mater. Trans. A **36A**, 3407 (2005)
19. K. Nagashio, K. Kuribayashi, Acta Mater. **53**, 3021 (2005)
20. K. Nagashio, K. Kuribayashi, J. Jpn. Assoc. Cryst. Growth **32**, 314 (2005) [in Japanese]
21. K. Fujiwara, K. Maeda, N. Usami, G. Sazaki, Y. Nose, K. Nakajima, Scripta Mater. **57**, 81 (2007)
22. K. Nagashio, M. Adachi, K. Higuchi, A. Mizuno, M. Watanabe, K. Kuribayashi, Y. Katayama, J. Appl. Phys. **100**, 033524 (2006)
23. K. Nagashio, K. Nozaki, K. Kuribayashi, Y. Katayama, Appl. Phys. Lett. **91**, 061916 (2007)
24. A. Karma, Int. J. Nonequilib. Process. **11**, 201 (1998)
25. T. Aoyama, K. Kuribayashi, Acta Mater. **51**, 2297 (2003)
26. K. Nagashio, H. Okamoto, H. Ando, K. Kuribayashi, I. Jimbo, Jpn. J. Appl. Phys. **45**, L623 (2006)
27. J.H. Perepezko, in *Proceedings of the 2nd Int. Conf. Rapid Solidification Processing*, ed. by R. Mehrabian et al. (Claitor's Pub. Division, Baton Rouge, LA, 1980), p. 56
28. Z. Liu, T. Nagai, A. Masuda, M. Kondo, J. Appl. Phys. **101**, 093505 (2007)

9

Liquid Phase Epitaxy

Alain Fave

Abstract. Liquid phase epitaxy (LPE) is a growth technique that can be suitable for photovoltaic applications regarding its simplicity and its capacity to produce high-quality thin film. The growth of Silicon proceeds from a molten solution (metal + Si), which is slowly cooled. Temperature range is typically 700–1,000°C and growth rate can be as high as 1 μm min^{-1}. In this chapter, we first introduce the fundamental principles of LPE and then we discuss on the influence of the metallic solvent (In, Sn, Cu, Al...), the temperature range on the growth rate and on the quality of the epitaxial layers. We also review epitaxial growth on polycrystalline silicon or foreign substrates. We finally discuss the development of high throughput LPE deposition equipment.

Liquid phase epitaxy (LPE) has been used for many years for the growth of semiconductors' multilayers, especially for optoelectronic applications with III–V based materials (GaAs, AlAs, InP, and related compounds) [1]. Efficient solar cells were also fabricated, especially for use under high-concentration from GaAs [2] and have reached up to 18% for Si.

The basis of LPE is the control of the liquid–solid phase equilibrium based on the solubility of silicon in a metallic solvent (In, Sn, Al, Ga, ...) [3]. The driving force for crystallisation of Silicon is the slow cooling of this saturated solvent. LPE have several advantages for solar cell production [4]: high deposition rates (0.1–2 μm min^{-1}) at moderate temperatures (typically 850–1,050°C); impurity segregation/rejection to the melt, thus, avoiding accumulation of electrically active impurities; since it is conducted close to thermodynamic equilibrium, it leads to epitaxial layers with low density of structural defects; low capital equipment and operating costs; and finally the selectivity of the growth, the feasibility of lateral overgrowth, and its ability to produce faceted crystals. Some disadvantages can be, however, pointed out, especially the difficulty in upscaling, and the layers' inhomogeneity and reproducibility.

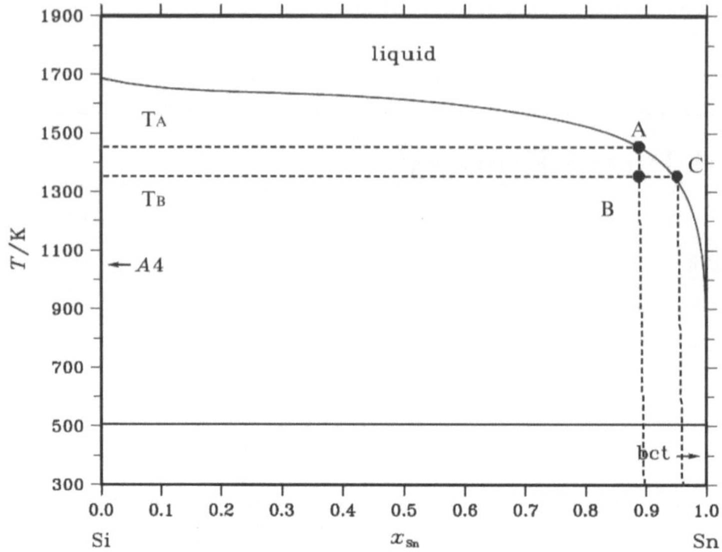

Fig. 9.1. Calculated phase diagram of Sn–Si (from [5])

9.1 Description

The principles of the epitaxial growth can be described using the phase diagram of the binary system Si–Sn as an example (Fig. 9.1). The upper curve represents the so-called "liquidus". At point A, the system is a liquid mixture of Sn solvent and Si, at temperature T_{A} with a composition X_{A}: the solution is saturated with silicon.

To obtain this composition, a metallic solvent is contacted with a silicon source at this temperature T_{A} during few hours. This leads to the saturation of the solvent in Si, at X_{A} concentration. Then, the silicon source is removed. If this liquid is brought into contact with a silicon substrate at this temperature, the equilibrium is not changed. However, when the solution is cooled down to T_{B}, the system presents two phases: silicon solid phase appears and the new concentration of the liquid phase in Si becomes X_{c}. If the cooling process is slow (few Kelvin per minute or less), "parasite" germination rate within the solvent is close to zero and epitaxial deposition of silicon occurs only on the substrate.

9.2 Kinetics of Growth

The phase diagram indicates the maximum fraction of silicon that can precipitate onto the substrate. But in practice, the kinetic growth of a layer from a solution is controlled by the following mechanisms:

1. Volumic transport of Si atoms by diffusion, convection or force flow
2. Transport through the boundary layer
3. Adsorption on the surface of the crystal seed
4. Surface diffusion to a growing step of Si
5. Linking to the step
6. Diffusion along the step
7. Attachment in the crystal on a site of the step

Mechanisms (1) and (2) are related to the mass transport, while the mechanisms (3–7) are related to the surface kinetics. The growth kinetics is, therefore, a combination of these two different processes. Depending on the experimental conditions, one of them will be preponderant and will control the kinetics of growth.

Three methods of growth have been developed and applied for semiconductors, which lead to different kinetics consideration:

- "Equilibrium cooling" or "ramp cooling" is a linear growth cooling from the equilibrium temperature. The temperature of the solution is lowered at a rate a (K min^{-1}) from the liquidus temperature (T_{A}) to the end temperature (T_{E}), while in contact with substrate. In this case, the kinetics of growth is limited by diffusion.
- "Step-cooling" growth: the solution is kept at a temperature ΔT below the T_{A} and then brought into contact with the substrate at constant temperature. The kinetics of growth is, thus, limited by the surface kinetics.
- "Super cooling" growth is a combination of equilibrium and step-cooling techniques. The temperature of the solution is decreased of ΔT below the T_{A}, brought into contact with the substrate and then the solution is cooled down at a rate a (K min^{-1}).

In order to determine the thickness of the resulting epitaxial layer, we will use the following assumptions:

1. The system is isothermal: There is no temperature gradient due to the furnace geometry or the holder. Cooling rate is low enough to neglect heat dissipation, due to crystallisation at the interface.
2. Diffusion of silicon into the solid phase is much lower than in the liquid phase. We neglect the solute diffusion from interface into the solid.
3. We can also neglect silicon flow due to the interface displacement compared to the silicon flow due to diffusion.

Under these conditions, the general equation of growth in solution controlled by the mass transport can be written as follows (in one dimension):

$$D\frac{\partial^2 C}{\partial x^2} + \nu\frac{\partial C}{\partial x} = \frac{\partial C}{\partial t} \tag{9.1}$$

where,

$C = C(x,t)$ is the solute (silicon) concentration. It is a function of growth time and is initially homogeneous,
$D = D(T)$ is the diffusion coefficient of silicon in the solvent, and
$\nu = \nu(t)$ is the growth rate.

Using the above assumptions, growth rate and thickness are calculated by solving the diffusion equation of silicon in the solution. If epitaxy takes place on a flat substrate with a weak supersaturation of the solution (this is generally the case for LPE), diffusion can be considered as one-dimension and stationary phenomena. Equation (9.1) becomes:

$$D_{\mathrm{Si}}\frac{\partial^2 C^{\mathrm{L}}_{\mathrm{Si}}(x,t)}{\partial x^2} = \frac{\partial C^{\mathrm{L}}_{\mathrm{Si}}}{\partial t}, \tag{9.2}$$

$$D_{\mathrm{Si}}\frac{\partial C^{\mathrm{L}}_{\mathrm{Si}}(0,t)}{\partial x} = \nu(t)\left[C^{\mathrm{S}}_{\mathrm{Si}}(0,\, t) - C^{\mathrm{L}}_{\mathrm{Si}}(0,\, t)\right], \tag{9.3}$$

$$\delta\mathrm{e} = \int_0^t \nu(t)\,dt = \int_0^t \frac{D_{\mathrm{Si}}}{C^{\mathrm{S}}_{\mathrm{Si}}(0,\, t) - C^{\mathrm{L}}_{\mathrm{Si}}(0,\, t)}\frac{\partial C^{\mathrm{L}}_{\mathrm{Si}}(0,\, t)}{\partial x}dt, \tag{9.4}$$

where t: time
D_{Si}: diffusion coefficient of Si in the liquid
C_{Si}: Si concentration
$\delta\mathrm{e}$: thickness of the epitaxial layer.

These equations lead to simple mathematical relationship between the film thickness and the time with the following assumptions:

4. The system is semi-infinite and duration of growth is short compared to the time of diffusion.
5. It is assumed that for small cooling intervals, the liquidus composition is a linear function of temperature:

$$T - T_{\mathrm{e}} = m(C - C_{\mathrm{e}}). \tag{9.5}$$

6. The diffusion coefficient of Si in the liquid phase is independent of temperature.

The equation giving $\delta\mathrm{e}$ depends on the boundary conditions associated with the different LPE growth techniques.

In the case of "equilibrium cooling":

$$C(x, t = 0) = C_0. \tag{9.6}$$

The temperature is reduced linearly with time, at a rate a, such that:

$$T(t) = T_0 - aT, \tag{9.7}$$

$$C(0,t) = C_0 - \left(\frac{a}{m_{\mathrm{Si}}}\right)t, \tag{9.8}$$

where m_{Si} is the slope of the liquidus curve and a is the cooling rate.

The solution was reported by Hsieh [6]:

$$\delta \mathrm{e} = \frac{4}{3} \frac{a}{m_{\mathrm{Si}} C_{\mathrm{Si}}^{\mathrm{S}}} \sqrt{\frac{D_{\mathrm{Si}}}{\pi}} t^{3/2}. \tag{9.9}$$

For the "step-cooling" growth, Boundary condition is:

$$C(0,t) = C_0 - \frac{\Delta T}{m_{\mathrm{Si}}} \tag{9.10}$$

and then:

$$\delta \mathrm{e} = \Delta T \frac{2}{m_{\mathrm{Si}} C_{\mathrm{Si}}^{\mathrm{S}}} \sqrt{\frac{D_{\mathrm{Si}}}{\pi}} t^{1/2}. \tag{9.11}$$

The "super cooling" growth is a combination of the equilibrium and step-cooling modes. Therefore, the solution (9.12) is given by a linear combination of (9.9) and (9.11) in a first approximation.

$$\delta \mathrm{e} = \frac{2}{m_{\mathrm{Si}} C_{\mathrm{Si}}^{\mathrm{S}}} \sqrt{\frac{D_{\mathrm{Si}}}{\pi}} \left[\Delta T t^{1/2} + (2/3)\, a t^{3/2} \right]. \tag{9.12}$$

In all cases, the amount of silicon available for the growth is limited by the solubility at the saturation temperature. For photovoltaic applications, there are also some specific requisites:

1. *Thickness*: The solar spectra absorption has to be maximised. About 50 μm thick film of crystalline silicon is necessary to absorb 80% of the solar spectra. If there is an efficient optical confinement, it can be reduced to 20 μm or even less
2. Good electrical properties of the epilayer (collection of the photogenerated carriers and generation of the current are linked to the diffusion length or lifetime of the minority carriers)

Moreover, taking into account the economic factors of solar cell production, one has to consider the throughput (growth rate), and the cost of the system (temperature, solvent, gas...). Having this in mind, we will discuss the different possibilities concerning the selection of the solvent.

9.3 Choice of the Solvent

The selection of the solvent is an important issue. One will take into account expectations concerning growth rate, temperature range and electrical properties of the epitaxial layers. The main criteria are listed below:

1. High solubility of silicon in the solvent: This parameter, or the evolution of solubility versus temperature during the growth has to be compatible with a reasonable time of obtainment of the active layer.
2. Low solubility of solvent into the silicon: During epitaxial process, atoms of the solvent are incorporated in the Si crystal. Their incorporation can modify the electronic properties of the layer. Actually, many metallic impurities act as recombination centre or as dopant and reduce the lifetime of minority carriers. Solvent purity is also an important parameter to avoid other impurities.
3. Low melting temperature and low vapour pressure: A low melting temperature allows better and faster homogenisation of the melt and, thus, a homogeneous growth. A low vapour pressure of the solvent is suitable to prevent loss of solvent during growth process.
4. Low toxicity and availability of the solvent: For instance, materials like antinomy (Sb), gold (Au) or silver (Ag) offer high solubility of silicon. However, the first one is very toxic and the others are quite expensive.

The "perfect" solvent does not exist. Each metal presents advantages and disadvantages. But among the different materials that have been proposed and studied, three of them can be selected: tin (Sn), indium (In) and copper (Cu). Gallium (Ga) or aluminium (Al), in spite of the high solubility of silicon (Table 9.1), are not good candidates; because of the high doping level, they will induce into the silicon epilayer. However, they can be added to the melt to dope the epitaxial layer. Lead (Pb) does not incorporate into the grown layer and is electrically inactive, but it has low silicon solubility and high vapour pressure [31]. Bismuth (Bi) and zinc (Zn) also have high vapour pressure.

A combination of two or more solvents can also be used. This is mainly the case, when lowering the growth temperature is the main objective (see Sect. 9.6).

9.3.1 Influence of the Substrate Surface

Epitaxial layer can be easily obtained on single crystal silicon substrate. (111) and (100) oriented substrates are commonly used. Due to the high affinity of

Table 9.1. Metal solvent melting points and silicon solubility

Metal solvent	Melting point °C	Atomic solubility of silicon	
		800°C(%)	1,000°C(%)
Indium (In)	156	0.38	1.9
Tin (Sn)	232	0.62	2.5
Copper (Cu)	1,083	30	30
Aluminium (Al)	660	27	44
Gallium (Ga)	30	3.9	12
Lead (Pb)	327	0.01	0.1

silicon with oxygen, formation of native oxide on the growth substrate has to be taken into consideration. In fact, it will hinder good wetting of the surface by the saturated melt. It can be prevented by using adequate cleaning procedure prior to the growth and the use of hydrogen as a growth ambient.

The cleaning procedure can be as follows:

1. De-greasing of the substrate, using solvent like acetone
2. Removal of native oxide with buffered HF
3. Formation of chemical silicon oxide with H_2SO_4–H_2O_2 (1:1)
4. Removal of this oxide using buffered HF

(Steps 1, 3 and 4 can be skipped if original substrate is already cleaned).

Moreover, since it is difficult (and expensive) to prevent the presence of oxygen and water vapour during the growth, a hydrogen ambient has to be used. It will reduce the silicon native oxide formed during the saturation step. The hydrogen has to be ultra-pure, which is commonly achieved by using Pd-diffusion membrane purification. Hydrogen presents potential hazards when exposed to air, and it is possible to reduce the risks by diluting hydrogen with a neutral gas like argon (H_2:15%–Ar:85%) [7].

Another possibility, which has to be used when temperature growth is lower than 900°C, is the addition of reducing agent in the melt like Al or Mg, which can lead to an oxide-free surface [8]. However, because Al is a p-type dopant for silicon, one has to reduce its amount or to use a two-melts process, one to remove native oxide, and one for the growth of the active layer [9].

Finally, meltback of the substrate is also suitable, when the surface of the substrate is not flat (for instance when using multicrystalline or metallurgical grade silicon). The substrate is contacted with the melt at a temperature above liquidus one. Then, return to equilibrium is achieved after dissolution of a thin region of the substrate, leaving the surface smoother and more accurate for the following growth.

9.4 Experimental Results

The thickness of the epitaxial layer depends on the temperature, solvent, cooling rate and amount of solvent and silicon, i.e. the weight of the melt compared to the surface of the substrate.

Figure 9.2 shows the dependence of the thickness versus the growth temperature. The LPE system is a laboratory scale horizontal graphite sliding boat (substrate area is 2 cm^2) with a palladium-purified hydrogen atmosphere. Graphite is used as crucible, since it presents: (1) no reactivity and low wettability with metallic solvent, (2) chemical inertness in growth atmosphere, (3) no contamination, especially electronically active impurities, such as P, B and (4) machinable [10]. Cleaning of the graphite can be done using HCl. Then, rinsing with ultrasonic and finally, backing for several hours.

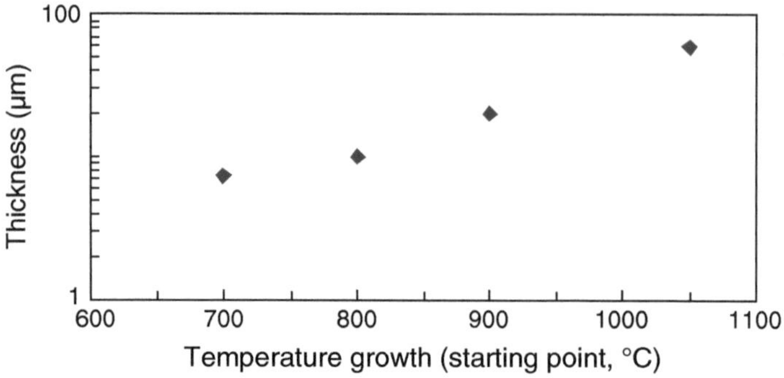

Fig. 9.2. Thickness versus temperature with Sn solvent, 0.5°C min^{-1} cooling rate and 2 h growth time

The experiments were done under the following conditions: 7 g of Sn (purity 5N), 2 h of growth, 0.5°C min^{-1} cooling rate. Since the amount of silicon incorporated in the melt increases with the temperature, it is possible to grow thick epitaxial layers. However, for low-temperatures (700–800°C), incorporation of silicon is low and limits the kinetics of growth. At this temperature range, thermodynamic properties of metallic alloys can increase the kinetics (see Sect. 9.6).

9.4.1 Growth with Sn and In Solvent in the 900–1,050°C Range

Thickness of the epitaxial layer depends mainly on the cooling rate and the duration of the growth. For a 0.5°C min^{-1} cooling rate, starting growth at 1,050°C, thickness increases linearly versus time for both solvents (Fig. 9.3a). Growth rate is twice higher with Sn than In (30 μm h^{-1} vs. 15 μm h^{-1}).

If we use moderate cooling rates (0.25°C min^{-1}, 0.5°C min^{-1} and 1°C min^{-1}), experimental results presented in Fig. 9.3b show that the final thickness is independent from the cooling rate and depends only on the temperature drop (undercooling).

These results confirm the observations of Baliga [11] stating that for a tin melt saturated with silicon in the 800–1,100°C temperature range, and for a cooling rate less than 1°C min^{-1}, the thickness does not depend on the duration of growth but only on the temperature drop ΔT. At these moderate cooling rates, the surface kinetics is rapid when compared to mass transport, then this last phenomena controls the kinetic of growth. However, for cooling rates above 2°C min^{-1}, the thickness is only related to the duration of epitaxy and neither to the temperature drop nor to the cooling rate. This transition, from mass transport limited growth to kinetically controlled growth, depends on the experimental set up (size, geometry of crucible, substrate holder...).

Moreover, the cooling rate also influences the quality of the epitaxial layer: Solvent inclusions and surface ripples appear when cooling rate is above

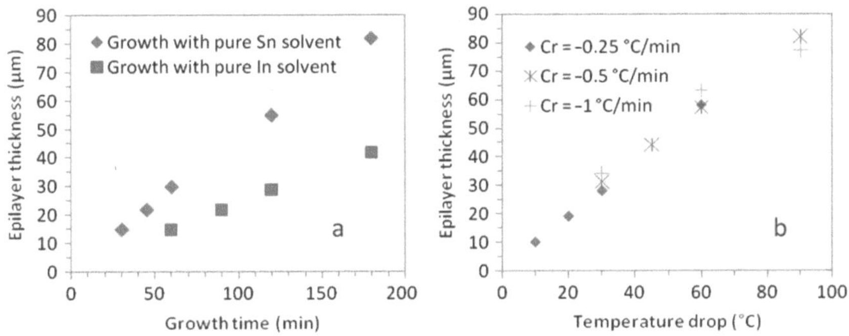

Fig. 9.3. (**a**) Linear increase of epitaxial layer thickness grown from 1,050°C on (111) substrate with 0.5°C min^{-1} cooling rate using tin or indium as solvent. (**b**) Thickness versus temperature drop for various cooling rate for tin solvent

1°C min^{-1}. Therefore, to obtain flat surface and thick epitaxial layer in a reasonable time, growth rates of 0.5–1°C min^{-1} are commonly used. We did not observe that addition of Ga in the melt modifies the growth rate noticeably.

The University of Konstanz [12] has also studied the influence of the cooling rate on the growth rate with In solvent at 920°C. They used (100) oriented Si substrates and cooling rate up to 0.33°C min^{-1}. The resulting growth rate reached 87.3 $\mu\text{m h}^{-1}$. Growth rate on (100) oriented surfaces is faster than on (111) surfaces, but the surface presented random pyramids distribution. Flat surface can be obtained only with low cooling rate/growth rate. In our case, we were able to grow flat layers on (100) substrates with high growth rate (2 $\mu\text{m min}^{-1}$) but only at 1,050°C. They have also developed a rapid LPE technique combined with a sonic agitation of the melt. The key element is the realisation of a strong temperature gradient between the substrate and the surface of the melt (a hole in the heating tube directly above the sample causes heat loss via radiation). Growth rate of 2 $\mu\text{m min}^{-1}$ (instead of 0.2 $\mu\text{m min}^{-1}$ without hole) have been demonstrated on (111) substrates from Sn melt with a cooling rate of 1°C min^{-1} [13]. In that case, a 25 μm thick solar cell absorber layer takes only 10 min to deposit.

9.4.2 Doping and Electrical Properties of Epitaxial Layers

The obtainment of p-type epitaxial layers with doping level in the range 10^{16}–$10^{17}\,\text{cm}^{-3}$ is of great interest for the fabrication of solar cells. The doping of the epitaxial layers can be obtained with the addition of an element in the solvent, for instance Ga or Al (Sb is a n type dopant). SIMS measurements have shown that incorporation of Ga is constant along the 30 μm thickness of the layers (at 1,050°C with cooling rate less than 1°C min^{-1}). Since pure Sn solvent leads to unintentional n doping (it is usually supposed that it is due to the Sb impurities present in the Sn source), one has to add more Ga in such solvent than in a pure In. For instance, at 1,050°C, 0.11% Wt of Ga is enough

Table 9.2. Electrical properties of epitaxial layers with Ga doping versus solvent and cooling rate. Growth temperature was 1,050°C [69]

	In/Ga (0.11%)		Sn/Ga (0.12%)	
Cooling rate	0.5°C min^{-1}	1°C min^{-1}	0.5°C min^{-1}	1°C min^{-1}
Doping (cm^{-3})	3.7×10^{16}	3.5×10^{16}	6×10^{16}	6.3×10^{16}
Mobility ($cm^2V^{-1}s^{-1}$)	263	226	232	198
Diffusion length (μm)	136	–	120	–
Defect density (cm^{-2})	1.1×10^4		1.3×10^4	

to obtain 3×10^{16} at cm^{-3} with In (6N), but not enough to compensate n type with Sn (5N) solvent. With 0.12% Wt of Ga, we can obtain p-type layer but with higher doping level (6×10^{16} at cm^{-3}) (Table 9.2).

Note that mobility is slightly decreased when higher cooling rate is applied. For lower cooling rate (0.25°C min^{-1}), we did not notice any improvement. Kopecek [12] also showed that the increase of the growth rate resulted in a higher resistivity and a lower quality of the LPE layer for In based melt.

Diffusion length (or lifetime) is a key parameter for the performance of solar cells. It is usually admitted that diffusion length of minority carriers has to be four times the thickness of the film to assure good photovoltaic efficiencies. At 1,050°C, appropriated values of 136 and 120 μm were obtained with In and Sn, respectively. The lower performance of epilayer grown from Sn melt can be explained by its high solid solubility (5×10^{19} cm^{-3} at 1,050°C). Incorporation of Sn atoms within the Si crystal could create a large stress and affect carrier transport. It is clearly related to the defects density of epitaxial film (measured by SECCO etching).

A comparison of Sn and In as solvents (and Ga and Al as dopants) has also been made at UNSW using single-crystal substrates [14]. They also obtained better results with In than with Sn and with Ga than with Al. It was found that layers grown with Sn solutions or doped with Al exhibit reduced mobility and lifetime. Presence of Al leads to the formation of an oxide on the surface of the melt and to the contamination of the epilayer, which exhibits higher density of shallow pits.

We summarise some experimental results concerning electrical properties of epitaxial layers presented by different authors in the following table (Table 9.3).

Boron is another possible p-type dopant. The incorporation of boron into silicon epitaxial layers grown from a tin melt has been studied by Baliga [15] and McCann [16]. Boron is provided via the silicon source wafer and its incorporation is a function of both time and temperature. Its segregation coefficient from liquid tin into solid silicon is temperature dependent and increases with temperature. Therefore, the content of boron into the layer will be maximum at the interface and minimum at the final surface. This doping gradient can be used to create a drift field in the base layer of the

Table 9.3. Electrical properties of epitaxial thin films grown from Sn and In solvents

Solvent	T (°C)	Dopant	Doping (cm^{-3})	Mobility ($cm^2V^{-1}s^{-1}$)	Diff. length (μm) Lifetime (μs)	References
In	920		9.5×10^{15}	259.7		[12]
In	920	Ga	4×10^{16}	207		[12]
In	900	Ga	5×10^{16}–10^{17}		50–300 μm	[70]
In	947	Ga	10^{17}		50–65 μm; 3–3.5 μs	[20]
In	1,050	Ga	3.7×10^{16}	263	135 μm	(This work)
In	930	Ga	2×10^{17}	100–200	25 μs	[14]
Sn	1,050	Ga	6×10^{16}	232	124 μm	(This work)
Sn	930	Al	1.5×10^{17}	100–200	8.5 μs	[14]

solar cell. Such behaviour was also proposed by Zheng [17] for Ga and In melt.

Although the majority of researchers use tin or indium, NREL (USA) has developed a technology using Cu based solvent [18]. Despite its reputation as a lifetime killer, they demonstrated that Cu can be a workable solvent for making silicon solar cells. The Cu–Si phase diagram is complex on the copper rich side, but it offers a wide temperature range for solution growth at compositions of ~30 at.% Si at the eutectic temperature of 802°C. It is possible to grow Si layer from pure Cu melt, but only at high-temperature (>802°C). They demonstrated that incorporation of Cu in the Si epitaxial layer is minimised when cooling rate is lower than 0.5°C min^{-1}, and it does not affect the quality of the junction. If the growth occurs on a (111) single crystal, when using cooling rate of 0.1°C min^{-1}, the growth rate can be 1 μm min^{-1}. If the cooling rate is significantly increased, it will lead to a morphological roughening of the grown layer with concomitant solvent entrapment. A minority carrier diffusion length of 109 μm was measured. They also used Cu/Al alloy, where the presence of Al allows the removal of the Si native oxide. With a 23%Si/28%Al/49%Cu, they obtained a mobility of 99.5 $cm^2 V^{-1}s^{-1}$ for a resistivity of 0.05 Ω cm. The corresponding diffusion length was 33.5 μm [19].

9.5 Growth on Multicrystalline Si Substrates

For solar cell applications, thin film LPE is economically viable only if it is combined with a low-cost multicrystalline Si substrate (high-throughput silicon ribbons, upgraded metallurgical grade silicon: MG-Si) or with a foreign substrate (glass, ceramic, metallic sheet...: see section 9–7).

Good-quality multicrystalline silicon (mc-Si) has been used as a model to develop techniques for depositing silicon on silicon. Grain boundaries are present but grains are large and impurity levels are not as high as in metallurgical grade silicon.

Wagner compared structural and electrical properties of polycrystalline Si layers grown by CVD or LPE (In melt, 947°C, 0.12 μm min^{-1}) with similar grain boundary structures [20]. The measured minority carrier lifetime was always higher and the recombination strength of the defects was smaller in the LPE layers than in the CVD layers. They attributed this to the higher purity of the LPE layer and its lower density of defects (rod-like defects).

The main limitation for high-quality layer is the presence of grain boundaries that retarded the epitaxial growth, resulting in a much rougher surface [21]. The higher energy of the grain boundary results in a reduced growth. The Si material diffusing towards the grain boundary is incorporated on either side of it. Once this surface roughness has been established, it is exacerbated during further growth by the phenomenon of constitutional supercooling. To reduce this effect, Australian National University (ANU) proposed to use a periodic meltback technique, where alternating cycles of heating and cooling are used (yoyo technique [22]). During heating cycle, the degree of undersaturation increases with distance away from the liquid–solid interface. Thus, protuberances are dissolved preferentially. These meltback cycles result in a much smoother surface. Morphology of the intra-grain region is also improved.

The left part of the Fig. 9.4 presents the growth by equilibrium cooling on mc-substrates and the resulting epitaxial layer and the right, with "yoyo" technique.

9.5.1 Photovoltaic Results Obtained with LPE Silicon Layers

Efficiencies of thin film solar cell grown by LPE can be as high as 18% on single crystal substrate and can display open circuit voltage above 660 mV [24], demonstrating their high electronic quality. ANU achieved 18.1% with a 35 μm thick epilayer grown on a lightly doped c-substrate (from In melt, at 950°C), combined with efficient light trapping system (texturation, oxide/Al back reflector). A similar process yielded 17% on heavily doped Si c-substrate (i.e. electrically inactive) [71, 73].

An efficiency of 16.4% has been achieved at UNSW on a 32 μm thick layer with an area of 4 cm^2 grown from In melt. A high-efficiency cell process was used with a micro-grooved surface texture [72], a ZnS/MgF_2 antireflection coating on passivating oxide, and a graded doping level in the active layer (Drift field). Note that they used an H_2/Ar forming gas mixture as the ambient gas rather than pure hydrogen. They also obtained 15.4% efficiency from a Sn melt [14]. Max-Planck-Institute produced a cell with an efficiency of 14.7% for a 16.8 μm thick layer (In solvent, 950°C) [26].

These high-performances on single crystal substrates have been confirmed on multicrystalline Si substrates. ANU fabricated a cell of 15.4% efficiency and V_{oc} of 639 mV on a lightly doped substrate and a cell of 15.2% efficiency and V_{oc} of 639 mV on a heavily doped substrate (with no texturing in either case,

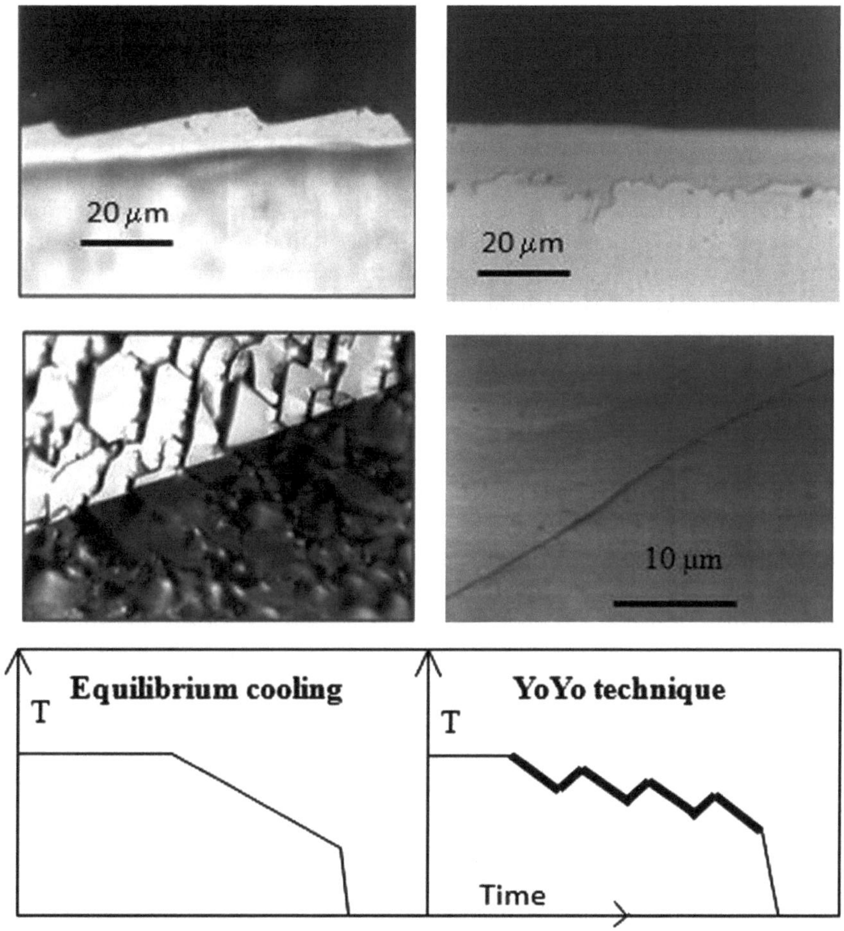

Fig. 9.4. LPE growth on multicrystalline silicon substrate at INL [23]: with equilibrium cooling (*left*) or yoyo technique (*right*). Before growth (950° with Sn), substrate roughness is about 2–3 μm. *From top to bottom* are cross section view, surface view and schematic temperature profile

but with a TiO_2 antireflective coating) [21]. Layers of 20–50 μm thick were grown and periodic meltback was used to obtain much smoother morphology at the grain boundaries, thereby to avoid emitter/substrate shunts.

To compensate the extra cost of the epitaxial growth of the Si thin film, one has to use a low-cost Si substrate like metallurgical grade Si (MG-Si). University of Konstanz [27] developed LPE on upgraded MG-Si (UMG-Si). An efficiency of 10% without surface texturation (Calculated potential: $n = 14\%$) was achieved. Thin layer of 30 μm thick was grown from In melt with 0.1 Wt% Ga at 990°C. The saturation of the melt was obtained with the meltback of the UMG-Si wafer: It is not necessary to add electronic grade Si to the solution.

The possible reason for cell efficiency not exceeding 10% is the diffusion of the impurities (Ca, Al, Fe, Ni and Cr) from the substrate into the LPE layer. Also, presence of small interruption in the layer leads to short-circuit in the cell. To improve the character of the layers, higher growth temperature above 1,000°C can be used. However, diffusion of impurities will be enhanced [28].

LPE on MG-Si substrates with a Cu/Al solvent has been investigated at NREL [25]. A diffusion length of 42 μm has been achieved for a layer thickness of 30 μm on a MG-Si substrate. The Cu was found to reduce Al incorporation into grown layers. An advantage of this approach is the rather high Si solubility in the Al/Cu solvent of 20–35 at.%. This high silicon concentration has been found to result in more isotropic growth rates on grains of different orientations than solvents with lower Si solubilities, such as In or Sn and also offers the potential for higher deposition rates.

9.6 Low-Temperature Silicon Liquid Phase Epitaxy

LPE has been widely used at high-temperature and has proved its capacity to grow high-quality epitaxial layers. But for low-temperature, it appears to be more difficult. However, it presents strong interest for low-cost applications (especially for solar cells) as it broadens the choice of the substrate. Actually, it will be possible to use glass substrate, if a suitable seed layer is previously deposited on the surface. But for such process, new solvents have to be defined. Main difficulties, compared to conventional LPE at 900–1,000°C, are the low solubility of silicon in usual solvent and the presence of native silicon oxide, which cannot be removed under H_2 flow.

A variety of solvents have been proposed like Au, Al, Ga [29,30], Pb [31], Sn [32] and others alloys as Au–Pb, Au–Zn, Al–Sn, Al–Zn [33], Au–Bi [32], Al–Ga [34], and Sn–Pb [35]. Resulting layers presented high doping level up to 10^{18} at. cm^{-3} (when using doping element like Al, Ga), uncompleted coverage of the surface (non-miscibility of Si in some alloys at low-temperature) or resulted in layers with electrical defects (Au).

Lee and Green [32] evaluated a number of binary or ternary alloys. The idea is to mix two metals, one (type X) with a low melting temperature (Sn, In, Pb, Bi) and one (type Y) with high solubility (Mg, Zn, Cu, Au). It is also important, to reduce the Si native oxide, to select some reducing agent like Al or Mg (type Z). Finally, the epitaxy will be realised with two melts. One to reduce SiO_2 (X–Z–Si or X–Y–Z–Si) and the other to grow epitaxial layer (X–Y–Si). Separation of cleaning and growth step avoids the diffusion of reducing agent like Al in the whole active epilayer. It is, then, necessary to study the ternary or quaternary diagrams to define the right composition. For instance, Abdo showed that the ternary Cu–Sn–Si and quaternary Cu–Sn–Al–Si systems are appropriate (Fig. 9.5) [36,37].

At 800°C, Si solubility increases up to 2 Wt% with this alloy (with pure Sn melt), Si solubility is only 0.16 Wt% at the same temperature. On a Si substrate, they were able to obtain flat and homogeneous 30 μm thick

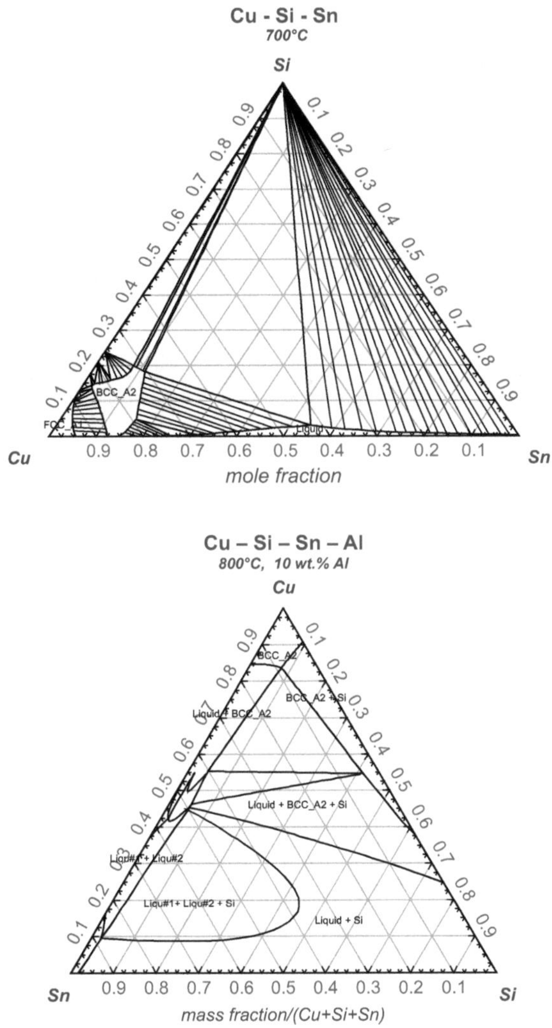

Fig. 9.5. Calculated phase diagram of Sn–Cu–Si (*left*) and Sn–Cu–Si with 10 Wt% Al (*right*) at 800°C [36]

layer (Fig. 9.6), starting the growth at 800°C and applying a cooling rate of 0.25°C min^{-1} during 2 h. The electrical properties of these layers have not been measured yet.

9.7 Liquid Phase Epitaxy on Foreign Substrates

As the cost of the silicon wafer is about half the final cost of the solar cell module, the use of crystalline thin film on non-silicon low-cost substrate is of great interest. The main issues are the compatibility of the substrate with

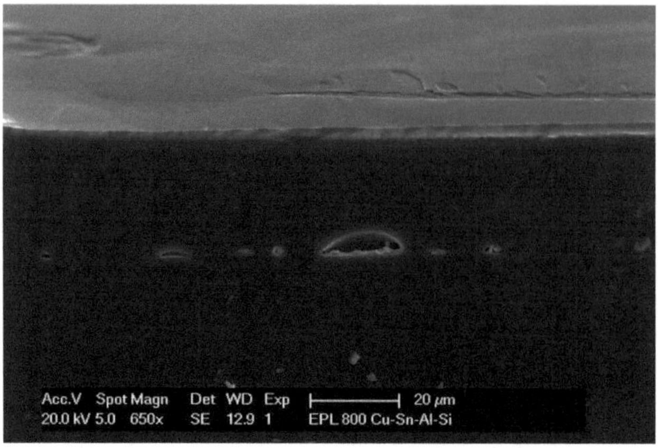

Fig. 9.6. SEM micrograph of Si layers grown at 800°C from a $Cu_{8.5}Al_{9.5}Sn_{80}Si_2$ solution [37]

growth temperature (thermal expansion coefficient close to silicon, presence of impurities able to diffuse into the Si layer) and the control of nucleation to obtain large grain size continuous layers.

Ceramics like Alumina or Mullite have been widely investigated. Alumina can be cheaply produced and Mullite, more expensive, has a thermal expansion coefficient close to Si for a wide range of temperature. They can sustain high-temperature annealing (1,000–1,200°C). Finally, they present high reflection coefficient, in the 80–90% range and then make an efficient back reflector for a solar cell. These substrates have to be initially coated with a seeding layer. For instance, PHASE and INSA combined seed layer deposition by RTCVD (with preferential (220) orientation) and growth of the active layer by LPE [38]. Grain size was increased from 1–10 μm (seeding layer) up to 100–200 μm, but continuous layers have not been achieved. It is conjectured that the (220) orientation is not suitable for subsequent LPE growth, because it leads to columnar growth. Higher aspect ratio of lateral versus vertical growth rate can be achieved when using (111) orientated seed layer and may lead to continuous layers.

Combination of seed layer and LPE was also studied at Max-Planck-Institute [39]. They used a glassy carbon substrate with a double seed layer a-Si/μc-Si (grain size is 1 μm). Low defect density and continuous layers over 4 cm2 have been obtained with Ga/Al melt. Resulting grain size was 10 μm.

ECN [40] developed silicon-enriched SiAlON tape-cast ceramic substrates using a plasma-sprayed Si seeding layer. Closed layers of 1 cm^2 have been achieved with grain sizes of 10–100 μm when using a melt of In with 1%Al at 960°C.

Some continuous layers have also be grown by UNSW with a Sn/Al melt on borosilicate glass using a seeding layer of a-Si on borosilicate glass at 750°C [68]. Grain sizes of 50 μm with a (111) preferential orientation were achieved with a layer thickness of 30 μm. They also reported the growth of continuous micrometers on unseeded glass substrates at around 750°C with an average grain size of 100 μm [41]. Al or Mg reduce SiO2 and lead to the formation of Si sites for the nucleation. The periodic melt-back and re-growth, suppressed growth normal to the plane of the substrate and lead to smoother silicon films.

As we can see, it is not easy to grow continuous Si thin films on foreign substrate with low defect density. To overcome this difficulty, the layer transfer process has been proposed [42–46]. This concept is based on the surface modification (usually a double porous layer) of a Si single crystal growth substrate that permits the transfer of the epitaxial layer on a low-cost substrate. The original growth substrate is reused several times. The porosification of the surface coupled with appropriate annealing under H_2 at 1,000–1,100°C allows epitaxial growth and further detachment. Majority of research groups involved in this concept used VPE as the growth process. Only ANU (see Sect. 9.8), Canon [47] and INL (former LPM) [48] proposed the use of LPE to grow the Si active layer. Re-structuration of the Si porous layer is a key point of the technique. Temperature and time have to be precisely adjusted to control the evolution of the surface roughness of the Si porous and also to hinder premature fragilisation of this layer [49]. The resulting electrical properties of the Si epilayer grown at 1,050°C using Sn/Ga melt were encouraging for solar cell application: mobility is 192 $cm^2\,V^-1\,s^-1$, diffusion length is 124 μm for a level of p doping of $7.1 \times 10^{16}\,cm^{-3}$.

9.8 Epitaxial Lateral Overgrowth

One of the specific features of LPE is its selectivity. It means that epitaxial growth can take place only in presence of Si nuclei seed. For instance, the growth on a Si wafer, partially covered with mask, will occur only in the exposed areas ("seeding windows") and not on the mask. These masking layers can be metallic or dielectric films (generally SiO_2, SiN_x). By using this specificity, it is possible to realise epitaxial lateral overgrowth (ELO) of Si on the mask. The growth starts in the seeding windows and proceeds vertically as well as laterally over the mask. In favourable conditions, adjacent stripes of Si tend to coalesce and form continuous epilayer and fully cover the mask [10, 50–53]. A silicon-on-insulator structure is then obtained. The properties of such structure can be of interest for PV applications. First, the mask acts as a defect filter and prevents the propagation of defects from the substrate. This can be interesting when considering growth of Si thin film on MG-Si. The mask could hinder diffusion of impurities into the active layer. Second, it can serve as a mirror layer to improve the optical confinement within the solar cells. Considering the objective of the fabrication of thin film crystalline Si solar cells, the presence of this dielectric mask can lower the requisite thickness and enhance the absorption [52]. Finally, the mask can be eventually etched and the epitaxial layer detached from the substrate for transfer and bonding to another (low-cost) substrate [54].

The epitaxial overgrown layer is characterised by the aspect ratio width versus thickness. When using (111) silicon substrates, this aspect ratio can be as high as 80 [55, 56]. It depends on the growth conditions and also on the directions of the seeding lines [57]. From this, Kraiem [58] used a grid shape (silicon lines were 35 μm width directed along directions <112> and <110> and silicon oxide was $90 \times 90\,\mu m^2$) to obtain a full and flat coalesced thin film after less than 1 h of growth (Sn melt, 1,050°C). Moreover, the resulted layer presented ten times less defects than usual LPE film.

Kinoshita [59] applied this technique to non-planar structured (111) silicon substrates. Despite the irregular topology of the substrate, epitaxial layers with flat top surfaces were always produced. Thanks to the formation of (111) facets in the plane of the epitaxial film (with Sn melt at 900°C).

When using (100) oriented substrate, since the vertical growth rate is much higher than lateral one, the behaviour of the grown layer is completely different. It is remarkably illustrated with the Epilift technique of the ANU. From a mesh-patterned silicon oxide layer (lines are along <110> directions), they were able to grow a 50–100 μm thick film presenting (111) facets along the lines (like a diamond shape). It gives a natural antireflection texture to this mesh. The interest of the technique rests on the further detachment of this film, thanks to a wet chemical or electrochemical etching. The masking layer and substrate may be reused as long as the masking layer is not attacked. Cells with any desired degree of transparency can be fabricated for specialist applications. They reported 13% efficiency on small area cell and a 50 cm^2 minimodule displayed a 10.9% efficiency [60].

Lateral growth of silicon sheet from Cu–Si melts was also reported [61]. This lateral growth occurs continuously by sliding the substrate toward a lower temperature region. The film thickness and its uniformity are dependent on the sliding speed. They assumed that it is possible to produce Si films with 1.5 $mm\,s^{-1}$ sliding speed. In addition, such lateral growth technique would enable the film growth on foreign substrates like ceramics or glasses with seeding Si nuclei only on the edge of the substrate.

9.9 High-Throughput LPE

The efficient results presented up to now were principally obtained by using a laboratory scale LPE apparatus (less than $5 \times 5\,cm^2$). They proved the efficiency of this technique to obtain high-quality epitaxial layers suitable for the fabrication of efficient solar cells.

The manufacturing of low-cost epitaxial thin film Si solar cells requires the development of high-throughput LPE deposition equipment to grow about 20–40 μm epitaxial layers at high-rate on a large surface [62] (Texturing step removes 10–20 μm of Si). Presently, LPE and CVD have not yet reached this objective, but some high-throughput reactors are recently emerging. These efforts are mainly focused on CVD, but LPE has some attractive features and can be adapted for this objective. It requires low capital investment and operating costs. There is no need for high vacuum and it does not consume or produce toxic gas. The only point to look out, concerning the safety, is the use of hydrogen (but a forming gas mixture can be considered [7]). The limitations for the transfer to a large scale production are the control of the temperature (related to the layer homogeneity), the adapted proportions between the surface of the wafers and the volume of the melt, and the reproducibility (thickness, doping). The horizontal sliding technique is generally used for the growth of III–V materials. It is necessary when multilayers are required. Many systems have been developed for these applications, since the early 1970s [63].

For solar cell production, the vertical dipping system seems to be the only suitable one because it can process large batches of wafers and grow a unique Si layer. Within the frame of the European project "Treasure", the University of Konstanz designed a reactor for 54, $10 \times 10\,\mathrm{cm}^2$ wafers in one batch [28]. Graphite was preferred than quartz as the crucible with surface treated by pyrolysis. They used a meltback step before each growth process to supply silicon to the melt from a UMG-Si source. Main problem on such large area wafers was the presence of pinholes in the grown epilayer. By using slight over-saturation and four serial meltback and re-growth steps, they obtained pinhole free LPE layers covering almost the whole $100\,\mathrm{cm}^2$ wafer area. Using In/Ga solvent and appropriate parameters, they can produce a gradient profile for the carrier concentration, from $3 \times 10^{18}\,\mathrm{cm}^{-3}$ at the interface with the substrate to $3 \times 10^{16}\,\mathrm{cm}^{-3}$ at the surface. With the help of this drift field, combined with the electrochemical macro-porous texturation, efficiencies up to 14% should be feasible using LPE growth on the low-cost UMG-Si wafer.

The Institute für Kristallzüchtung in Berlin [64] proposed the use of a temperature difference method on large area polycrystalline substrates ($10 \times 10\,\mathrm{cm}^2$). The thermodynamic driving force is generated by a temperature gradient perpendicular to the substrate surface. It favours the diffusion of solute from the Si source material towards the Si substrate. $30\,\mu\mathrm{m}$ thick Si films have been grown from In/Ga melt at 980°C using a gradient of $10\,\mathrm{K\,cm}^{-1}$. The growth rate was $0.3\,\mu\mathrm{m\,min}^{-1}$ and the doping concentration was adjusted in the range of 10^{16}–$2 \times 10^{18}\,\mathrm{cm}^{-3}$. The minority charge carrier lifetime of typically 5–10 μs was measured. Grain boundaries did not show significant effects on the life time in the TDM-grown layers. This is a promising technique for the deposition of Si thin film for PV applications.

Canon [65] patented a silicon LPE dipping method for large metallurgical grade polycrystalline wafers ($5\,\mathrm{in.}^2$). As mentioned before in Sect. 9.5, grooves may be formed at the grain boundaries during the growth, leading to short-circuits of the solar cell. Canon patented a method with an optimised time–temperature program (yoyo technique combined with two independent heaters), to reduce the effects of grain boundaries in solar cells. The main idea is to control the ratio of the depth of the groove regarding the thickness of the epitaxial layer by adjusting the heating and cooling temperature ranges of the periodic meltback. By limiting this ratio lower than 0.25, they were able to produce solar cells with 15% efficiency (In melt, temperature range is 950–865°C). The effective area of these solar cells is not specified.

Canon [66] demonstrated LPE growth also on 5 in. Si wafers with porous silicon for further detachment. They also used vertical dipping system with In solvent (4N) with a quartz carrier. Growth rate was in the range of 0.1–1 μm min^{-1} depending on saturation temperature, cooling rate and direction or situation of the substrate in the solvent.

9.10 Conclusion

LPE has shown its capacity to produce high-quality Si epitaxial layer for solar cell applications. However, it is not currently used for commercial production. Shortage in solar-grade Si feedstock will continue to motivate research for Si thin film on low-grade Si or foreign substrates approaches. The "simplicity", the low capital equipment and operating costs, the high growth rate of LPE may promote its development to find out solutions to overcome the difficulties to upscale this technique.

To know completely about LPE, some specific books can be of great interest, especially [4] and [67].

Acknowledgments

The author acknowledges Dr. Marc Gavand and Pr. André Laugier, who introduced him the LPE technique. He also thanks his present and former colleagues and students, especially, Dr. Jed Kraiem, Dr. Fatima Abdo and Dr. Sylvain Joblot.

References

1. M.G. Astles, *Liquid Phase Epitaxial Growth of III–V Compound Semiconductor Materials and Their Device Applications* (Adam Hilger, Bristol, 1990)
2. J.C. Maroto, A. Marti, C. Algora, G.L. Araujo, in *Proceedings of the 13th European Photovoltaic Solar Energy Conference*, Nice, 1995, p. 343
3. B.J. Baliga, J. Electrochem. Soc. **133**, 5C (1986)
4. P. Capper, M. Mauk, (eds.), *Liquid Phase Epitaxy of Electronic, Optical, and Optoelectronic Materials* (Wiley, Chichester, 2007)
5. P. Franke, D. Neuschütz, *SGTE, Binary Systems. Part 4: Binary Systems from Mn-Mo to Y-Zr,, Volume 19 'Thermodynamic Properties of Inorganic Materials' of Landolt-Börnstein – Group IV 'Physical Chemistry'*, (Springer, Berlin, 2006)
6. J.J. Hsieh, J. Cryst. Growth **27**, 49 (1974)
7. R. Bergmann, J. Kurianski, Mater. Lett. **17**, 137 (1993)
8. Z. Shi, J. Mater. Sci. Electron. **5**, 305 (1994)

9. F. Abdo, A. Fave, M. Lemiti, A. Laugier, C. Bernard, A. Pish, Phys. Stat. Sol. (c) **4**, 1397 (2007)
10. M. Konuma, ed. by Y. Pauleau. *Feature and Mechanisms of Layer Growth in Liquid Phase Epitaxy of Semiconductor Materials, in Chemical Physics of Thin Film Deposition Processes for Micro- and Nano-Technologies* (Springer, Berlin, 2002), p. 384
11. B.J. Baliga, J. Electrochem. Soc. **124**, 1627 (1977)
12. R. Kopecek, K. Peter, J. Hötzel, E. Bucher, J. Cryst. Growth **208**, 289 (2000)
13. K. Peter, G. Willeke, E. Bucher, in *Proceedings of the 13th European Photovoltaic Solar Energy Conference*, Nice, 1995, p. 379
14. Z. Shi, W. Zhang, G.F. Zheng, V.L. Chin, A. Stephens, M.A. Green, R. Bergmann, Sol. Energy Mater. Sol. Cells **41–42**, 53 (1996)
15. B.J. Baliga, J. Electrochem. Soc. **128**, 161 (1981)
16. M.J. McCann, K.J. Weber, M. Petravic, A.W. Blakers, J. Cryst. Growth **241**, 45 (2002)
17. G.F. Zheng, W. Zhang, Z. Shi, D. Thorp, R.B. Bergmann, M.A. Green, Sol. Energy Mater. Sol. Cells **51**, 95 (1998)
18. T.F. Ciszek, T.H. Wang, X. Wu, R.W. Burrows, J. Alleman, C.R. Schwertfeger, T. Bekkedahl, in *Proceedings of the 23rd IEEE Photovoltaic Specialist Conference*, Louisville, 1993, p. 65
19. T.H. Wang, T.F. Ciszek, in *Proceedings from the 1st World Conf. on Photovoltaic Solar Energy Conversion IEEE*, Hawai, USA, 1994, p. 1250
20. G. Wagner, H. Wawra, W. Dorsch, M. Albrecht, R. Krome, H.P. Strunk, S. Reidel, H.J. Möller, W. Appel, J. Cryst. Growth **174**, 680 (1997)
21. G. Ballhorn, K.J. Weber, S. Armand, M.J. Stocks, A.W. Blakers, Sol. Energy Mater. Sol. Cells **52**, 61 (1998)
22. T. Sukegawa, M. Kimura, A. Tanaka, J. Cryst. Growth **108**, 598 (1991)
23. A. Fave, B. Semmache, E. Rauf, A. Laugier, in *Proceedings from "Matériaux et procédés pour la conversion photovoltaïque de l'énergie solaire"*, Ademe, Sophia Antipolis, 1998, p. 35
24. A.W. Blakers, J.H. Werner, E. Bauser, H.J. Queisser, Appl. Phys. Lett. **60**, 2998 (1992)
25. T.H. Wang, T.F. Ciszek, C.R. Schwerdtfeger, H. Moutinho, R. Matson, Sol. Energy Mater. Sol. Cells **41/42**, 19 (1996)
26. J.H. Werner, S. Kolodinski, U. Rau, J.K. Arch, E. Bauser, Appl. Phys. Lett. **62**, 2998 (1993)
27. K. Peter, R. Kopecek, P. Fath, E. Bucher, C. Zahedi, Sol. Energy Mater. Sol. Cells **74**, 219 (2002)
28. C. Zahedi, E. Enebakk, M. Mueller, D. Kunz, R. Kopecek, K. Peter, C. Lévy-Clément, S. Bastide, M. Mamor, T.H. Bergstrom, in *Proceedings of the 19th European Photovoltaic Solar Energy Conference*, Paris, 2004, p. 1273
29. E. Sumner, R.T. Foley, J. Electrochem. Soc. **125**, 1817 (1978)
30. H. Ogawa, Q. Guo, K. Ohta, J. Cryst. Growth **155**, 193 (1995)
31. M. Konuma, G. Cristiana, E. Czech, I. Silier, J. Cryst. Growth **198/199**, 1045 (1999)
32. S.H. Lee, M.A. Green, J. Electron. Mater. **20**, 635 (1991)
33. Z. Shi, T.L. Young, M.A. Green, Mater. Lett. **12**, 339 (1991)
34. B. Girault, F. Chevrier, A. Joullie, G. Bougnot, J. Cryst. Growth **37**, 169 (1977)
35. H.J. Kim, J. Electrochem. Soc. **119**, 1394 (1992)

36. F. Abdo, PhD thesis, INSA de Lyon, 2007
37. F. Abdo, A. Fave, M. Lemiti, A. Laugier, C. Bernard, A. Pish, Arch. Metall. Mater. **51**, 533 (2006)
38. A. Fave, S. Bourdais, A. Slaoui, B. Semmache, J.M. Olchowik, A. Laugier, F. Mazel, G. Fantozzi, in *Proceedings of the 11th International Photovoltaic Science and Engineering Conference*, Sapporo, Japan, 1999, p. 733
39. A. Gutjahr, I. Silier, G. Cristiani, M. Konuma, F. Banhart, V. Schöllkopf, H. Frey, in *Proceedings of the 14th European PV Solar Energy Conference, Barcelona*, 1997, p. 1460
40. S.E. Schiermeier, C.J. Tool, J.A. van Roosmalen, L.J. Laas, A. von Keitz, W.C. Sinke, *2nd World Conference on PV Solar Energy Conversion*, Vienna, 1998, p. 1673
41. Z. Shi, T.L. Young, M.A. Green, in *Proceedings of the 1st World Conf. on Photovoltaic Solar Energy Conversion*, Hawai, 1994, p. 1579
42. T. Yonehara, K. Sakaguchi, N. Sato, Appl. Phys. Lett. **64**, 2108 (1994)
43. H. Morikawa, Y. Nichimoto, H. Naomoto, Y. Kawama, A. Takami, S. Arimoto, T. Ishihara, K. Namba, Sol. Energy Mater. Sol. Cells **53**, 23(1998)
44. R. Brendel, in *Proceedings of the 14th European Photovoltaic Solar Energy Conference*, Barcelona, Spain, 1997, p. 1354
45. H. Tayanaka, K. Yamauchi, T. Matsushita, in *Proceedings of the 2nd World Conference and Exhibition on Photovoltaic Solar Energy Conversion*, Vienna, Austria, 1998, p. 1272
46. J. Kraiem, S. Amtablian, O. Nichiporuk, P. Papet, J.-F. Lelievre, A. Fave, A. Kaminski, P.-J. Ribeyron, M. Lemiti, in *Proceedings of the 21st European PVSEC*, Dresden, 2006, p 1268
47. S. Nishida, K. Nakagawa, M. Iwane, Y. Iwasaki, N. Ukijo, M. Mizutani, in *Technical Digest of the 11th International Photovoltaic Science and Engineering Conference*, Kyoto, Japan, 1999, p. 537
48. S. Berger, A. Fave, S. Quoizola, A. Kaminski, A. Laugier, A. OuldAbdes, N.-E. Chabane-Sari, in *Proceedings from the 17th European Photovoltaic Solar Energy Conference and Exhibition*, Munich, 2001, p. 1772
49. J. Kraiem, O. Nichiporuk, E. Tranvouez, S. Quoizola, A. Fave, A. Descamps, G. Bremond, M. Lemiti, in *Proceedings of the 20th European photovoltaic solar Energy Conference*, Barcelona, Spain, 2005
50. H. Raidt, R. Köhler, F. Banhart, B. Jenichen, A. Gutjahr, M. Konuma, I. Silier, E. Bauser, J. Appl. Phys. **80**, 4101 (1996)
51. I. Silier, A. Gutjahr, N. Nagel, P.O. Hansson, E. Czech, M. Konuma, E. Bauser, F. Banhart, R. Köhler, H. Raidt, B. Jenichen, J. Cryst. Growth **166**, 727 (1996)
52. M.G. Mauk, P.A. Burch, S.W. Johnson, T.A. Goodwin, A.M. Barnett, in *Proceedings of the 25th IEEE Photovoltaic Specialists Conference*, IEEE, Washington, 1996, p. 147
53. R. Bergmann, J. Cryst. Growth **110**, 823 (1991)
54. K.J. Weber, K. Catchpole, M. Stocks, A.W. Blakers, *26th Photovoltaic Solar Conference*, Anaheim, 1997, p. 107
55. Y. Suzuki, T. Nishinaga, Jpn. J. Appl. Phys. I **29**, 2685 (1990)
56. Y. Suzuki, T. Nishinaga, T. Sanada, J. Cryst. Growth **99**, 229 (1990)
57. I. Jozwik, J.M. Olchowik, J. Cryst. Growth **294**, 367 (2006)
58. J. Kraiem, A. Fave, A. Kaminski, M. Lemiti, I. Jozwik, J.M. Olchowik, in *Proceedings of the 19th European Photovoltaic Solar Energy Conference*, Paris, 2004, p. 1158

59. S. Kinoshita, Y. Suzuki, T. Nishinaga, J. Cryst. Growth **115**, 561 (1991)
60. M.J. Stocks, K.J. Weber, A.W. Blakers, in *Proceedings from 3rd World Conference of Photovoltaic Solar Energy Conversion*, Osaka, 2003, p. 1268
61. K. Kita, C.-J. Wen, J. Otomo, K. Yamada, H. Komiyama, H. Takahashi, J. Cryst. Growth **234**, 153 (2002)
62. J. Poortmans, V. Arkipov, *Thin Film Solar Cells* (Wiley, New York, 2006), p. 471
63. M.G. Mauk, J.B. McNeely, in *Equipment and Instrumentation for Liquid Phase Epitaxy*, ed. by P. Capper, M. Mauk. Liquid Phase Epitaxy of Electronic, Optical, and Optoelectronic Materials (Wiley, New York, 2007), p. 85
64. B. Thomas, G. Muller, P.-M. Wilde, H. Wawra, in *Proceedings of the 26th IEEE Photovoltaic Specialist Energy Conference*, Anaheim, 1997, p. 771
65. K. Nakagawa, S. Ishihara, H. Sato, S. Nishida, Y. Takai, US Patent 6951585
66. S. Nishida, K. Nakagawa, M. Iwane, Y. Iwasaki, N. Ukiyo, M. Mizutani, T. Shoji, Sol. Energy Mater. Sol. Cells **65**, 525 (2001)
67. S. Dost, B. Lent, *Single Crystal Growth of Semiconductors from Metallic Solutions* (Elsevier, Amsterdam, 2006)
68. Z. Shi, T.L. Young, G.F. Zheng, M.A. Green, Sol. Energy Mater. Sol. Cells **31**, 51 (1993)
69. A. Fave, S. Quoizola, J. Kraiem, A. Kaminski, M. Lemiti, A. Laugier, Thin Solid Films **451–452**, 308 (2004)
70. S. Kolodinski, J.H. Werner, U. Rau, J.K. Arch, E. Bauser, in *Proceedings of the 11th European Photovoltaic Solar Energy Conf.*, Harwood, Chur, Switzerland, 1992, p. 53
71. A.W. Blakers, K.J. Weber, M.F. Stuckings, S. Armand, G. Matlakowski, A.J. Carr, M.J. Stocks, A. Cuevas, T. Brammer, Progr. Photovoltaics **3**, 193 (1995)
72. G.F. Zheng, W. Zhang, Z. Shi, M. Gross, A.B. Sproul, S.R. Wenham, M.A. Green, Sol. Energy Mater. Sol. Cells **40**, 231 (1996)
73. A.W. Blakers, K.J. Weber, M.F. Stuckings, S. Armand, G. Matlakowski, M.J. Stocks, A. Cuevas, in Proceedings of the 13th European Photovoltaic Solar Energy Conference, Nice, 1995, p. 33.

10

Vapor Phase Epitaxy

Mustapha Lemiti

Abstract. The main advantages of the vapor phase epitaxy (VPE) are the ability to grow very good quality layers, with high growth rate (higher than μm min^{-1}). Its principle is relatively simple and allows great flexibility (change in doping level or type of doping…). In addition, the VPE can handle several large wafers, which is particularly desirable for photovoltaic applications. In this chapter, we introduce the principle of this method before discussing the theories and modeling for understanding the mechanisms governing the kinetics of crystal growth. It is followed by a detailed description of SiH_2Cl_2/H_2 system, well adapted to the growth of films for photovoltaic applications.

10.1 Introduction

Epitaxy is the regularly oriented growth of a crystalline material on a crystalline substrate. The epitaxial layer builds up on the substrate with the same crystallographic orientation, the substrate acting as a seed for the growth. If an amorphous/polycrystalline substrate surface is used, the film will also be amorphous or polycrystalline. Thus, when the substrate is multicrystalline, the growth is also multicrystalline and follows the orientation of the substrate grains.

Epitaxial films frequently have superior characteristics than either polycrystalline or amorphous films. The epitaxial growth concerns a large number of materials: silicon, silicon–germanium alloys, III–V compounds, binary and ternary composites, metals, etc.

When an epitaxial film grows on a substrate of the same nature, one deals with homoepitaxy. If the growth occurs on a different substrate, one deals with heteroepitaxy.

In the case of two different materials, the agreement of lattice parameters becomes a key issue. For a disagreement of lattice parameter, which is too large, crystal growth is rendered impossible by the appearance of many defects like dislocations, which relax the mechanical constraints at the interface. The maximum difference of lattice constant allowed is such that $\frac{a_{\mathrm{epi}}-a_{\mathrm{sub}}}{a_{\mathrm{sub}}} \leq 10^{-3}$, where a_{epi} is the lattice parameter of the epitaxial layer and a_{sub} the lattice parameter of the substrate.

For silicon, the process can be used to grow films with thicknesses of ~1 μm to >100 μm. Some processes require high substrate temperature, whereas others do not require significant heating of the substrate. For photovoltaic applications, epitaxial silicon is usually grown using liquid-phase epitaxy (LPE) [1–3] and vapor-phase epitaxy (VPE) [4–6], which is a modification of chemical vapor deposition (CVD).

The vapor phase epitaxy (VPE) is a technique widely used in the microelectronics industry for the growth of thin films on silicon substrate (or other compound semiconductors) [7]. Consequently, many studies have been done on this technique. Its main advantages are the ability to grow epitaxial layers with very good quality and with high growth rates (above the μm/min). Its principle is relatively simple and allows great flexibility during the process (variation of doping level or type of doping...). Moreover, the VPE can handle multiple large wafers, which is very beneficial for photovoltaic applications.

The precursors are in a gaseous form. The idea is to provide enough energy to dissociate the gas species above the substrate and, thus, cause the deposit and arrangement of atoms on the surface. There are several types of reactor depending on the method of supply of energy and also on the pressure in the chamber. Thus, in APCVD (Atmospheric pressure chemical vapor deposition), the growth takes place at atmospheric pressure. The gas filling speed is higher than in a reactor working at low pressure called LPCVD (Low pressure CVD). In both types of reactor, heating is provided mainly by magnetic induction or by joule effect through a resistance. In a PECVD reactor (Plasma enhancement CVD), additional energy is provided by a plasma discharge. This type of machine can work at low temperatures. A last type of process to be mentioned is RTCVD (Rapid thermal CVD), where heating is provided by halogen lamps.

The growth rates can be as high as tens of microns per minute and this explains why the vapor phase technology is the most widespread in industry.

Silicon epitaxy involves different types of chemical reactions according to the considered precursors. The gas precursors are SiH_4 (silane), or chlorinated compounds like $SiCl_4$ (silicon tetrachloride), $SiHCl_3$ (trichlorosilane), or SiH_2Cl_2 (dichlorosilane). The choice of precursor inevitably directs the epitaxy technique and the working temperature [8]:

- From silane (SiH_4):

$$\mathrm{SiH_{4(gas)}} \rightarrow \mathrm{Si_{(solid)}} + \mathrm{2H_{2(gas)}}.$$

The use of silane permits the lowest growth temperature (up to 700°C) but the growth rate remains low. Silane, which is inflammable in air requires special care and imposes a reaction chamber at low pressure to avoid leaks.

– From silicon tetrachloride ($SiCl_4$):

$$SiCl_{4(gas)} + 2H_{2(gas)} \longleftrightarrow Si_{(solid)} + 4HCl_{(gas)}.$$

This reaction produces a high quality material but takes place at high temperature (1,250°C) causing the redistribution of dopants.

– From trichlorosilane ($SiHCl_3$):

$$SiHCl_{3(gas)} + H_{2(gas)} \longleftrightarrow Si_{(solid)} + 3HCl_{(gas)}.$$

Reduction of trichlorosilane is the most used industrial method. It takes place at around 1,100°C. It is inexpensive and the precursor is available in large quantities.

– From dichlorosilane (SiH_2Cl_2):

$$SiH_2Cl_{2(gas)} \longleftrightarrow Si_{(solid)} + 2HCl_{(gas)}.$$

Pyrolysis of dichlorosilane provides a good quality crystal with a growth rate relatively high. The dichlorosilane is more expensive than the previous one but stays in a gas state at atmospheric pressure. Moreover, the deposition temperature is around 1,100°C.

10.2 Theoretical Aspects of VPE

The technique of growth by gas-phase epitaxy is naturally sensitive to the hydrodynamics of gases. The film quality and growth rate depend on the nature of the deposited material. In this section, we review briefly the concepts of laminar and turbulent flows, before describing the various cases of kinetic growth. Finally, we will detail some aspects of the experimental growth of silicon from dichlorosilane. The purpose of the following paragraphs is not to make a comprehensive study but rather to provide the reader with the elements enabling him to understand the various mechanisms involved in the growth process.

10.2.1 Notions of Hydrodynamics

The flow of precursor gases to the surface of the substrate plays an important role in the kinetics of growth and also on the uniformity and quality of the crystalline films.

The parameters affecting the dynamics of gases are mainly:

– The temperature gradient: gas relaxes in the hot zones, while its density remains high in the coldest parts.
– The speed of gas, which depends on the effects of convection related to the viscosity and to the forces of friction.

Different flow regimes (natural convection and forced convection) have to be considered. In the first case, the flow is guided by gradients of temperature and gas concentration. However, despite the presence of a temperature gradient, a forced convection regime is privileged. This regime occurs for higher flow rates of gas for which the flow is governed by the action of mechanical forces. In this case, the forces of friction and pressure determine the movement of the gas particles.

The type of flow is characterized by the Reynolds number:

$$Re = \frac{v_\mathrm{g} L}{\nu} = \rho \frac{1}{\eta} v_\mathrm{g} L$$

where, ν is the cinematic viscosity of the gas $(\mathrm{m^2\,s^{-1}})$, ρ is the gas density $(\mathrm{g\,m^{-3}})$, v_g is the gas velocity $(\mathrm{m\,s^{-1}})$ and η is the dynamic viscosity (Pa s). L is the geometrical characteristics of the reactor (m) (the length if the reactor is horizontal or the diameter if the reactor is cylindrical).

The gas flow is laminar when $Re < 5{,}400$: the forces of friction stabilize the flow.

The gas flow is turbulent when $Re > 5{,}400$: the forces of inertia overtake the forces of friction, creating turbulence.

In a cold wall reactor, the convection regime is mixed. On the one hand, the temperature gradient between the substrate and the walls of the reactor tends to establish a system of natural convection (laminar flow). On the other hand, the flow of gas induces a forced convection (turbulent flow). A laminar flow is necessary to ensure a good uniformity of the epitaxial film thickness.

10.2.2 Kinetics and Growth Regimes

The growth in vapor phase can be decomposed into several stages:

- Transport and distribution of precursor gases through the carrier gas from its entering in the reactor until the surface of substrate.
- Adsorption of species on the surface of the substrate.
- Diffusion of species to preferential sites, chemical decomposition.
- Incorporation into the lattice of constituents of the epitaxial film.
- Desorption of secondary products.
- Diffusion of secondary products away from the surface.

These steps can be separated into two categories: those consisting of transport of matter and those corresponding to surface chemical reactions. From these two categories follows the two ways to control the growth kinetics, i.e. limited by either providing the material or the reaction of the species on the surface of the substrate. Experimentally, the growth regime is determined by the conditions of temperature and gas flow rate.

10.2.2.1 Theoretical Approach

The theoretical framework of the growth kinetics takes into account the physics of species distribution within the gas phase. The key parameters are, therefore, the flow of reactive species from the gas phase to the surface of the substrate and the flow caused by the consumption of these species during growth.

The following formalism comes from the work of Grove [9]. Figure 10.1 is a schematic view of the gas flows during the epitaxial growth. G_c is the concentration of silicon in the volume of the gas chamber that depends on the partial pressure of gas precursor and C_s is the concentration of silicon atoms at the interface between gas and crystal.

We can then express the gas flows by:

$F_g = h_g (C_g - C_s)$, where h_g is the mass transport coefficient $\left(\text{in cm s}^{-1}\right)$.
$F_s = k_s\, C_s$ and k_s is the surface kinetic coefficient $\left(\text{in cm s}^{-1}\right)$.
F_g represents the flow of silicon atoms from the gas phase to the substrate and F_s represents the flow of silicon atoms consumed by chemical reactions during the crystal growth.

At equilibrium, these flows are equal, then it comes:

$$C_s = C_g \frac{h_g}{h_g + k_s} = \frac{C_g}{1 + \frac{k_s}{h_g}}.$$

The growth rate can be written, according to N, as the number of silicon atoms per unit volume in the crystal that is equal to $5 \times 10^{22}\,\text{cm}^{-3}$ and $v = \frac{F_s}{N}$. This leads to:

$$v = \frac{h_g k_s}{h_g + k_s} \frac{C_g}{N}.$$

The growth rate is proportional to the concentration of silicon C_g in the gas phase and is determined by the relative values of the coefficients h_g or k_s. Two growth regimes are possible.

If $h_g > k_s$, the surface kinetics determines the growth rate. On the contrary, if $h_g < k_s$, the transport of silicon from the gas phase to the surface of the substrate controls the growth rate.

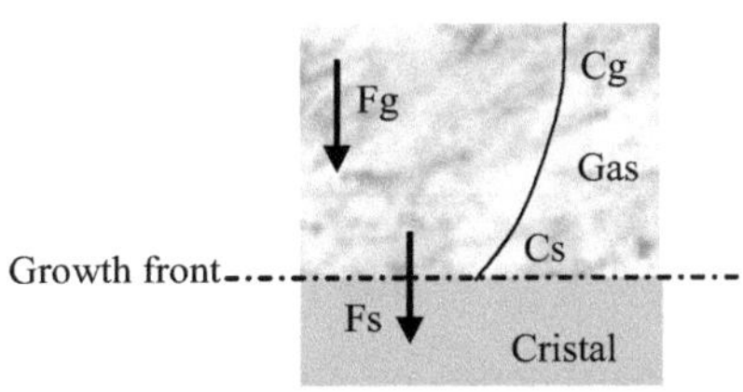

Fig. 10.1. Schematic view of the gas flows during the epitaxial growth

10.2.2.2 Limitations Due to the Surface Kinetics or Mass Transfer

When the surface kinetic reactions are the limiting factor, the growth rate can be written:

$$v = k_s \frac{C_g}{N}.$$

It does depend on the concentration of the reactive gases and of the coefficient of surface kinetics k_s. As often, the activation of the chemical reactions follows an Arrhenius law and k_s is expressed by $k_s = B \exp\left(-\frac{E_A}{kT}\right)$. In this growth regime, the growth rate depends on the temperature via the exponential factor.

At high temperatures, the limiting factor is not the surface kinetics but the contribution of matter by the gas phase. The growth rate depends, therefore, on the coefficient h_g whose variation with temperature is slow $\left(h_g \propto T^{3/2}\right)$. It is then considered that the growth rate in the regime of mass transfer is proportional to the concentration C_g.

Experimentally, there are two growth regimes as shown in Fig. 10.2 for each silicon precursor.

Moreover, Fig. 10.2 shows that more a precursor is chlorine-rich, more the growth rate is low. This phenomenon is explained by the etching of silicon by

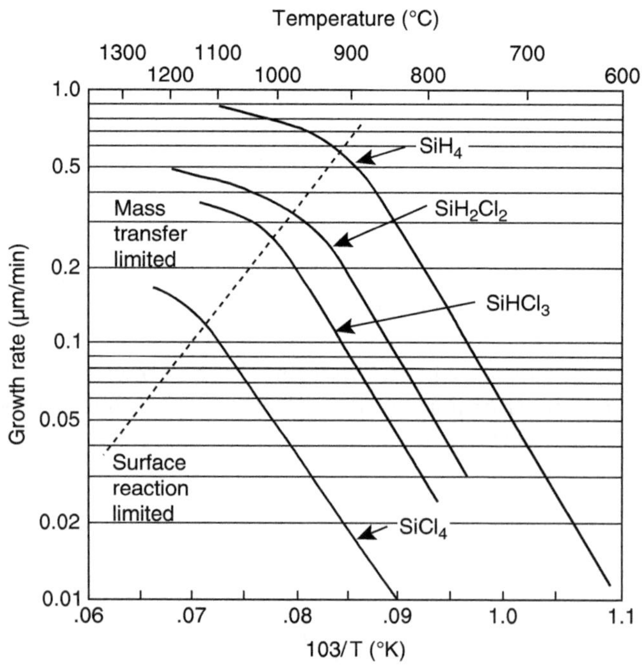

Fig. 10.2. Temperature dependence of growth rates of silicon at atmospheric pressure and for a concentration of 0.1% mol. in H_2 [9]

hydrogen chloride which is formed during the decomposition of chlorinated precursors.

The presence of hydrogen chloride in the gas phase implies that the epitaxial process results from a balance between growth and etching.

10.2.2.3 Model of the Boundary Layer

In fluid mechanics, the flow of gas along a wall is modeled by the formation of boundary layers with gradients of temperature, reactant species concentration, and gas flow speed. These various boundary layers are generally stacked. The boundary layer related to the gas flow speed gradient models the transition from the null speed at the surface of the substrate to the full speed of gas. It is noted δ and expressed in terms of parameters that reflect the forces of inertia and viscosity by:

$$\delta_{\mathrm{s}} = a\sqrt{\frac{\eta x}{\rho v}} = a\sqrt{\left(\frac{\eta RT}{MPv}x\right)} = a\sqrt{x}\sqrt{\frac{L}{Re}}$$

where,

- a is a coefficient of proportionality, and
- $\rho = \frac{nM}{V} = \frac{MP}{RT}$ is the gas density, P is the gas pressure, M is the mass molar of the gas
- v: gas velocity
- x: position on the susceptor
- $\eta = f(T)$: viscosity
- L: length of the reactor

The diffusion of reactive species within the boundary layer will be the limiting factor in mass transfer limited regime. Applying the laws of diffusion and considering a constant average thickness of the boundary layer, leads to the expression of the growth rate as follows:

$$v = H_{\mathrm{g}}\frac{C_{\mathrm{g}}}{N} = \frac{D_{\mathrm{g}}}{\delta}\frac{C_{\mathrm{g}}}{N} = C_{\mathrm{ste}}D_{\mathrm{g}}\sqrt{\frac{\rho v_{\mathrm{g}}}{\eta L}}\frac{C_{\mathrm{g}}}{N}.$$

The growth rate depends on the reactant concentration in the gas phase and varies as the square root of the carrier gas velocity (equivalent to flow).

10.3 Experimental Aspects of VPE

10.3.1 Experimental Approach of the Kinetics and Mechanisms of Silicon Growth in a SiH_2Cl_2/H_2 System

The pyrolysis of dichlorosilane creates many chlorinated compounds in more or less important quantities. The work of Morosanu [10] shows that the molecule dichlorosilane ($SiCl_2$) is the most important intermediate product of the reaction in the formation of silicon from dichlorosilane (DCS).

All the reactions of DCS decomposition involved in the gas phase are given by Claassen and Bloem [11]:

Reactions in the gas phase:

$$SiH_2Cl_{2(g)} \leftrightarrow SiCl_{2(g)} + H_{2(g)}$$
$$SiCl_{2(g)} + HCl_{(g)} \leftrightarrow SiHCl_{3(g)}$$

Adsorption at the surface (* represents surface free site, X* are species adsorbed):

$$SiCl_{2(g)} + * \leftrightarrow SiCl_2^*$$
$$H_{2(g)} + 2* \leftrightarrow 2H^*$$

Chemical reactions at the surface:

$$SiCl_2^* + H^* \leftrightarrow SiCl^* + HCl_{(g)}$$
$$SiCl^* + H^* \leftrightarrow Si + HCl_{(g)}$$
$$SiCl_2^* + H_{2(g)} \leftrightarrow Si^* + 2HCl_{(g)}$$
$$SiCl_2^* + SiCl_{2(g)} \leftrightarrow SiCl_{4(g)} + Si$$
$$Si^* \leftrightarrow Si$$

High temperature epitaxy of silicon from dichlorosilane precursor is the sum of many chemical reactions. The whole process can be summarized by the following equation:

$$SiH_2Cl_{2(g)} \leftrightarrow Si + 2HCl_{(g)}$$

In fact, other reactions are added as a result of the introduction of dopant gas like boron or phosphorus precursors that are used to modify the electrical properties of silicon.

The experimental approaches [11–13] on the growth rate of silicon on silicon substrate confirm the presence of two growth regimes depending on the temperature: a regime controlled by chemical kinetics for temperatures below 1,000°C and a regime controlled by mass transfer regime for temperatures above 1,000°C (Fig. 10.3).

Figure 10.4 shows that the growth rate varies linearly with the concentration of DCS only from a temperature of 1,000°C onwards.

The work of Baliga [14] and Lekholm [15] shows the linear variation of the growth rate with the concentration of SiH_2Cl_2 (Fig. 10.5).

These results are consistent with earlier models. They also highlight the possibility of substantial growth rates of several μm per minute. According to Coon [13], the desorption kinetics of hydrogen chloride HCl is the limiting factor for the growth of silicon in the regime of low temperature between 650 and 900°C. At low temperature, calculating the speed of HCl desorption shows

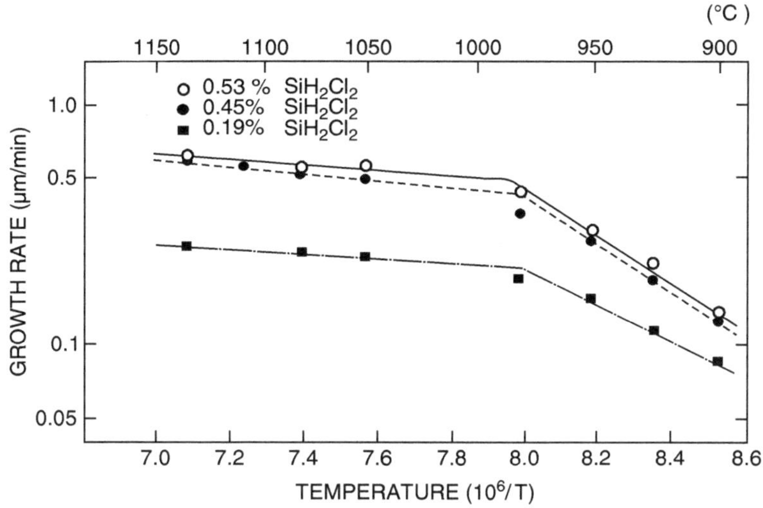

Fig. 10.3. Growth rate as a function of substrate temperature [12]

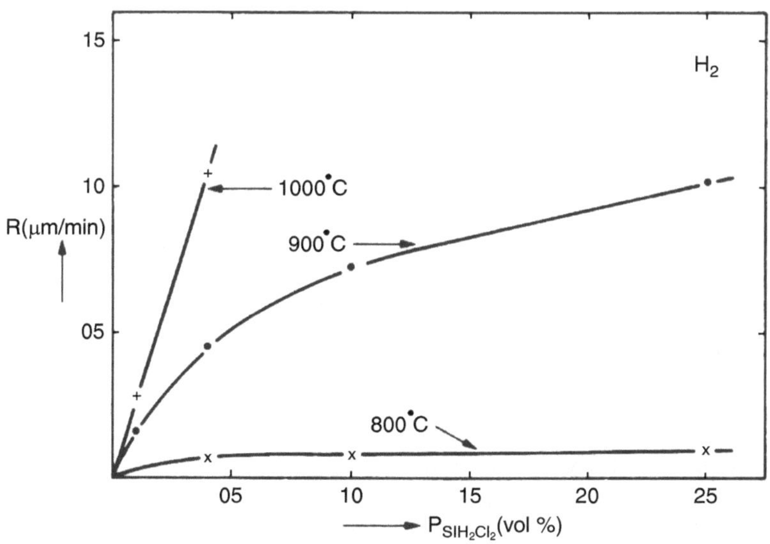

Fig. 10.4. Growth rate as a function of temperature and percentage of dichlorosilane in H_2 ambient [11]

that the surface of the silicon substrate is saturated and that the desorption is essential to free up vacant sites for adsorption of SiH_2Cl_2. In contrast, at high temperature, the surface is mainly composed of free sites. The influence of hydrogen chloride is also discussed in the work of Claassen and Bloem [11], who observed a change in the growth rate of silicon when adding HCl in the gas phase.

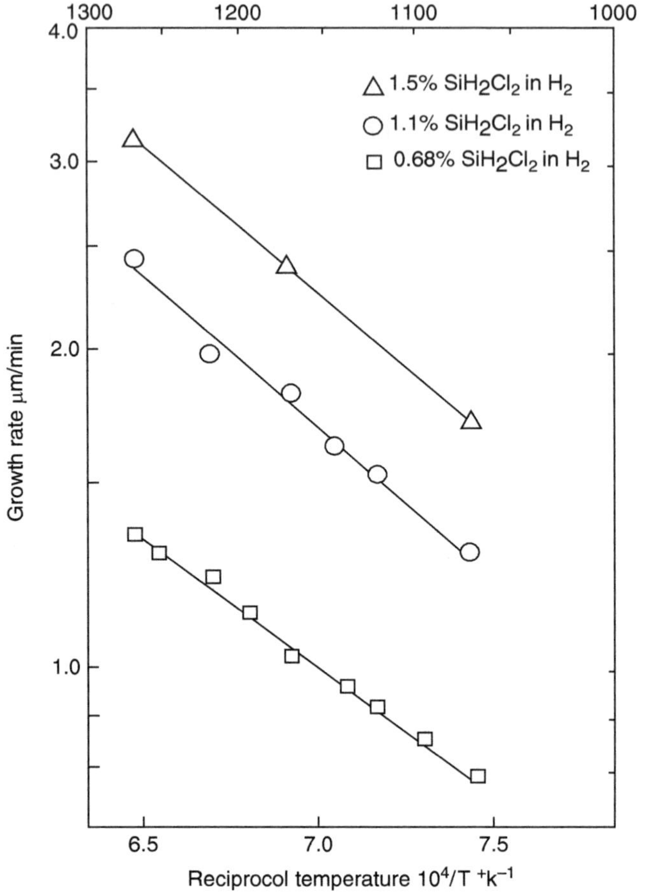

Fig. 10.5. Growth rate for various concentration of dichlorosilane [7]

10.3.2 Doping of Epitaxial Films

The process of doping consists in introducing the semiconductor n-type or p-type impurities that are necessary for the electrical characteristics required for material applications. This operation demands a perfect control of the quantity of dopants and of their profile in the epitaxial layer. For doping of silicon, arsine (AsH_3) and phosphine (PH_3) are gases commonly used for n-type doping, while the diborane (B_2H_6) is used for p-type doping.

10.3.2.1 Dopant Incorporation

In the case of in-situ doping during epitaxy, a small amount of dopant gas is introduced into the reactor at the same time as the silicon precursor. The dopant concentration can vary in a broad range, by controlling the partial pressure of dopant gas. The lower doping levels are fixed by the purity and dilution of used gases. The upper limit of doping depends on the maximum solubility of the dopant element in silicon at the deposition temperature.

At high temperature, the molecules of dopant are dissociated into different volatile species. The equilibrium partial pressures of these species can be calculated thermodynamically. In the case of diborane, equilibrium partial pressures of gaseous species in the system Si–H–Cl–B have been calculated by Bloem and Giling [8]. The data can provide information on dominant dopant species for a given partial pressure. For example, for low partial pressure of PH_3, the dominant species in the gas phase are PH_3 and PH_2. For high partial pressure PH_3, P_2 becomes the dominant species.

The actual incorporation of dopants in silicon is determined by a factor of segregation K_{eff}. It is given by the concentration of dopants in silicon divided by the ratio of partial pressure of dopant gas (P^0_{dopant}) and silicon precursor (P^0_{Si}):

$$K_{\mathrm{eff}} = \frac{[\mathrm{dopant}]_{\mathrm{Si}}/(5 \times 10^{22})}{P^0_{\mathrm{dopant}}/P^0_{\mathrm{Si}}}.$$

On the one hand, when K_{eff} is less than 1, part of the incorporated dopant atoms are rejected from the silicon layer. On the other hand, when K_{eff} is equal to 1, the atoms of the dopant gas are completely incorporated in the silicon.

The consequences of the introduction of dopants on the growth kinetics have been studied by Agnello et al. [16] and Lengyel et al. [17]. Figure 10.6 illustrates the change in growth rates as a function of temperature for different dopant gases. For a low partial pressure (10^{10} atm. to 10^5 atm.), the incorporation is proportional to the partial pressure of diborane. For higher pressure ($>10^4$ atm.), there is a decrease of incorporation due to the condensation of species containing boron. Indeed, the values of equilibrium partial pressures show that these species begin to condense for a high partial pressure of diborane and, therefore, do not participate in the doping process.

In the regime limited by the surface kinetics, adding a dopant gas to induce n-type or p-type doping causes an increase in the growth rate. On the contrary, in the regime limited by mass transfer, adding dopants in the reactive gas does not affect the silicon growth rate [16–18].

10.3.2.2 Profile of the Concentration of Dopants

Doping occurs simultaneously with the growth of silicon and can induce an abrupt concentration profile in p–n junction elaboration. Nevertheless, there are certain limitations due to the phenomena of exodiffusion and autodoping.

Exodiffusion: At high temperature, the distribution of dopant atoms extends both from the gas phase to the substrate and in the opposite direction. The diffusion from the solid phase to the gas phase aims at restoring the

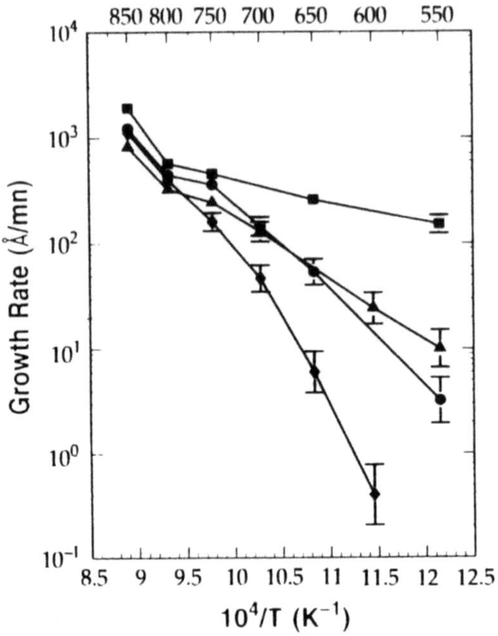

Fig. 10.6. Temperature dependence of growth rates of silicon (from. precursor: 1% of dichlorosilane in H_2). *Filled diamond*: no gas doping, *filled square*: B_2H_6, *filled circle*: PH_3, *filled triangle*: AsH_3 [16]

thermodynamic equilibrium between the two phases by increasing the partial pressure until a saturation value [8]. The exodiffusion is, therefore, a loss of dopant concentration at the surface of the substrate.

During growth, this phenomenon known as exodiffusion does not take place because the gas phase is deliberately saturated to dope the layer. However, it starts at the end of the growth during the cooling process. The importance of the phenomenon depends on the dopant species: it varies with its diffusion coefficient, which is a function of temperature and with its saturation pressure in the gas phase. The nature of the carrier gas also affects the value of the diffusion coefficient [20].

Experimentally, the loss of dopant atoms on the surface of the sample depends on the initial doped layer and on the time/temperature parameters during the cooling step.

A method for reducing the dopant concentration loss on the surface is to cover the surface with an oxide layer, the latter acting as diffusion barrier [19]. Another method is to carry out the cooling step under a dopant gas flow, in such a way that the gas phase is saturated with boron or phosphorus. This trick helps ensuring a uniform doping along the volume of the epitaxial layer.

Autodoping phenomenon: The autodoping refers to the incorporation of dopant atoms in the epitaxial layer from unwanted external sources.

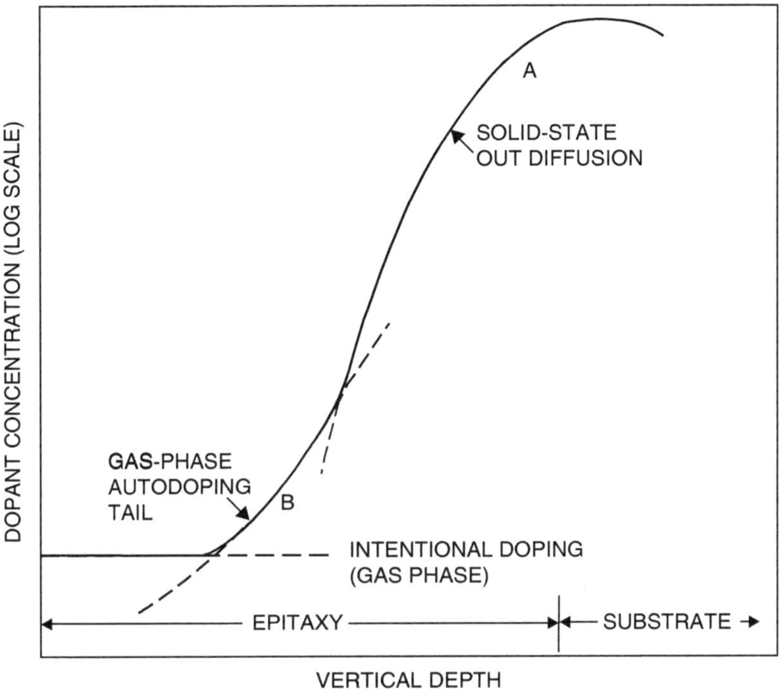

Fig. 10.7. Dopant profile at the epitaxial layer/substrate interface [21]

It has several origins:

- Prior deposits on the susceptor and walls of the reaction chamber.
- The substrate itself [20, 21], from its front or rear surface via an exodiffusion.

This phenomenon is again guided by the diffusion parameters: temperature and diffusion coefficient. The autodoping phenomenon combined with the exodiffusion, which occurs during the rise in temperature before the growth step, explains that the profile of the dopant concentration between the epitaxial layer and the substrate cannot be abrupt. The concentration profile at the junction shows a subdoping of the substrate and an overdoping of the epitaxial layer as shown in Fig. 10.7.

The level of autodoping also depends on the initial doping of the substrate. The diffusion lengths, for boron at 1,100°C can be about 0.5 μm.

10.4 Epitaxial Growth Equipments

The modern microelectronics and semiconductor industries have imposed severe demands on the quality of films produced by the silicon epitaxy process and the epitaxial film deposition techniques need to fulfill several general requirements, such as high growth rate good epitaxial thickness uniformity minimum particulate generation and economic use of reactants.

A basic epitaxial reactor should consist of at least the following items: (a) a reactor tube or chamber to isolate the epitaxial growth environment; (b) a system that distributes the various chemical species for epitaxial growth in a very controlled manner; (c) a system for heating the wafers; and (d) a system for scrubbing the effluent gases.

Epitaxial reactors are high-temperature chemical vapor deposition systems. High temperature reactors can be divided into hot-wall and cold-wall reactors. The former is predominantly used in systems where the deposition reaction is exothermic in nature, since the high wall temperature minimizes or even prevents undesirable deposition on the reactor walls. Hot-wall reactors are frequently tubular in form; heating is most often accomplished by resistance elements or RF coupling with a graphite sleeve surrounding the reactor tube. The cold-wall reactors can be used for growth of silicon from halides or hydride. Since these are endothermic reactions, they will proceed most readily on the hottest surfaces of the system.

The simplest configuration is the horizontal reactor, which consists of a horizontal quartz tube. Wafers are placed horizontally on a graphite susceptor in the tube. The wafers are heated by the susceptor that is RF power coupled. Gases used for growing epitaxial silicon enter at one end of the tube and are exhausted from the other end. The flow of gas is parallel to the wafer surface and the reactant species are supplied to the growth interface via diffusion through the boundary layer on the surface. This kind of reactor offers lower construction cost, but controlling the growth over the entire susceptor is a problem. It is difficult to get good temperature, thickness, and doping uniformities within a wafer and from wafer to wafer.

In the vertical pancake reactor working at atmospheric pressure [24], the wafers are placed on the silicon carbide coated graphite susceptor, which is heated by the underlying RF coils (Fig. 10.8). High frequencies (few kHz to few hundred kHz) are used for RF heating to temperature typically 1,000–1,200°C. The susceptor is near the bottom of the quartz bell-jar. The reactant gases enter from the center of the circular susceptor, rise to the top of the bell-jar and then spread downward. Some gas exits at the bottom while some flows over the susceptor. The gases are distributed evenly across all wafers and the susceptor rotates to further smooth out any nonuniformity in flow. Thus, good thickness and doping uniformities are obtained. The vertical pancake system is capable of running at reduced pressure as well as at atmospheric pressure to minimize autodoping effects and pattern shift.

For photovoltaic applications, several reports illustrate the growing interest in research to develop techniques of epitaxy on large surfaces. The major challenge is to get a device at a lower cost.

Thus, at the University of Madrid, Rodriguez et al. [22] have designed a reactor for the recycling of gases that are not consumed. This innovation appears of paramount importance, because the gas consumption efficiency

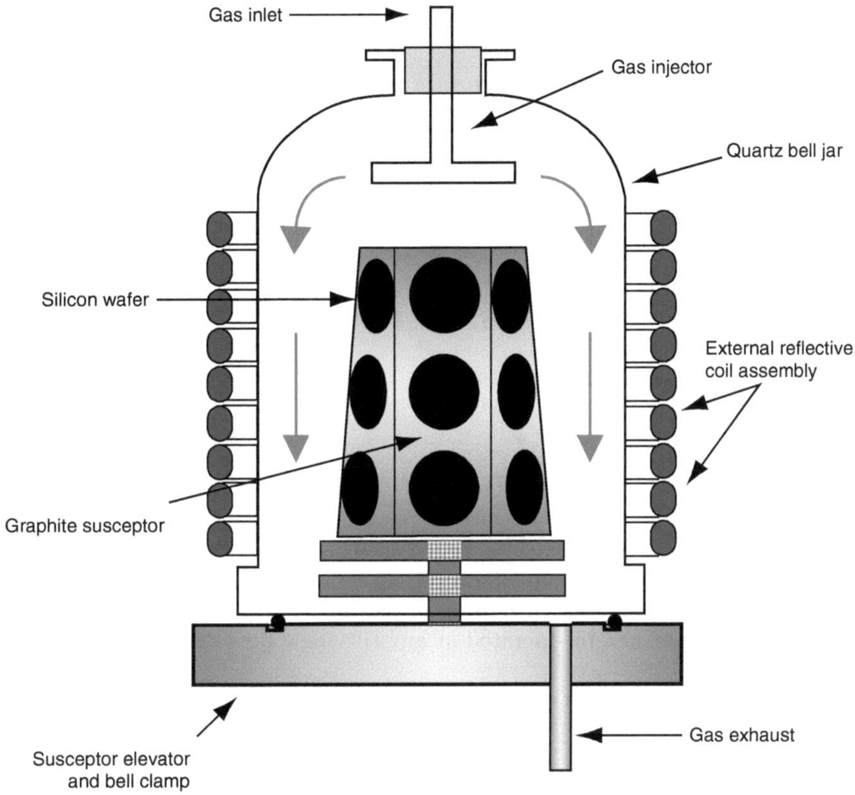

Fig. 10.8. Schematic view of vertical barrel epitaxial reactor [24]

during the epitaxy process rarely exceeds 30% and the amount of high-purity hydrogen is significant. Heating by Joule effect coupled to a system of reflecting mirrors provides a greater energy efficiency.

In the European project SWEET [26], financed between 2002 and 2005, several partners (IMEC, Fraunhofer ISE, ATERSA, Crystalox Ltd, PV Silicon AG) were pursuing the goal to lead industrial maturity to the fabrication of epitaxial solar cells on cheap silicon substrates. Part of their work has resulted in the design of a high productivity epitaxial reactor, called Con-CVD, that can reach continuous operation speed of $1.2\,\mathrm{m}^2\,\mathrm{h}^{-1}$ with a growth rate of $2.9\,\mu\mathrm{m}\,\mathrm{min}^{-1}$. Given the large volume of the reaction chamber, argon is used as carrier gas as it presents best guarantees of safety compared to hydrogen.

Another example is the work of Kunz et al. [23] at ZAE Bayern, on the design of an epitaxial reactor that uses the internal convection of gases to obtain homogeneous layers on large surfaces of more than $(40 \times 40)\,\mathrm{cm}^2$ (Fig. 10.9). This system, called CoCVD (Convection-assisted CVD), uses

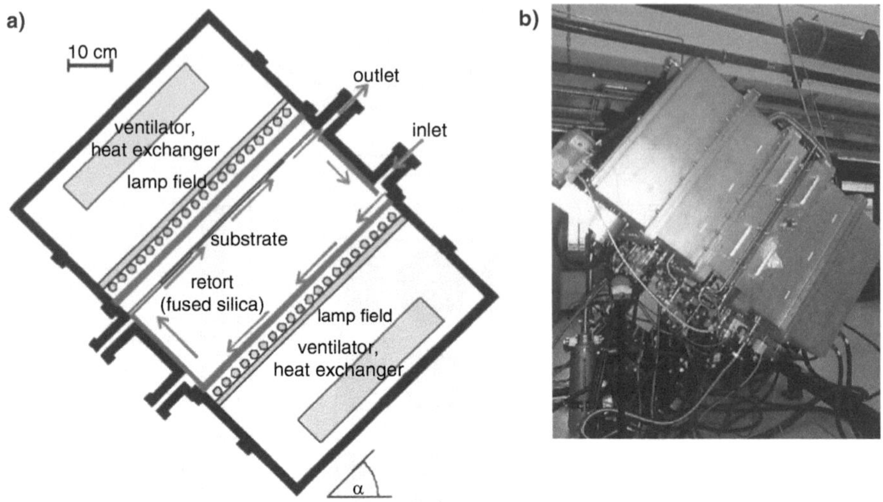

Fig. 10.9. Drawing (**a**) and photograph (**b**) of the new CoCVD System. The flow of gas is governed by thermal convection resulting in a uniform laminar flow along the bottom side of the substrate. An important parameter is the substrate inclination, therefore the machine can be operated at any tilt angle between 0° and 90° [25]

trichlorosilane ($SiHCl_3$) and boron trichloride (BCl_3) as gas precursors. A heating lamp can reach a temperature greater than 1,100°C. The authors reported growth rates ranging from 0.4 to 1.2 μm min^{-1}. The defect density is less than 5×10^3 cm^{-2}. The quality of the epitaxial films is confirmed by the realization of cells on monocrystalline substrate p$^+$-type (10^{19} cm^{-3}). Conversion efficiencies of 11.5% and 12.8% were reported for layer thicknesses ranging between 8.5 and 20 μm.

In solar cell technology, nowadays, the epitaxy technique is used to grow thin, highly doped back-surface field layers (BSF) and active layers on top of highly doped substrates. However, besides the growth of the base, the emitter can also be produced, by only adding a few more minutes to the standard epitaxy process. This epitaxial emitter concept is ideally suited for an in-line continuous epitaxy reactor with several sequential deposition chambers. Therefore, a process can be designed to grow a complete solar cell structure in-situ.

It is clear that the development of vapor phase epitaxy reactor at atmospheric pressure is booming. The process leads to good quality material and is compatible with the photovoltaic industry criteria of efficiency and reduced production costs. Indeed, the active layer deposition step represents the main cost of a thin cell process. Any innovation to reduce the thermal budget of this operation is of major interest.

References

1. A.W. Blakers, J.H. Werner, E. Bauser, H.J. Queisser, Appl. Phys. Lett. **60**, 2752 (1992)
2. J. Kraiem, E. Tranvouez, S. Quoizola, M. Lemiti, in *Proceedings of the 31st IEEE Photovoltaic Specialists Conference*, 2005, p. 1107
3. K. Snoeckx, G. Beaucarne, F. Duerinckx, I. Gordon, J. Poortmans. Semicon. Int. **30**, 45 (2007)
4. A. Fave, S. Quoizola, J. Kraiem, A. Kaminski, M. Lemiti, A. Laugier, Thin Solid Films **451–452**, 308 (2004)
5. G.F. Zheng, W. Zhang, Z. Shi, M. Gross, A.B. Sproul, S.R. Wenham, M.A. Green, Solar Energy Mater. Solar Cells **40**, 231 (1996)
6. K.J. Weber, A.W. Blakers, M.J. Stocks, M.J. Stuckings, A. Cuevas, A.J. Carr, T. Brammer, G. Matlakowski, *1st WCPEC*, Hawaii, 1994, p. 1391
7. B.J. Baliga, in *Epitaxial Silicon Technology*, ed. by B.J. Baliga, Reference Book (Academic, London, 1986), p. 328
8. J. Bloem, L.J. Giling, in *Current Topics in Materials Science*, vol. 1, ed. by E. Kaldis (North-Holland, The Netherlands, 1978), p. 147
9. A.S. Grove, A. Roder, T. Sah, J. Appl. Phys. **36**, 802 (1965)
10. C.E. Morosanu, D. Iosif, E. Segal, J. Cryst. Growth **61**, 102 (1983)
11. W.A.P. Claassen, J. Bloem, J. Cryst. Growth **50**, 807 (1980)
12. K.I. Cho, J.W. Yang, C.S. Park, S.C. Park, in *Proceedings of the 10th International Conference on Chemical Vapor Deposition*, vol. **87–88** (The Electrochemical Society, New Jersey, 1987), p. 379
13. P.A. Coon, M.L. Wise, S.M. George, J. Cryst. Growth **130**, 162 (1993)
14. B.J. Baliga, J. Electrochem. Soc. **129**, 1078 (1982)
15. A. Lekholm, J. Electrochem. Soc. **119**, 1122 (1972)
16. P.D. Agnello, T.O. Sedgwick, J. Electrochem. Soc. **140**, 2703 (1993)
17. I. Lengyel, K.F. Jensen, Thin Solid Films **365**, 231 (2000)
18. L.J. Giling, Mater. Chem. Phys. **9**, 117 (1983)
19. S.P. Murarka, J. Appl. Phys. **56**, 2225 (1984)
20. K. Suzuki, H. Yamawaki, Y. Tada, Solid-state Electronics **41**, 1095 (1997)
21. G.R. Srinivasan, J. Electrochem. Soc. **127**, 1334 (1980)
22. H.J. Rodriguez, J.C. Zamorano, I. Tobias, *22nd European Photovoltaic Solar Energy Conference*, Milan, Italy, 2007, 1091
23. T. Kunz, I. Burkert, R. Auer, A.A. Lovtsus, R.A. Talalaev, Y.N. Makarov, J. Cryst. Growth **310**, 1112 (2008)
24. M.C. Nguyen, J.P. Tower, A. Danel, *196th Meeting of the Electrochemical Society*, Honolulu, 1999, 99–16
25. T. Kunz, I. Burkert, R. Auer, R. Brendel, W. Buss, H.v. Campe, M. Schulz, *19th European Photovoltaic Solar Energy Conference*, Paris, France, 2004, p. 1241
26. S. Reber, F. Duerinckx, M. Alvarez, B. Garrard, F.-W. Schulze, EU project SWEET on epitaxial wafers equivalents: Results and future topics of interest. in *21st European Photovoltaic Solar Energy Conference*, Dresden, Germany, 2006

11

Thin-Film Poly-Si Formed by Flash Lamp Annealing

Keisuke Ohdaira

Abstract. Flash lamp annealing (FLA) has attracted attentions as a technique of rapidly crystallizing precursor a-Si films to form poly-Si films with high crystallinity on low-cost glass substrates. In this chapter, a brief explanation on fundamental physics typically seen in nonthermal equilibrium annealing and in utilization of metastable a-Si as precursor films have been given. Recent findings concerning FLA-triggered crystallization of micrometer-order-thick a-Si films and microstructures of the poly-Si films are also introduced.

11.1 Introduction

Thin-film polycrystalline Si (poly-Si) with high crystallinity formed through postcrystallization of precursor amorphous Si (a-Si) films is an attractive material for high-efficiency solar cells, because of the formation with less amount of material than bulk crystalline Si, higher carrier mobility and resulting longer carrier diffusion length than a-Si-containing thin films, and no light-induced degradation [1–8]. To improve the throughput of the crystallization process, rapid crystallization techniques are expected to be applied instead of conventional furnace annealing. Flash lamp annealing (FLA) is a millisecond-order rapid annealing technique, which is expected to be a high-throughput crystallization process to form poly-Si films. In this section, crystallization of precursor a-Si films on glass substrates by FLA, particularly aiming at solar cell application, are briefly introduced.

11.2 FLA Equipment

Figure 11.1 shows an example of the schematic diagram of FLA equipment [9]. The flash lamps, e.g., xenon lamps, are energized by discharging a capacitor/inductor bank flash power supply [10], resulting in millisecond-order radiation onto wafers, to increase the temperature of target material by light

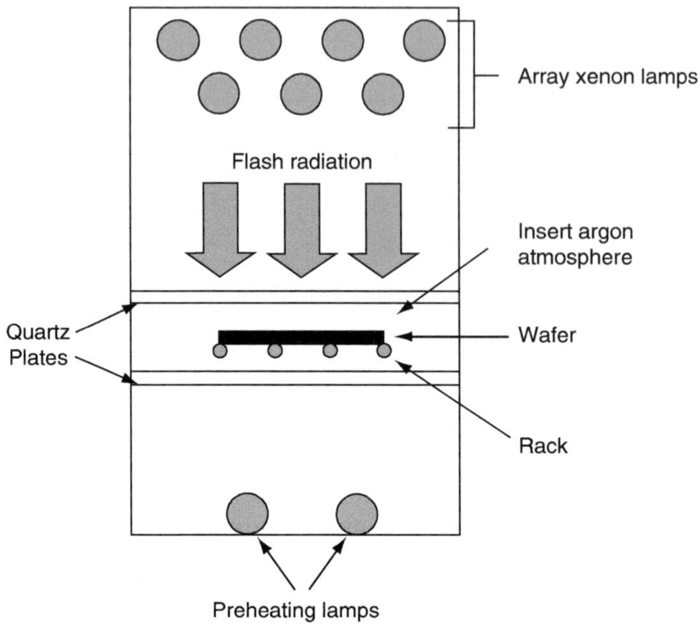

Fig. 11.1. Example of the schematic diagram of FLA equipment [9]. The flash lamps are energized by discharging a capacitor/inductor bank flash power supply [10], which can systematically control lamp irradiance

absorption. The lamp irradiance, typically on the order of several tens of $\mathrm{J\,cm^{-2}}$, can be systematically controlled by varying the capacitor charge [10]. Since irradiation from xenon lamps has broad spectrum in the visible range [9], precursor a-Si films as well as crystalline Si (c-Si) wafers can be effectively heated by optical absorption. The samples are often put in an inert gas atmosphere, or sometimes in a vacuumed space, during FLA to prevent reaction of films to gases, such as oxygen. Preheating is often applied to control an in-depth temperature profile of wafers or films during FLA. Although this technique was originally developed for activation of ion-implanted dopants keeping their shallow profiles [10–13], it has recently drawn renewed attention as a rapid crystallization technique [9, 14–16]. Unlike laser annealing, large-area irradiation can be realized by only one flash pulse without scanning, which will lead to a highly productive annealing process.

11.3 Thermal Diffusion Length

Degree of thermal diffusion is an important factor in treating rapid annealing, since it will govern in-depth crystallinity of poly-Si films formed and thermal damage to glass substrates. One-dimensional thermal diffusion coefficient can be expressed as

$$\frac{\partial}{\partial t}(\rho h) = \frac{\partial}{\partial x}\left(\kappa \frac{\partial T}{\partial x}\right) + S, \tag{11.1}$$

where T is the temperature, h is enthalpy per unit weight, ρ is the density, κ is the thermal conductivity and S is a source term for the absorption of the flash lamp light, respectively. For simplicity, assuming that ρ and κ are independent of T and of depth x, and only the surface is heated at a constant temperature T_s, (11.1) can be solved as

$$\frac{T - T_0}{T_s - T_0} = 1 - \text{erf}\left(\frac{x}{2\sqrt{Dt}}\right), \tag{11.2}$$

where T_0 is the temperature at $x = \infty$, T_s is the temperature at the surface ($x = 0$), "erf" represents the error function, and D is the thermal diffusion coefficient expressed using the specific heat capacity c as

$$D = \frac{\kappa}{c\rho}. \tag{11.3}$$

The factor $\sqrt{Dt}$ contained in (11.2), referred to as the thermal diffusion length (L_T), indicates degree of in-depth thermal diffusion. Figure 11.2 shows L_T of a-Si, quartz, and soda lime glass as a function of duration t [9,17–22]. Annealing durations realized by some of the rapid annealing techniques and corresponding L_T are particularly indicated in the graph. In the case of conventional rapid thermal annealing (RTA), pulse duration exceeds 1 s, which results in

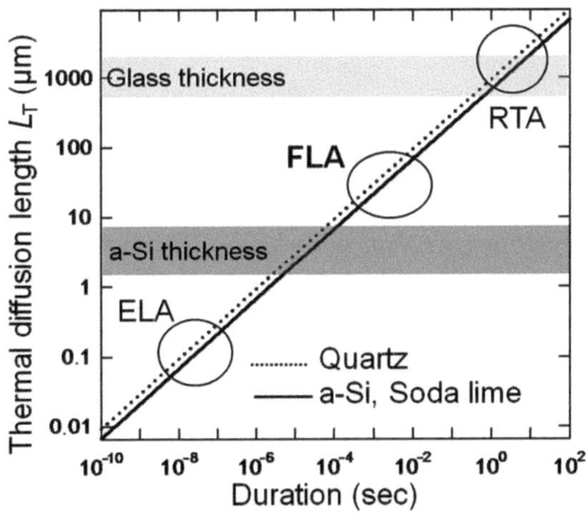

Fig. 11.2. L_T of a-Si, quartz, and soda lime glass as a function of pulse duration. Durations and corresponding L_T in the cases of conventional RTA and ELA as well as of FLA are indicated as *circles.* Typical thicknesses of a precursor a-Si film aiming at solar cell application and glass substrates are also illustrated. To crystallize a micrometer-order-thick a-Si film, avoiding thermal damage to a whole glass substrate, *the circle* should be between the two areas, which can be realized by FLA

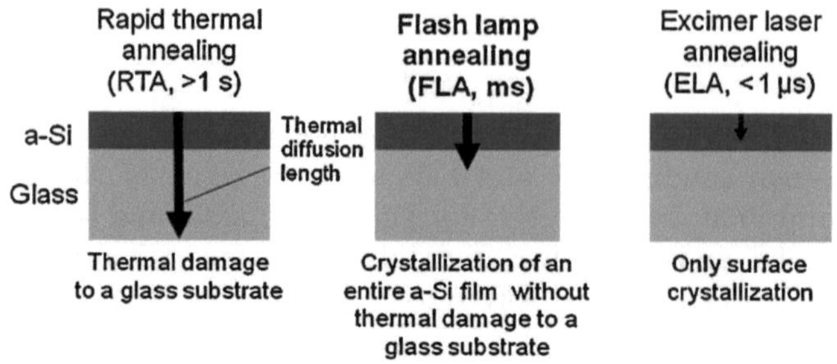

Fig. 11.3. Schematic diagrams of rapid crystallization by RTA, FLA, and ELA. RTA, with pulse duration over 1 s, will completely heat a glass substrate, resulting in thermal damage to an entire substrate. ELA cannot crystallize a micrometer-order-thick a-Si film because of too short pulse duration (<1 μs). FLA, with millisecond-order duration, is most proper to crystallize a micrometer-order-thick a-Si film without thermal damage to an entire glass substrate

complete heating of a glass substrate, even if heat generation occurs only in an a-Si film. However, excimer laser annealing (ELA), which has been frequently used in forming thin (<100 nm) poly-Si films [23, 24], provides pulse lights with duration less than 1 μs. This duration corresponds to L_T of a-Si less than 1 μm, meaning incomplete crystallization of micrometer-order-thick a-Si films. Judging from the relation between t and L_T shown in Fig. 11.2, FLA with millisecond-order duration is most proper to crystallize micrometer-order-thick a-Si films without thermal damage to entire glass substrates, which leads to utilization of low-cost glass substrates with poor thermal resistivity. The annealing features using these three techniques are schematically shown in Fig. 11.3.

11.4 Thermal Model of FLA

The actual FLA system contains more complicated factors, to determine in-depth temperature profiles, such as temperature and material dependence of c, κ, and ρ, time-dependent lamp irradiance, and an in-depth profile of light absorption because of finite absorption coefficient and of broad spectrum of the flash lamp light. Phase transition must also be considered to discuss FLA-triggered crystallization. Based on a finite difference method, (11.1) can be rewritten as

$$\begin{aligned} h_{i,n+1} = h_{i,n} + \frac{\Delta t}{\rho(\Delta x)^2}\Bigg[&\frac{2\kappa_i}{v_i(1+v_i)}\left[v_i T_{i+1,n} - (1+v_i)T_{i,n} + T_{i-1,n}\right] \\ &+ \frac{1}{2}(\kappa_{i+1}-\kappa_i)(T_{i+1,n}-T_{i,n}) + \frac{1}{2v_i^2}(\kappa_i-\kappa_{i-1})(T_{i,n}-T_{i-1,n})\Bigg] \\ &+ S_{i,n}\Delta t, \end{aligned} \tag{11.4}$$

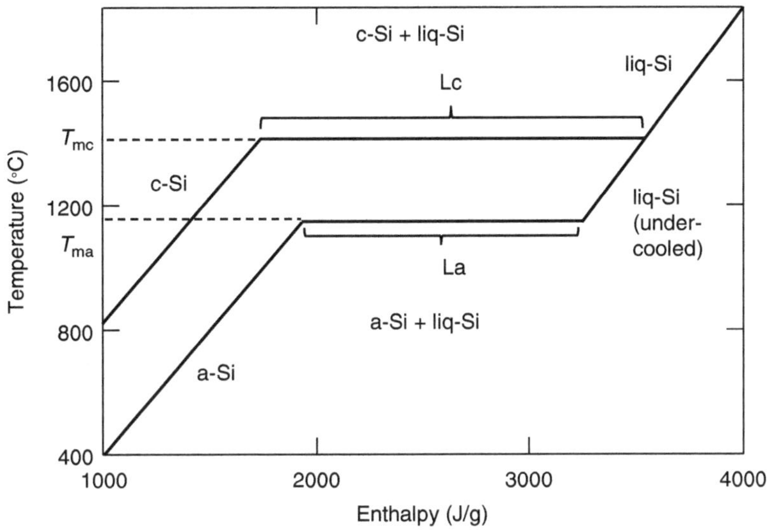

Fig. 11.4. State diagram for Si. a-Si and c-Si have different melting points of T_{ma}(1,414°C) and T_{mc}(1,145°C), and have latent heats of L_{a} $\left(1{,}148\,\mathrm{J\,g^{-1}}\right)$ and L_{c} $\left(1803.75\,\mathrm{J\,g^{-1}}\right)$, respectively [9]

where Δx is the cell width, i is the depth index, n is the time step, v_i is the ratio of distance between the center of the previous and next finite element cell [25]. T is related to h by the state diagram for Si shown in Fig. 11.4, and calculated from the relationship

$$T = \begin{cases} h/c & \text{for } h < cT_{\mathrm{m}} \\ T_{\mathrm{m}} & \text{for } cT_{\mathrm{m}} < h < cT_{\mathrm{m}} + L\,, \\ (h-L)/c & \text{for } cT_{\mathrm{m}} + L < h \end{cases} \tag{11.5}$$

where L is the latent heat and T_{m} is the melting point of Si [25]. Phase transition through melting growth and solid-phase crystallization (SPC) can be predicted by applying classical nucleation theory [26], nucleation rate from undercooled Si melt, and liquid-phase epitaxy for each cell and time step. Both melting growth and SPC are found to be possible in FLA-triggered crystallization [25]. Since precursor metastable a-Si has a higher enthalpy than stable c-Si, crystallization of a-Si leads to thermal emission, which sometimes plays an important role during FLA-triggered lateral crystallization, as shown in Sect. 11.6.

11.5 Control of Lamp Irradiance

Figure 11.5 shows the Raman spectra of the Si films with different thicknesses on quartz substrates after FLA, under the same irradiance with 5 ms duration [27]. No c-Si phase can be observed, and the broad signals centered at

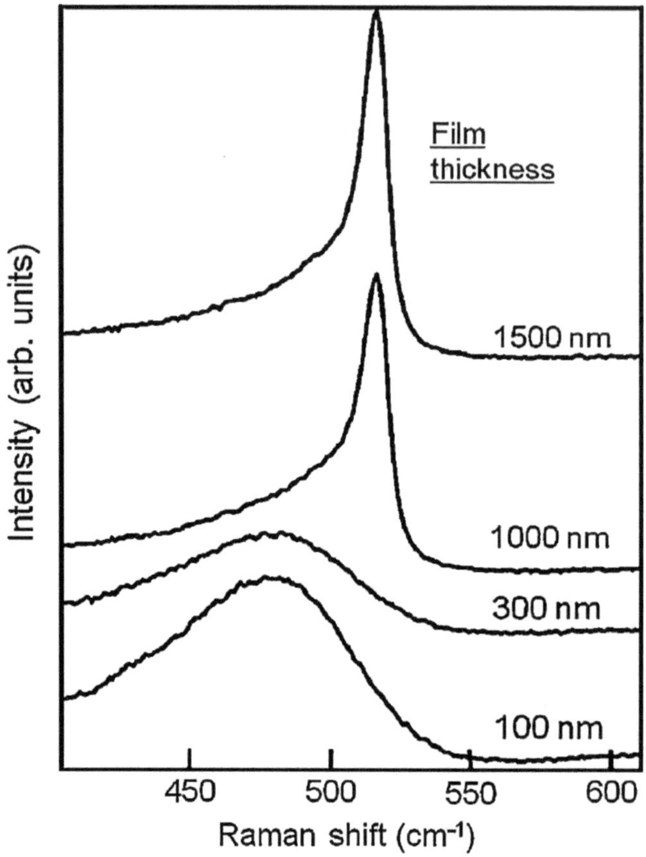

Fig. 11.5. Raman spectra of Si films with various thicknesses formed on quartz substrates after FLA with a same lamp irradiance [27]. Thicker a-Si films tend to be crystallized under a lower lamp irradiance

480 cm^{-1} related to the a-Si phase are dominant in the spectra of the films with a film thickness less than 300 nm. However, the c-Si phase can clearly be seen in the case of thicker films. These results indicate that thick a-Si films can be crystallized under a low lamp irradiance compared with thin films. This phenomenon can be understood as a result of differences in the total generated heat in a-Si films caused by optical absorption and in L_T of the generated heat. L_T of a-Si and quartz in 5 ms are on the order of several tens of μm, as shown in Fig. 11.2, indicating that the heat generated in the a-Si film during FLA almost homogeneously diffuses entirely through the Si film and also into the quartz surface with a length of at least several tens of μm. Because a broad spectrum of flash lamp light, mainly in visible range, the total heat generated in thick a-Si films is larger than that generated in thin a-Si films. Further, the total L_T from the surface is relatively small in

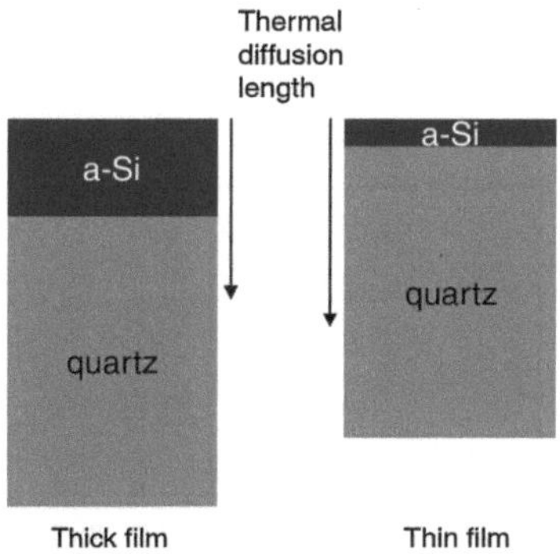

Fig. 11.6. Schematic diagrams of thermal diffusion in cases of thick and thin a-Si films [27]. The total generated heat in thick a-Si films is larger than that in thin a-Si films, whereas the total thermal diffusion length from the surface is relatively small in case of thick a-Si, resulting in crystallization at lower lamp irradiance

case of thick a-Si because of the lower thermal diffusivity of a-Si than that of quartz, as shown schematically in Fig. 11.6 [27]. The thicker film, therefore, holds more heat per unit thickness, resulting in crystallization at lower lamp irradiance. As understood from the discussion above, the threshold irradiance for crystallization can also be affected by thermal constants of substrates. a-Si films formed on soda lime glass substrates can actually be crystallized by FLA under lower lamp irradiance than the films with same film thickness on quartz substrates, which have longer L_T than soda lime glass [28]. The lamp irradiance must, therefore, be tuned according not only to a-Si film thickness but also to substrate materials and structures.

11.6 FLA for Solar Cell Fabrication

Peeling of Si films from glass substrates during FLA becomes serious problem, particularly when the precursor a-Si films are more than 1 μm thick, and thus, adhesion between Si and glass substrates must be improved. Figure 11.7 shows the surface images of lamp-annealed 4.5-μm-thick Si films on $20 \times 20\,\mathrm{mm}^2$ quartz and soda lime glass substrates with and without Cr film insertion [28,29]. The Si films are completely peeled off without Cr insertion, and there are no suitable conditions for the crystallization of a-Si without peeling in this structure, although lamp irradiance is systematically varied. However, the polycrystallization of a-Si without peeling can be seen because of excellent adhesiveness of Cr to glass and to Si [30–32]. The inserted Cr films can also be utilized as back contacts and back electrodes in actual solar cell structures.

Peeling of Si films tends to be suppressed also by using precursor a-Si films with less amount of hydrogen content [33], as in the case of ELA. From this point of view, catalytic chemical vapor deposition (Cat-CVD), which can

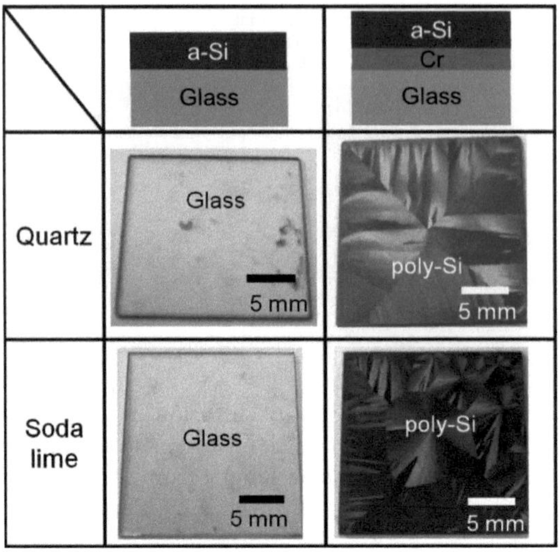

Fig. 11.7. Surface images of lamp-annealed 4.5-μm-thick Si films on $20 \times 20\,mm^2$ quartz and soda lime glass substrates with and without Cr film insertion [28, 29]. The Si films are completely peeled off without Cr insertion, whereas the peeling of Si films during FLA can be suppressed with insertion of Cr films because of high adhesiveness of Cr

easily yield a-Si films with hydrogen content of as low as approximately 3%, is more suitable than conventional plasma-enhanced CVD (PECVD) for the deposition technique of precursor a-Si films.

Keeping abrupt profiles of dopants is also required to apply the FLA technique to solar cell fabrication process. Figure 11.8 shows secondary ion mass spectroscopy (SIMS) profiles of Cr and P atoms in the bottom layers and of B in the surface doping layer of p–i–n Si stacked films before and after FLA, the structure of which is also schematically shown [34]. The abrupt profiles of the surface B atoms as well as of the bottom Cr and P atoms are maintained after FLA, which results from millisecond-order rapid annealing, and shows the possibility of immediate formation of p–i–n poly-Si structure with only one irradiation of flash lamp for p–i–n stacked a-Si layers.

11.7 Microstructure of the Poly-Si Films

Polycrystallization of micrometer-order-thick a-Si films by FLA exhibits curious phenomena different from the case of crystallization of thin (<1 μm) a-Si films, and consequently forms poly-Si films with characteristic microstructures. Figure 11.9 shows surface images of the Si films after FLA under various lamp irradiances [35]. Only the edges are crystallized in the case of the lowest lamp irradiance, and the crystallized area expands towards the center of the

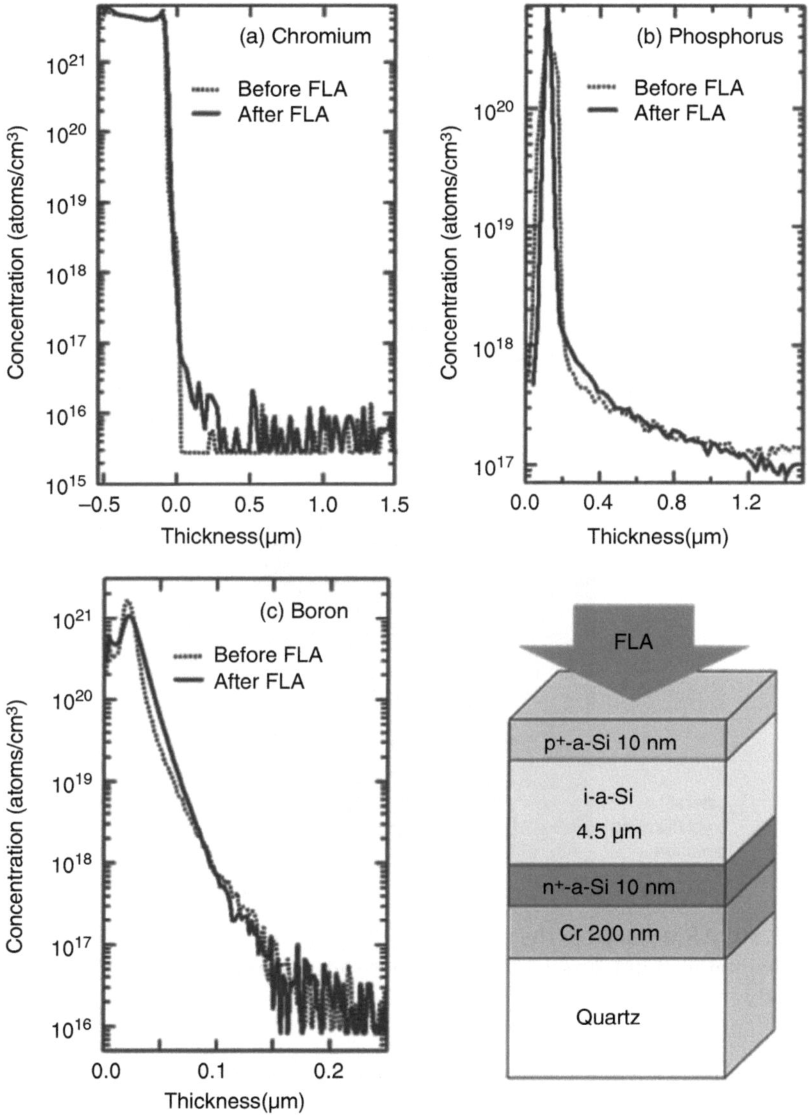

Fig. 11.8. SIMS profiles of (**a**) Cr, (**b**) P, and (**c**) B atoms before and after FLA [34]. Schematic diagram of the p–i–n stacked poly-Si films formed by FLA is also shown. The abrupt profiles of the surface B atoms as well as of the bottom Cr and P atoms are maintained after FLA

Si films with increase in lamp irradiance, showing that the crystallization of the a-Si films by FLA takes place laterally from the edges towards the center. To start the crystallization, the temperature of a-Si has to reach a threshold value, such as the melting point of a-Si in the case of melting growth, or the temperature required for a sufficient nucleation rate in the case of SPC. Thus,

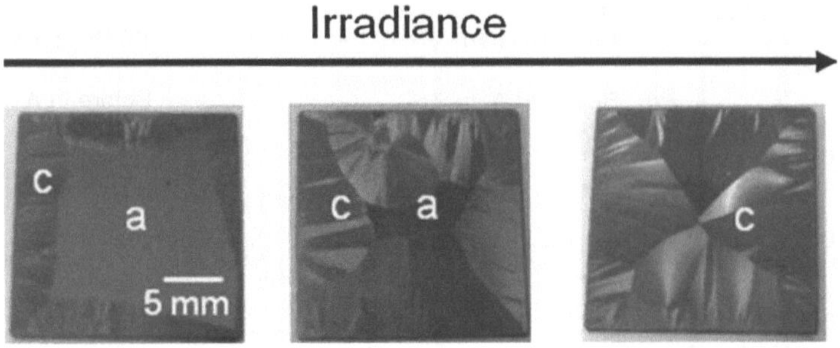

Fig. 11.9. Surface images of the Si films after FLA under various flash lamp irradiance [35]. The characters "a" and "c" in the images represent amorphous and crystallized areas, respectively. Only edge areas are crystallized under lower lamp irradiance, while crystallized area expands toward center with increasing lamp irradiance

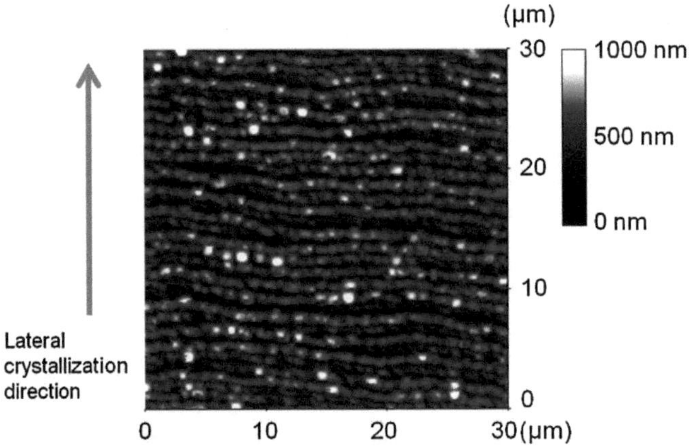

Fig. 11.10. AFM image of the poly-Si surface [36]. Periodic structures with a pitch of approximately 1 μm, formed perpendicular to the lateral crystallization direction, are clearly seen

the experimental results indicate that the edges of Si have higher temperature than the interior areas, which is well explained by the additional heating of the Si film edges by the lamp irradiation. The a-Si film is heated by the absorption of the irradiated light with high homogeneity in the lateral direction, and the edges of the Si films also receive additional heating because flash lamp light expands from lamps and the angle of incidence must be considered. As a consequence, the edges reach higher temperature, and become starting point of lateral crystallization.

Figure 11.10 shows an AFM image of the poly-Si surface, showing stripe structures with a pitch of approximately 1 μm perpendicular to the lateral growth direction [36]. Root mean square (RMS) roughness of the surface is estimated to be as large as 120 nm. Such a significant structural change from

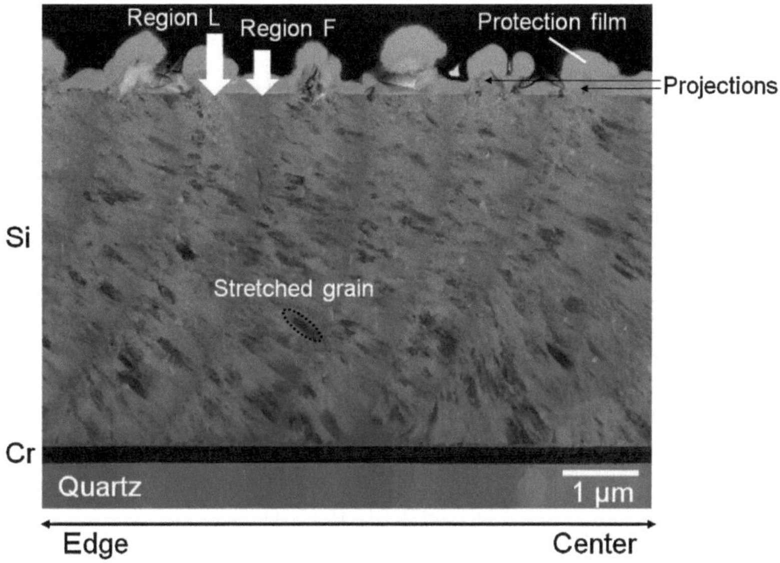

Fig. 11.11. Cross-sectional TEM image of the poly-Si film formed by FLA [35]. The region L (containing large stretched grains) and the region F (consisting of 100-nm-sized fine grains) alternatively appear in lateral crystallization direction

a flat a-Si surface probably indicates the melting of the surface Si during FLA. This large surface roughness and periodic stripe pattern behave like an optical grating, resulting in a rainbow-colored surface of the poly-Si. The naturally formed surface roughness also acts to reduce optical reflectance [36]. Figure 11.11 shows a cross-sectional TEM image of the poly-Si film formed by FLA, indicating that the film consists of small grains with sizes less than 1 μm. The periodic projecting parts can be seen on the surface of the poly-Si film, as seen in the AFM image. There exist two kinds of regions in the poly-Si; one contains relatively large grains, with a size in the order of several hundreds of nm, represented as black and light gray contrasts in the TEM image (region L), and the other consists only of 10-nm-sized fine grains with almost uniform dark gray color (region F). The region L connects to the surface projecting regions, while the region F lies below the flat region. The relatively large grains tend to be stretched in the lateral crystallization direction, which is also a clear indication of lateral crystallization. Figure 11.12 shows typical Raman spectra of the poly-Si films [28]. A clear c-Si peak located at 520 cm^{-1} can be observed, whereas no significant signal relating to a-Si can be observed at approximately 480 cm^{-1}. A full width at half maximum (FWHM) of the c-Si peaks is estimated to be 7–9 cm^{-1}, which indicates the existence of grains of less than 10 nm in size [37], and is consistent with the TEM image.

Although the surface Si seems to be melted during FLA, crystallization after complete melting of the whole Si is unlikely in this case. This is because the dopant profiles for boron (B) and phosphorus (P) show no significant

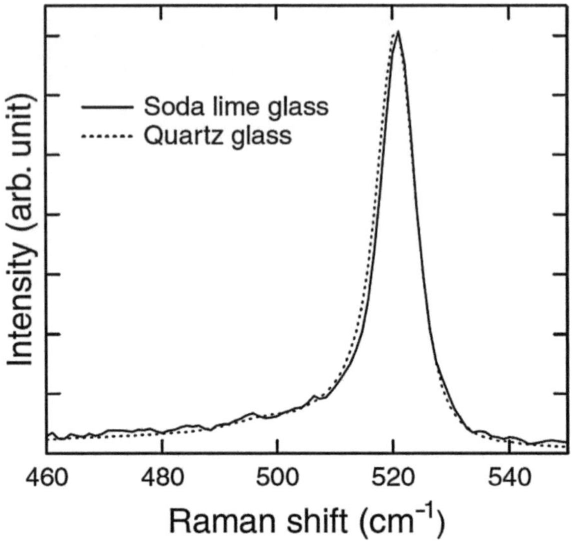

Fig. 11.12. Raman spectra of poly-Si films formed on Cr-coated soda lime and quartz glass substrates [28]. A clear c-Si peak located at 520 cm^{-1} can be observed, whereas no significant signal relating to a-Si can be observed at approximately 480 cm^{-1}, indicating high crystallinity. A FWHM of the c-Si peaks of 7–9 cm^{-1} indicates the existence of fine grains less than 10 nm in size, consistent with the TEM image

change after FLA, as shown in Fig. 11.8. The diffusion coefficients of B and P in Si melt are 2.4×10^{-4} and $5.1 \times 10^{-4}\,cm^2\,s^{-1}$ [38], corresponding to the diffusion lengths of 11 and 16 μm in 5 ms, respectively. The profiles of the dopants would, therefore, be completely broken, if the whole Si film is melted during the annealing time of 5 ms.

A possible mechanism to describe the lateral crystallization is heating due to differences of enthalpy between the a-Si and c-Si states and diffusion of the generated heat to neighboring a-Si. a-Si is a metastable state, and hence, has higher enthalpy than c-Si, resulting in thermal generation due to crystallization. The generated heat diffuses into the neighboring area, and can induce further crystallization. This lateral crystallization based on heat generation due to crystallization has been known as "explosive crystallization" (EC). The velocity of the lateral crystallization discussed here, in the order of m/s, can be fully explained by EC [39], and surface periodic structure has also been observed in some cases [40,41]. Geiler et al. have proposed the following four types of EC: (1) explosive solid-phase nucleation (ESPN), governed by nucleation directly from a-Si to c-Si phase, (2) explosive solid-phase epitaxy (ESPE), in which epitaxial growth in solid phase is dominant, (3) explosive liquid-phase nucleation (ELPN), dominated by nucleation from Si melt, and (4) explosive liquid-phase epitaxy (ELPE), in which liquid-phase epitaxy governs the crystallization [39]. Of these types, the crystallization shown here is probably governed by ESPN, because there exist a large number of small grains both in regions L and F, and their nucleation rate must be so high that it cannot be explained by nucleation from undercooled Si melt [9]. The region L is accompanied by partial melting of a-Si, because of the significant change of the surface morphology. The large-sized grains extending in the lateral crystallization directions can be explained by epitaxial growth from partially melted Si onto grains. The surface projections are probably formed through volume expansion from Si melt to c-Si. Auvert et al. also observed the mixed feature of liquid-phase and solid-phase EC in cw laser crystallization, and found that the poly-Si formed through liquid-phase crystallization has a rough surface, whereas SPC region shows a smooth surface [42]. These results are consistent with those shown here. The reason of keeping abrupt profiles of B atoms even after melting of surface Si is that the melting duration is estimated to be less than 0.5 μs, because 10,000 steps of 1-μm-long crystallization and heat generation processes must occur within 5 ms for 1-cm-long lateral crystallization, within which diffusion length of B atoms is less than 0.11 μm.

11.8 Summary

In this chapter, polycrystallization of precursor a-Si films by FLA is briefly introduced, together with physics in nonthermal equilibrium annealing. The poly-Si films formed by FLA have also been found to have high potential as solar cell material. Minority carrier lifetimes of the poly-Si films can be improved to be as long as 10 μs after defect termination [43], and actual solar cell operation has been demonstrated using p–i–n stacked poly-Si films formed by simultaneous crystallization of p–i–n a-Si films [34]. Although the study to fabricate thin-film solar cells using poly-Si films formed by FLA is just getting started, individual fundamental characteristics of the poly-Si films are sufficient as solar cell material, and therefore, future developments in this technique for highly productive solar cell process are expected.

Acknowledgements

The author acknowledges Prof. Matsumura and the members of the laboratory in JAIST for their support to part of the study introduced here, and Prof. Usami of Tohoku University for giving me an opportunity to write this chapter.

References

1. A.G. Aberle, in *Proceedings of the IEEE 4th WCPEC, Hawaii*, 2006, p. 1481
2. A.H. Mahan, S.P. Ahrenkiel, B. Roy, R.E.I. Schropp, H. Li, D.S. Ginley, in *Proceedings of the IEEE 4th WCPEC, Hawaii*, 2006, p. 1612

3. T. Matsuyama, M. Tanaka, S. Tsuda, S. Nakano, Y. Kuwano, Jpn. J. Appl. Phys. **32**, 3720 (1993)
4. R. Morimoto, A. Izumi, A. Masuda, H. Matsumura, Jpn. J. Appl. Phys. **41**, 501 (2002)
5. M.J. Keevers, T.L. Young, U. Schubert, M.A. Green, in *Proceedings of the 22nd EU PVSEC, Milan*, 2007, p. 1783
6. I. Gordon, D. Van Gestel, L. Carnel, G. Beaucarne, J. Poortmans, K.Y. Lee, P. Dogan, B. Gorka, C. Becker, F. Fenske, B. Rau, S. Gall, B. Rech, J. Plentz, F. Falk, D. Le Bellac, in *Proceedings of the 22nd EU PVSEC, Milan*, 2007, p. 1890
7. L. Carnel, I. Gordon, D. Van Gestel, G. Beaucarne, J. Poortmans, in *Proceedings of the* 22nd *EU PVSEC, Milan*, 2007, p. 1880
8. S. Janz, M. Kuenle, S. Lindekugel, E.J. Mitchell, S. Reber, in *Proceedings of the 33rd IEEE PVSC, San Diego*, 2008, p. 168
9. M. Smith, R. McMahon, M. Voelskow, D. Panknin, W. Skorupa, J. Cryst. Growth **285**, 249 (2005)
10. W. Skorupa, R.A. Yankov, W. Anward, M. Voelskow, T. Gebel, D. Downey, E.A. Arevalo, Mater. Sci. Eng. B **114–115**, 358 (2004)
11. R.L. Cohen, J.S. Williams, L.C. Feldman, K.W. West, Appl. Phys. Lett. **33**, 751 (1978)
12. H.A. Bomke, H.L. Berkowitz, M. Harmatz, S. Kronenberg, R. Lux, Appl. Phys. Lett. **33**, 955 (1978)
13. L. Correra, L. Pedulli, Appl. Phys. Lett. **37**, 55 (1980)
14. B. Pécz, L. Dobos, D. Panknin, W. Skorupa, C. Lioutas, N. Vouroutzis, Appl. Surf. Sci. **242**, 185 (2005)
15. M. Smith, R.A. McMahon, M. Voelskow, W. Skorupa, J. Stoemenos, J. Cryst. Growth **277**, 162 (2005)
16. M.P. Smith, R.A. McMahon, K.A. Seffen, D. Panknin, M. Voelskow, W. Skorupa, Mater. Res. Soc. Symp. Proc. **910**, 571 (2007)
17. H. Kiyohashi, N. Hayakawa, S. Aratani, H. Masuda, High Temp. High Pressure **34**, 167 (2002)
18. J. Huang, P.K. Gupta, J. Noncryst. Solids **139**, 239 (1992)
19. I. Sawai, S. Inoue, J. Soc. Chem. Ind. Jpn. **41**, 663 (1938) [in Japanese]
20. V.K. Bityukov, V.A. Petrov, High Temp. High Pressure **38**, 293 (2000)
21. S. Inaba, S. Oda, K. Morinaga, J. Non-Cryst. Solids **325**, 258 (2003)
22. Y. Kikuchi, H. Sudo, N. Kuzuu, J. Appl. Phys. **82**, 4121 (1997)
23. A.A.D.T. Adikaari, N.K. Mudugamuwa, S.R.P. Silva, Appl. Phys. Lett. **90**, 171912 (2007)
24. I. Steinbach, M. Apel, Mater. Sci. Eng. A **449–451**, 95 (2007)
25. M. Smith, R.A. McMahon, M. Voelskow, W. Skorupa, J. Appl. Phys. **96**, 4843 (2004)
26. C. Spinella, S. Lombardo, J. Appl. Phys. **84**, 5383 (1998)
27. K. Ohdaira, S. Nishizaki, Y. Endo, T. Fujiwara, N. Usami, K. Nakajima, H. Matsumura, Jpn. J. Appl. Phys. **46**, 7198 (2007)
28. K. Ohdaira, T. Fujiwara, Y. Endo, S. Nishizaki, H. Matsumura, Jpn. J. Appl. Phys. **47**, 8239 (2008)
29. K. Ohdaira, Y. Endo, T. Fujiwara, S. Nishizaki, H. Matsumura, Jpn. J. Appl. Phys. **46**, 7603 (2007)
30. N.M. Poley, H.L. Whitaker, J. Vac. Sci. Technol. **11**, 114 (1974)

31. R. Berriche, D.L. Kohlstedt, Mater. Dev. Microelectron. Packag.: Perform. Reliab., Proc. Electron. Mater. Process. Congr., 4th 1991, p. 47
32. N. Jiang, J. Silcox, J. Appl. Phys. **87**, 3768 (2000)
33. K. Ohdaira, K. Shiba, H. Takemoto, T. Fujiwara, Y. Endo, S. Nishizaki, Y.R. Jang, H. Matsumura, Thin Solid Films 517, 3472 (2009)
34. K. Ohdaira, T. Fujiwara, Y. Endo, K. Shiba, H. Takemoto, S. Nishizaki, Y.R. Jang, K. Nishioka, H. Matsumura, in *Proceedings of the 33rd IEEE PVSC, San Diego*, 2008, p. 418
35. K. Ohdaira, T. Fujiwara, Y. Endo, S. Nishizaki, H. Matsumura, J. Appl. Phys. (in press)
36. K. Ohdaira, T. Fujiwara, Y. Endo, S. Nishizaki, K. Nishioka, H. Matsumura, in *Tech. Digest 17th PVSEC, Fukuoka*, 2007, p. 1326
37. H. Campbell, P.M. Fauchet, Solid State Commun. **58**, 739 (1986)
38. H. Kodera, Jpn. J. Appl. Phys. **2**, 212 (1963)
39. H.D. Geiler, E. Glaser, G. Götz, W. Wagner, J. Appl. Phys. **59**, 3091 (1986)
40. C. Grigoropoulos, M. Rogers, S.H. Ko, A.A. Golovin, B.J. Matkowsky, Phys. Rev. B **73**, 184125 (2006)
41. T. Takamori, R. Messier, R. Roy, Appl. Phys. Lett. **20**, 201 (1972)
42. G. Auvert, D. Bensahel, A. Perio, V.T. Nguyen, G.A. Rozgonyi, Appl. Phys. Lett. **39**, 724 (1981)
43. K. Ohdaira, Y. Endo, T. Fujiwara, S. Nishizaki, K. Nishioka, H. Matsumura, T. Karasawa, T. Torikai, in *Proceedings of the 22nd EU PVSEC, Milan*, 2007, p. 1961

12

Polycrystalline Silicon Thin-Films Formed by the Aluminum-Induced Layer Exchange (ALILE) Process

Stefan Gall

Abstract. Thin, large-grained polycrystalline Si (poly-Si) films can be formed on foreign substrates (e.g., glass) by the aluminum-induced layer exchange (ALILE) process, which is based on the aluminum-induced crystallization (AIC) of amorphous Si (a-Si). During an annealing step, below the eutectic temperature of the Al/Si system (577°C), the initial substrate/Al/a-Si stack is transformed into a substrate/poly-Si/Al(+Si) stack. In this chapter, the ALILE process itself and the properties of the resulting poly-Si films are discussed in detail from the scientific as well as technological point of view.

12.1 Introduction

During the last decade, the production of solar cells has grown dramatically. With a share of about 90%, wafer-based crystalline silicon solar cells are still dominating the market. In order to maintain high growth rates in the future, significant cost reductions are necessary. The reduction of the silicon thickness is an appealing way to bring down costs because even relatively thin crystalline Si solar cells feature the potential for high efficiencies. The potential has already been demonstrated by the preparation of a solar cell with an efficiency of 21.5% on a thinned-down monocrystalline Si wafer with a thickness of 47 μm [1]. Unfortunately, this is not a real Si thin-film technology but still a Si wafer technology. In order to bring down costs substantially, high efficiencies have to be reached with a real Si thin-film technology utilizing competitive production techniques for large-area low-cost foreign substrates (e.g., glass).

The Si thin-film solar cells on glass, available on the market today, are mainly based on hydrogenated amorphous Si(a-Si:H) and hydrogenated microcrystalline Si(μc-Si:H). Both are usually prepared by plasma enhanced chemical vapor deposition (PECVD). A big advantage of both technologies is that the Si films can be prepared at very low temperatures (below 300°C). Although stabilized single-junction solar cell efficiencies of about 10% have already been reached on small areas in the laboratory (e.g., 9.5% on 1 cm^2 for a-Si:H [2,3]), both materials probably do not have the potential for very high single-junction efficiencies as the structural quality is relatively poor. To reach higher efficiencies, a-Si:H/μc-Si:H tandem solar cells have been developed. So far, stabilized mini-module efficiencies of about 10% have been obtained with this tandem (double junction) structure (e.g., 10.1% on 64 cm^2 [4]).

To overcome the current single-junction efficiency limits of both a-Si:H and μc-Si:H, the Si material quality has to be improved substantially. Large-grained polycrystalline Si(poly-Si) thin-films, characterized by both (1) a grain size much larger than the film thickness and (2) an intragrain quality comparable to wafer-based Si, seem to be a suitable material for high efficiency Si thin-film solar cells on glass. The preparation of such large-grained poly-Si films poses a big challenge because the glass substrate limits the process temperatures to about 600°C (the precise temperature limit strongly depends on the type of glass substrate). Several concepts have been studied to prepare large-grained poly-Si films on glass. Due to the fact that the direct deposition of Si on glass below 600°C always results in amorphous Si and/or finecrystalline Si (i.e., with a grain size much smaller than the film thickness), the techniques investigated so far are usually based on a two step process. In the first step, an amorphous Si(a-Si) film is deposited and in the second step this a-Si film is crystallized. For example, the a-Si layer can be crystallized thermally at about 600°C (solid phase crystallization – SPC) [5]. At such low temperatures, the process is relatively slow and the Si films formed by SPC feature a grain size comparable to the film thickness (for a film thickness of 1–2 μm). Based on SPC, mini-module efficiencies of up to 10.4% have already been obtained [6]. However, it remains questionable whether the Si films prepared by SPC are suitable for thin-film solar cells with very high efficiencies. Beside SPC, techniques, such as laser crystallization (LC) [7] and metal-induced crystallization (MIC) [8] have been investigated to form suitable large-grained poly-Si films on glass. In this chapter, a very specific MIC technique, which is based on aluminum-induced crystallization (AIC) of amorphous Si, is discussed. The associated process is called aluminum-induced layer exchange (ALILE). Here, the resulting poly-Si films are called ALILE films (in the literature also the expression AIC films is used).

12.2 General Aspects of the ALILE Process

It is a well known phenomenon that metals can significantly influence the crystallization of amorphous Si. The temperature required to crystallize bare a-Si is about 600°C. The contact of a-Si with metals usually leads to a strong reduction of this crystallization temperature (metal-induced crystallization). Metal/Si systems can be divided into (1) compound forming systems, which form stable metal silicide phases in thermodynamic equilibrium (e.g., Ni/Si, Pd/Si, Pt/Si) and (2) simple eutectic systems, which do not form stable metal silicides (e.g., Al/Si, Ag/Si, Au/Si). For simple eutectic systems in a layer stack configuration, the MIC process that takes place below the eutectic temperature as a solid phase transition, can be divided into three steps: (1) dissociation of Si atoms from the a-Si into the metal, (2) diffusion of Si atoms through the metal and (3) nucleation and incorporation of Si atoms into already existing Si crystals [9]. In this chapter, we will focus on aluminum-induced crystallization of amorphous Si. Due to the fact that Al is a shallow acceptor in Si with an energy level of 67 mV above the valence band edge [10], aluminum-induced crystallization of a-Si always leads to p-type material. Over the past decades, AIC has been investigated in different sample structures. Here we will focus on some important results, which are relevant for the ALILE process on glass substrates.

In 1977, Majni and Ottaviani investigated a Si-wafer/Al/a-Si structure [11]. The layers (Al and a-Si) were deposited by electron-beam evaporation onto the Si(100) wafer with a thickness of about 700 nm each. After annealing at 530°C for 12 h, an exchange of the two layer positions was observed resulting in a Si-wafer/Si/Al structure. The Si layer had grown epitaxially on the Si(100) wafer and, therefore, the related process is called solid phase epitaxy (SPE). Due to doping with Al, the hole concentration of the epitaxially grown p-type Si layer was $2 \times 10^{18}\text{cm}^{-3}$. In 1979, the same authors reported on a barrier layer at the initial Al/a-Si interface, which plays a fundamental role for the process by both controlling the diffusion during the layer exchange and limiting the thickness of the epitaxially grown Si film to the thickness of the initial Al layer [12]. In 1981, Tsaur et al. used this process to form solar cells featuring a p-type SPE-grown emitter with a thickness of about 200 nm on a n-type Si wafer [13]. Without antireflection coating (ARC) and back surface field (BSF), they obtained efficiencies at AM1 of up to 10.4% and 8.5% on monocrystalline Si(100) wafers and multicrystalline Si wafers, respectively. In 1998, it was shown that the layer exchange process does not occur on Si wafers only but also on foreign substrates: Koschier et al. demonstrated the process on oxide-covered Si wafers [14] and Nast et al. on glass substrates [15]. Nast et al. showed that the utilization of glass substrates leads to the formation of poly-Si films, which are continuous and feature a uniform thickness. In the following years, different aspects of the ALILE process were investigated [16–22].

After this short historical overview, the ALILE process on glass substrates will be introduced in more detail. Starting point for the aluminum-induced layer exchange (ALILE) process on glass is usually a glass/Al/a-Si stack (left hand side of Fig. 12.1). Different glass substrates have been used for the process (e.g., Corning 1737, Schott Borofloat 33). The preparation of the initial layer stack usually starts with a cleaning process of the glass substrate. Then, the layers (Al and a-Si) are deposited by physical vapor deposition (PVD) (i.e., thermal evaporation, electron-beam evaporation or sputtering). For the deposition of the a-Si layer PECVD has also been used [23]. A typical thickness is about 300 and 375 nm for the Al layer and the a-Si layer, respectively.

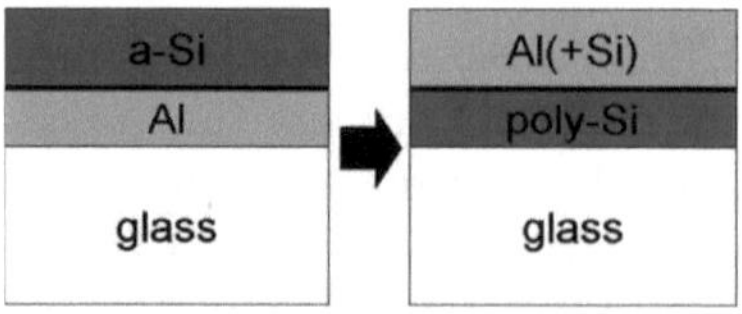

Fig. 12.1. Schematic illustration of the aluminum-induced layer exchange (ALILE) process. During an annealing step below the eutectic temperature of the Al/Si system, the initial glass/Al/a-Si stack is transformed into a glass/poly-Si/Al(+Si) stack. The permeable membrane between the layers (*black line*) stays in place during the ALILE process

The excess of Si compared to Al is necessary for the preparation of continuous poly-Si films on the glass substrate. As already mentioned above, the ALILE process requires a thin permeable membrane (barrier layer) between the Al and the a-Si layer, which controls the diffusion of Al and Si. Usually, the permeable membrane consists of an Al oxide layer formed by exposure to air (e.g., for 2 h) of the Al-coated glass substrate prior to the a-Si deposition. Annealing of the initial glass/Al/a-Si stack at temperatures below the eutectic temperature of the Al/Si system ($T_{\mathrm{eu}} = 577°\mathrm{C}$) leads to a layer exchange and a concurrent crystallization of Si resulting in a glass/poly-Si/Al(+Si) stack (right hand side of Fig. 12.1). The permeable membrane stays in place during the entire ALILE process (indicated by a black line in Fig. 12.1). Thus, the thickness of the resulting poly-Si film is determined by the thickness of the initial Al layer (in this example about 300 nm). It was shown that also ultra-thin poly-Si films with a thickness down to 10 nm [24] and poly-Si layers with a thickness of up to 1 μm [25] can be prepared. Due to the excess of Si, which is necessary to form continuous poly-Si films, the Al layer on top of the poly-Si film contains some Si inclusions, also referred to as "Si islands." It was found that the crystal quality of the "Si islands" can be similar to the crystal quality of the underlying poly-Si film, although there was no obvious relation between the crystal orientations [26]. The amount of Si within the final Al(+Si) layer is determined by the ratio of the thickness of the initial a-Si layer to the thickness of the initial Al layer.

The main process steps of the layer exchange are schematically depicted in Fig. 12.2. (a glass/Al/a-Si stack is shown with the initial Al layer marked in light gray, the initial a-Si layer in dark gray and the permeable membrane in black). The process starts with the dissociation of a-Si and the subsequent diffusion of Si atoms across the permeable membrane into the initial Al layer (process step 1). This leads to an increase in the Si concentration C_{si} within the initial Al layer, until the critical concentration for nucleation is reached. Then, Si nuclei are formed locally within the initial Al layer (process step 2). These nuclei grow in all directions, until they are confined vertically between the glass substrate and the permeable membrane. The Si growth is fed by lateral diffusion of Si atoms within the initial Al layer towards existing grains

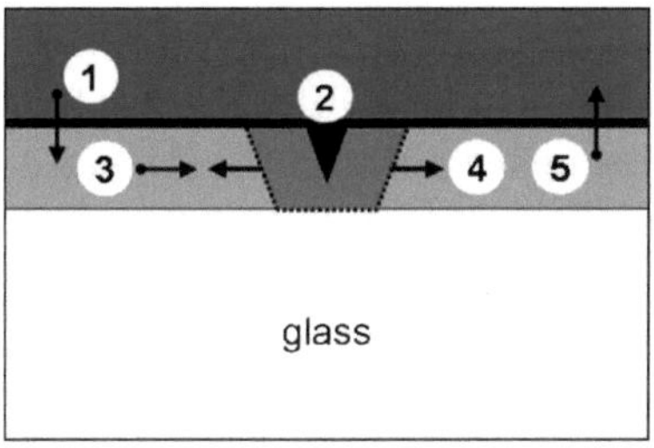

Fig. 12.2. Main steps of the aluminum-induced layer exchange (ALILIE) process (a glass/Al/a-Si stack is shown): (1) Dissociation of a-Si and diffusion of Si into the Al layer; (2) Nucleation of Si within the Al layer; (3) Lateral diffusion of Si within the Al layer towards existing Si grains; (4) Lateral Si grain growth; (5) Displacement of Al

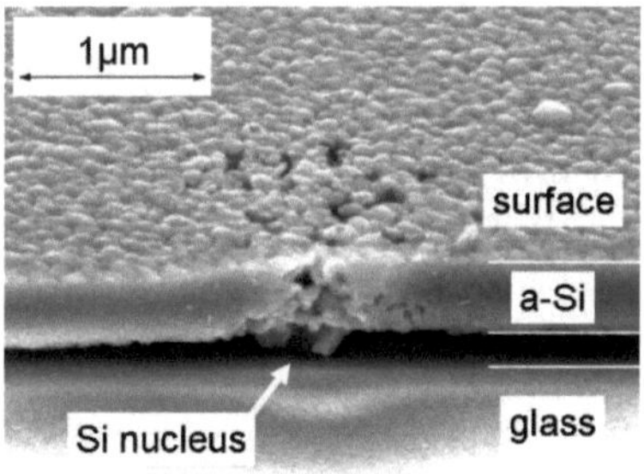

Fig. 12.3. Scanning electron microscopy (SEM) image of a Si nucleus formed between the initial glass/Al and Al/a-Si interfaces (the Al was etched off chemically). The sample was tilted by 30° to show the cross section as well as the surface. (from [27])

(process step 3). The Si growth continues only laterally (process step 4), until adjacent grains coalesce and finally, form a continuous poly-Si film on the glass substrate. Due to the growth of Si grains within the initial Al layer, the Al is displaced to the initial a-Si layer (process step 5) resulting finally in an Al(+Si) layer on top of the poly-Si film. However, the Al is not completely displaced from the initial Al layer. Some local Al inclusions remain along the grain boundaries of the final poly-Si film. More details of the ALILE process are discussed later.

The local formation of a Si nucleus within the initial Al layer (process step 2) is shown in Fig. 12.3 [27]. The sample investigated here was annealed at 420°C for 3 h. Due to the rather low annealing temperature nucleation had just started (the lower the annealing temperature the longer it takes to form first Si nuclei). In the vicinity of the Si nucleus, Al had segregated into the a-Si layer (process step 5). After the annealing step, the Al was etched off chemically. Not only the segregated Al within the a-Si layer was etched off, but also parts of the initial Al layer in the vicinity of the Si nucleus. Figure 12.3

shows a scanning electron microscopy (SEM) image of such a Si nucleus. The Si nucleus, which is confined between the glass substrate and the a-Si layer, is surrounded by a hollow space because the Al was etched off in this region.

For technological applications, it is important that the ALILE process is not limited to small areas. This was demonstrated by the preparation of a continuous poly-Si film on a 3-in. glass substrate (see Fig. 12.4) [28]. To obtain this figure, the Al on top of the poly-Si film was selectively removed by wet chemical etching. The sample is transparent due to the small thickness of the continuous poly-Si film (about 300 nm). If the Al is etched off selectively (like in Fig. 12.4), the "Si islands" are still on top of the continuous poly-Si film. This can be clearly seen in Fig. 12.5, which shows a SEM image of a poly-Si surface after selective removal of Al by wet chemical etching (main image) [29]. To get smooth continuous poly-Si films on glass, techniques, such as chemical mechanical polishing (CMP) can be applied. The inset of Fig. 12.5 shows a SEM image of a smooth poly-Si surface after removal of the complete Al(+Si) top layer by CMP. Unfortunately, CMP is probably not up-scalable at low costs for a large-area thin-film solar cell production. An appealing method to remove the "Si islands" (after etching off the Al) is based on wet chemical

Fig. 12.4. Continuous poly-Si film formed by the ALILE process on a 3-in. glass substrate (To obtain this figure the Al on top of the poly-Si film was selectively removed by wet chemical etching). (from [28])

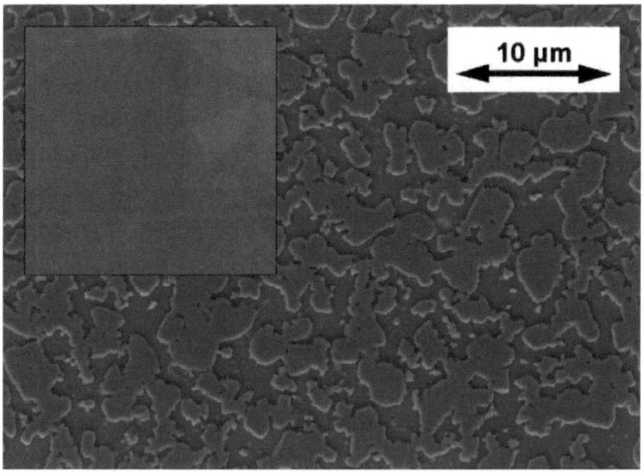

Fig. 12.5. SEM image of a poly-Si surface after selective removal of Al by wet chemical etching (*main image*) and after removal of the complete Al(+Si) top layer by chemical mechanical polishing (CMP) (*inset*). The scale applies for both main image and inset. In the main image the remaining "Si islands" are clearly visible. (from [29])

etching of the Al oxide layer (permeable membrane) between the continuous poly-Si film and the "Si islands" (lift-off process) [30]. Also selective etching techniques, which remove the "Si islands" at least as fast as the Al, were investigated (e.g., by reactive ion etching [31]).

The layer exchange not only occurs for the above mentioned layer sequence ("normal structure"), but also for the "inverse structure" [32, 33]. In this case, an initial glass/a-Si/Al stack is transformed into a glass/Al(+Si)/poly-Si stack. The "inverse structure" requires a permeable membrane, too. This membrane can also be prepared by exposure to air. But this time, a Si oxide layer is formed. In comparison with the creation of an Al oxide layer, the Si oxide layer formation takes much longer duration (days instead of hours), because a-Si is not as reactive as Al, regarding oxide formation. It is very remarkable that the layer exchange takes place even if completely different permeable membranes are used. On oxidized wafers, the "inverse structure" was already investigated in 1996 [34]. The "inverse structure" has some advantages compared to the "normal structure" described before: (1) The poly-Si film features a smooth surface and is directly accessible for subsequent process steps (e.g., epitaxial thickening). (2) The Al layer between the poly-Si film and the glass substrate can be used as a contact layer in device configurations. The disadvantage is related to the fact that for the "inverse structure" all subsequent process steps are limited to the eutectic temperature of the Al/Si system (577°C), to prevent the formation of a liquid phase. In the following sections, only the "normal structure" is discussed, but most of the results can more or less be transferred to the "inverse structure."

12.3 Kinetics of the ALILE Process

This section deals with the kinetics of the ALILE process. The nucleation and the subsequent growth can be observed directly during the layer exchange process, using an optical microscope equipped with a heating stage (in-situ investigation). The samples under investigation are placed upside down onto the heater. Hence, the initial glass/Al interface can be studied through the transparent glass substrate. Due to the different reflectivity of Al and Si, it is possible to distinguish between these two materials (dark and bright areas correspond to poly-Si and Al, respectively).

Figure 12.6 shows three optical micrographs of the initial glass/Al interface observed after various annealing times during an ALILE process at 510°C [35]. Figure 12.6a shows the initial glass/Al interface, 7 min after the first dark spots were observed, with the optical microscope. The time when the first dark spots appear is referred to as nucleation time t_{N}, which in this case amounts to 12 min. The time difference between the real t_{N} (i.e., the formation of first stable nuclei within the Al layer) and $\mathrm{t_N}$ is relatively small and can, therefore, be neglected. Almost all grains, which correspond to the isolated dark areas visible in Fig. 12.6a, appeared at the same time. With increasing annealing time t_{A} (Fig. 12.6a–c), the size of the Si grains increases, but there is only very little additional nucleation. This self-limitation of the nucleation is an important feature of the ALILE process. The Si grains grow uniformly in all lateral directions until they touch each other. Finally, a continuous poly-Si is formed (i.e., the whole optical micrograph appears dark).

The crystallized fraction, which was defined as the ratio of the dark area to the total area in the optical micrographs, is shown in Fig. 12.7 as a function of the t_{A} for different annealing temperatures T (from 500 to 530°C). The t_{A} necessary to form a continuous poly-Si layer (i.e., to reach a crystallized fraction of 100%) becomes shorter with increasing annealing temperature. For example, at 530°C a continuous poly-Si film was formed after 19 min. Whereas, it took 90 min to form a continuous poly-Si film at 500°C. The time necessary to finalize the ALILE process can be separated into the above

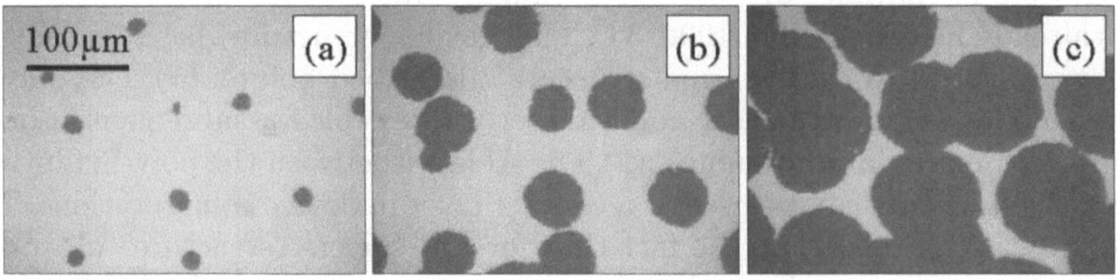

Fig. 12.6. Optical micrographs of the initial glass/Al interface taken during a crystallization process at 510°C. The corresponding annealing times are: (**a**) 19 min., (**b**) 26 min., (**c**) 33 min. Each optical micrograph shows a total area of $350 \times 260\,\mu\mathrm{m}^2$. *Dark* and *bright* areas correspond to poly-Si and Al, respectively. (from [35])

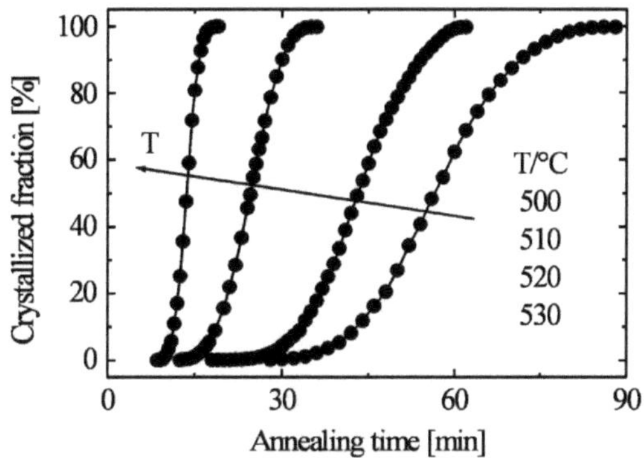

Fig. 12.7. Crystallized fraction versus annealing time for different annealing temperatures T (from 500 to 530°C). (from [35])

introduced t_N and the growth time t_G, which was defined as the time after t_N necessary to form a continuous poly-Si film. Both, t_N and t_G decrease with increasing annealing temperature. Using the temperature dependence of the t_N shown in Fig. 12.7, an activation energy E_A of 1.8 eV was determined for the nucleation process [35]. The corresponding prefactor of the exponential function amounted to 4×10^{-11} min. The determined activation energy is much lower than the activation energy of normal solid phase crystallization of a-Si (4.9 eV [36]). It was suggested that the activation energy for the nucleation process is determined by the barrier layer at the initial Al/a-Si interface [35].

For a thick barrier layer, which was prepared by thermal oxidation of the Al surface at 560°C prior to the a-Si deposition, a similar activation energy for the nucleation process was found ($E_A = 1.9$ eV) [37]. In addition to the nucleation process, the grain growth of this sample was also investigated. In order to determine the grain growth velocity (v_g) of a single grain the grain radius (r_g) was measured as the function of the t_A by analyzing the optical micrographs. After the t_N, the grain growth velocity ($v_g = dr_g/dt_A$) increases, until it reaches a final constant value, which was used for further analysis. By investigating several grains the average grain growth velocity was determined. At 450°C, the average grain growth velocity was found to be 0.076 μm min^{-1}, which is almost the same as for a sample where the barrier layer was formed just by exposure to air (0.077 μm min^{-1}). This means that the barrier layer does not influence the grain growth velocity. Therefore, the completely different t_G of the two samples are based on completely different numbers of grains. From the temperature dependence of the grain growth velocity of the sample with the thick barrier layer, an activation energy of 1.8 eV was determined. Thus, the activation energies for nucleation (1.9 eV) and grain growth velocity (1.8 eV) are almost identical. This activation energy may be associated with the energy barrier for the transport of Si across the barrier layer (permeable membrane).

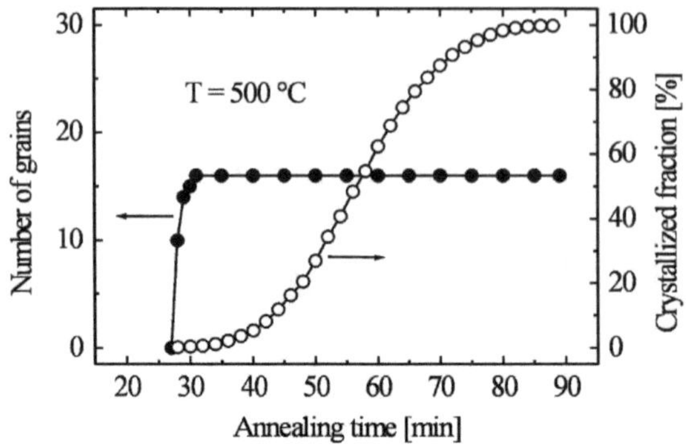

Fig. 12.8. Number of grains (*solid circles*) and crystallized fraction (*open circles*) versus annealing time at an annealing temperature T of 500°C. (from [35])

Most of the time, the number of grains does not increase monotonously with time after t_N, but reaches the final value in an early state of the growth process. Figure 12.8 shows an extreme case of a sample where the number of grains in the area under investigation increases from 0 to the final value of 16 within a few minutes after the t_N (solid circles). During the rest of the t_G, no new nucleation occurs. For comparison, the crystallized fraction of this sample is also shown in Fig. 12.8 (open circles). This curve is identical with the 500°C curve of Fig. 12.7. It is obvious that the suppression of the nucleation allows for the growth of large grains. This will be discussed in the next section.

12.4 Structural and Electrical Properties of the Poly-Si Films

In this section, the most important structural (grain size, grain orientation and intragrain defects) and electrical properties of the poly-Si films prepared by the ALILE process are discussed.

From the final number of grains (N_∞), i.e., the total number of grains, which appear in the optical microscope within the area under investigation (A_{total}), the average grain size of the resulting poly-Si film can be estimated by the following consideration. The area of every single grain equals the total area A_{total} divided by N_∞. Assuming a square shape for the grains, the edge length of the squares d can be calculated which is then called the estimated average grain size ($d = (A_{\text{total}}/N_\infty)^{1/2}$). The lower the T_{A}, the lower is the final number of grains N_∞, hence, the larger is the estimated average grain size. For the samples shown in Fig. 12.7, the estimated average grain sizes range from 51 μm at 530°C to 75 μm at 500°C. It is important to keep in mind that this analysis offers just a rough estimate of the grain size but cannot replace a real crystallographic investigation (although it is a relatively good approximation). The estimated average grain sizes given here are rather large compared to what is usually observed. For example, Nast et al. investigated the grain size by electron backscatter diffraction (EBSD) and found an average grain size of 6.2 μm [17]. Some grains showed sizes exceeding 10 μm but there were also grains, which were considerably smaller than the average grain size. The different results (e.g., 51–75 μm vs. 6.2 μm) are attributed to the strong influence of the sample preparation on the grain size of the final poly-Si film. Besides the T_A, the grain size is influenced by the way the initial sample structure is prepared (e.g., choice of the foreign substrate, deposition of both Al and a-Si, and the preparation of the barrier layer). For example, Pihan et al. studied the influence of the foreign substrate and they determined average grain sizes based on EBSD ranging from about 8 μm to about 21 μm, always coming along with a relatively broad grain size distribution [38]. Due to the fact that the grain size depends strongly on the sample preparation, it is not possible to define a standard average grain size for the poly-Si films prepared by the ALILE process, but usually an average grain size of about 10 μm can be taken as a good measure. In general, the grain size is much larger than the film thickness, which is in the order of a few hundred nanometers. Therefore, this material is referred to as large-grained poly-Si.

A typical EBSD map of the surface of a poly-Si film on glass prepared by the ALILE process is shown in Fig. 12.9 (left) [39]. The sample was annealed at 425°C for 16 h. Afterwards, the Al(+Si) top layer was removed by CMP. The poly-Si film shown here features an average grain size of 7 μm and a maximum grain size of 18 μm. From such EBSD measurements, not only the grain size but also the crystallographic orientation of the grains can be determined.

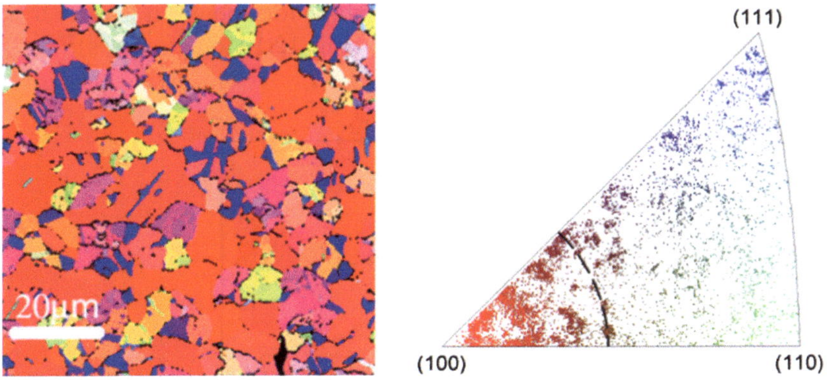

Fig. 12.9. Electron backscatter diffraction (EBSD) map showing the grain structure of a poly-Si film on glass prepared by the ALILE process (*left*) and the corresponding inverse pole figure showing the preferential (100) orientation of the poly-Si surface (*right*). The region used for the definition of the preferential (100) orientation $R_{(100)}$ is indicated by a *dashed line* (20° tilt with respect to the perfect (100) orientation). The sample was annealed at 425°C for 16 h. Afterwards the Al(+Si) top layer was removed by CMP. The area under investigation was 80 × 80 μm^2. *Red*, *green* and *blue* correspond to (100), (110) and (111), respectively. (from [39])

Figure 12.9 (right) shows the inverse pole figure of the EBSD map shown in Fig. 12.9 (left). The surface of the poly-Si film is preferentially (100) orientated. This can be seen by the agglomeration of EBSD measurement points close to the (100) corner of the inverse pole figure. To quantify this behavior, the percentage of the measurable area under investigation, which is tilted by less than 20° with respect to the perfect (100) orientation is called the preferential (100) orientation $R_{(100)}$. The corresponding region of the inverse pole figure is indicated by a dashed line (see Fig. 12.9 (right)). This definition can also be applied to the (110) and the (111) orientation. The poly-Si film shown in Fig. 12.9. features a preferential orientation of 66%, 4%, and 10% for $R_{(100)}$, $R_{(110)}$, and $R_{(111)}$, respectively. This shows clearly the strong preferential (100) orientation of the poly-Si surface. Recently also, a preferential (103) orientation of the poly-Si surface was found [40].

A high preferential (100) orientation of the poly-Si surface is favorable for subsequent epitaxial growth at low temperatures [41–43]. Due to the preferential (100) orientation, the utilization of the poly-Si films formed by the ALILE process as a template (seed layer) for subsequent epitaxial thickening at low temperatures is quite attractive [44].

Kim et al. found that the T_A influences the preferential (100) orientation of the poly-Si surface [45]. With decreasing T_A the preferential (100) orientation is enhanced. Figure 12.10 shows an example where the preferential (100) orientation $R_{(100)}$ increases from about 40% to about 70%, when the T_A decreases from 550 to 450°C (solid circles) [46]. The preferential (100) orientation is not only influenced by the T_A but also by the way the permeable membrane is formed. Samples with a barrier layer formed in a furnace at 560°C by exposure to an oxygen atmosphere for about 2 h feature a much lower preferential (100) orientation $R_{(100)}$ (Fig. 12.10, open circles): $R_{(100)}$ is

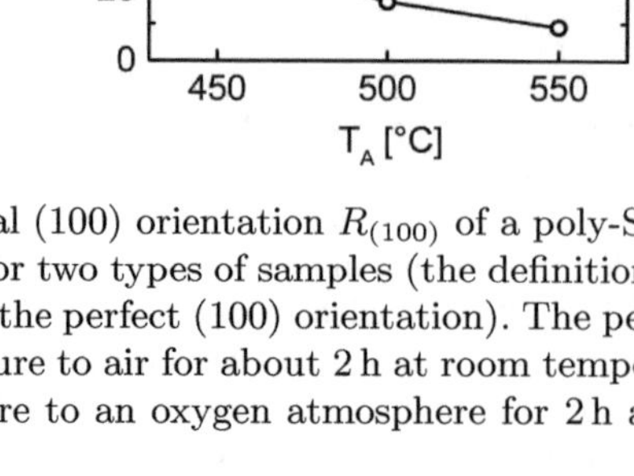

Fig. 12.10. Preferential (100) orientation $R_{(100)}$ of a poly-Si surface versus annealing temperature (T_A) for two types of samples (the definition of $R_{(100)}$ is based on a 20° tilt with respect to the perfect (100) orientation). The permeable membrane was formed either by exposure to air for about 2 h at room temperature (*solid circles*) or in a furnace by exposure to an oxygen atmosphere for 2 h at 560°C (*open circles*). (from [46])

reduced from about 70% to about 30% at $T_A = 450°C$ and from about 40% to about 10% at $T_A = 550°C$.

From the EBSD measurements the grain boundaries can also be classified ... the grain boundaries can be classified by the determination of the orientation of neighboring grains (in this way the actual microscopic structure of the grain boundaries remains unknown). Coincidence site lattice (CSL) grain boundaries (twin boundaries of first order ($\Sigma 3$) and second order ($\Sigma 9$)) as well as other types of grain boundaries were found [17]. In small-grained areas, mainly $\Sigma 3$ and $\Sigma 9$ grain boundaries were detected. These grain boundaries are probably formed during the growth of the individual grains. In silicon, $\Sigma 3$ and $\Sigma 9$ grain boundaries are considered to be electronically inactive.

Regarding the structural properties of the poly-Si films prepared by the ALILE process, research has mainly focused on the grain size and the grain orientation. But to obtain high quality poly-Si films the intragrain defect density plays a key role. To benefit from a large grain size, the intragrain defect density has to be quite low. Nast et al. showed that the grains of the poly-Si films formed by the ALILE process contain a large number of intragrain defects [17]. Pihan et al. found that the main defects are twins and low-angle grain boundaries, but no twins were found in grains with a (100) surface orientation [47]. Due to the fact that the ALILE poly-Si films are mainly used as seed layer for subsequent epitaxial thickening, intragrain defects were analyzed quite often in the epitaxially grown films. By defect etching, very large intragrain defect densities of about $10^9\,cm^{-2}$ were found in (100)-oriented grains of Si layers obtained by epitaxial thickening of ALILE seed layers (high temperature epitaxy was used) [48]. These intragrain defects, which are mainly stacking faults formed in the seed layer or at the interface between the seed layer and the epitaxial layer, are electronically active. Another investigation showed near defect-free regions in (100)-oriented grains and highly defective regions in (111)-oriented grains of Si layers obtained by epitaxial thickening of ALILE seed layers (here low-temperature epitaxy was used) [49]. In the highly defective regions of (111)-oriented grains, twins in the seed layer and stacking faults in the epitaxial layer originating from the seed layer surface are the dominant defects. The knowledge about intragrain defects in poly-Si films formed by the ALILE process is still in an early state. More investigations are needed to understand the formation and the influence of these intragrain defects.

Hall effect measurements were used to investigate the electrical properties of the poly-Si films formed by the ALILE process. Due to the incorporated Al, the poly-Si films are always p-type. At room temperature, a hole concentration of $2.6 \times 10^{18}\,cm^{-3}$ and a hole mobility of $56.3\,cm^2\,V^{-1}\,s^{-1}$ were determined [16]. Temperature dependent Hall measurements revealed both valence band conduction and defect band conduction (two-band conduction). For such highly doped material, the presence of a defect band conduction is expected. The Al concentration in the poly-Si films was measured by secondary ion mass spectroscopy (SIMS). An Al concentration of about $3 \times 10^{19}\,cm^{-3}$ was found, which is about a factor of 10 larger than the measured hole concentration. The concentration of Al is determined thermodynamically by the solubility of Al in Si.

12.5 Influence of the Permeable Membrane

The permeable membrane (barrier layer) plays a crucial role for the ALILE process. Usually, the barrier layer is formed by exposure of the Al-coated glass substrate to air. The thickness of this barrier layer (Al oxide), which is on a nanometer scale, can be influenced by the variation of the exposure time. With increasing exposure time, the thickness of the barrier layer increases. The thicker the barrier layer, the lower is the nucleation density and the longer is the process time necessary to form a continuous poly-Si film [19]. By X-ray photoelectron spectroscopy (XPS) measurements it was shown that the barrier layer stays in place during the whole ALILE process.

In order to demonstrate the importance of the barrier layer for the layer exchange process, an initial glass/Al/a-Si stack without a barrier layer was prepared [50]. To prevent the formation of a barrier layer the a-Si was deposited directly onto the Al (without vacuum break). The SEM images in Fig. 12.11 make the influence of the barrier layer clear. Two samples are shown – one with barrier layer (by exposure to air for 2 h) (left hand side) and one without barrier layer (right hand side). Both samples were annealed for 45 min at 480°C. After the annealing step, the Al was etched off chemically. To show the cross section as well as the surface, the samples were tilted by 30° for the SEM measurements. The sample with barrier layer (left) shows a continuous poly-Si film with "Si islands" on top (similar to what is shown in Fig. 12.5), whereas the sample without barrier layer (right) does not show a continuous poly-Si film but a porous film structure. The former interface between the Al layer and the a-Si layer is not visible. In some parts, the

Fig. 12.11. SEM images of a sample with barrier layer (*left*) and a sample without barrier layer (*right*). The samples were annealed at 480°C for 45 min. After the annealing step, the Al was etched off. The samples were tilted by 30° to show both cross section and surface. The sample with barrier layer (*left*) features a continuous poly-Si film with "Si islands" on top. The sample without barrier layer (*right*) features a porous structure. (from [50])

surface of the initial glass/Al/a-Si stack can be identified (smooth areas). This experiment shows clearly that a barrier layer between the Al layer and the a-Si layer is absolutely necessary for the formation of a continuous poly-Si film. Kim et al. carried out similar experiments by preparing samples without intended oxidation of the Al surface [45]. In contrast to what has been shown before, these authors observed a layer exchange forming a poly-Si film. This might be due to an unintentional oxidation during the preparation of the initial glass/Al/a-Si stack caused by the very high reactivity of the Al surface. This shows clearly that very small differences in the sample preparation can cause a significant impact on the ALILE process. This was demonstrated by well controlled oxidation experiments inside the vacuum system. Exposure of the Al surface to an oxygen atmosphere of about 6.5×10^{-3} mbar for only 2 min leads already to the formation of a continuous poly-Si film [50].

In addition to the experiments with very thin barrier layers, rather thick barrier layers have also been investigated [51]. The thick barrier layers were formed by thermal oxidation of the Al surface (e.g., for 2 h at 560°C). With such a thick barrier layer, an estimated average grain size of above 200 μm was reached. But at the corresponding T_{A} of 450°C, it took days to form a continuous poly-Si film. The influence of a barrier layer formed by thermal oxidation on the preferential orientation of the poly-Si surface was already described (see Fig. 12.10).

12.6 Model of the ALILE Process

In this section, a model of the ALILE process is described. The five main process steps of the ALILE process have already been introduced in Sect. 12.2 (see Fig. 12.2). The silicon, which is required for the growth of existing Si grains, is supplied by lateral diffusion of Si within the Al layer (process step 3). Such a diffusion-limited growth process is based on the fact that the incorporation of dissolved Si atoms into existing Si grains is fast when compared to the diffusion of Si towards existing grains. The diffusion of Si towards existing grains is driven by a gradient of the Si concentration C_{Si}. The C_{Si} in the direct vicinity of existing grains is lower than far away from the grains. This means that there are Si depletion regions around existing grains [18].

In the first phase of the ALILE process, no stable nuclei are formed (stage I). The formation of stable nuclei starts at the t_{N}. As already discussed in the Sect. 12.3 about the kinetics of the ALILE process, nucleation takes usually place only in a short time period (see Fig. 12.8). This short time period is called nucleation phase (stage II). After the nucleation phase, no new stable nuclei are formed but the growth of the existing grains continues until the grains coalesce and finally form a continuous poly-Si film on the glass substrate (stage III). The corresponding self-regulated suppression of nucleation, which allows for the growth of large grains, is a characteristic feature of the ALILE process. Nast et al. suggested that the self-regulated suppression of nucleation is due to overlapping Si depletion regions around existing grains [18].

The experimentally observed results regarding nucleation and growth are discussed with the help of the phase diagram of the Al/Si system [52–54]. The Al-rich part of the phase diagram is shown in the left part of Fig. 12.12. In the phase diagram the temperature T is shown versus the C_{Si}. The equilibrium lines are indicated as solid lines. At the eutectic temperature of the Al/Si system ($T_{\mathrm{eu}} = 577°\mathrm{C} = 850\,\mathrm{K}$), a maximum of about 1.5at.% Si can be dissolved in solid Al. The ALILE process takes place below the eutectic temperature (e.g., at 750 K). This means that the transition occurs directly from the (Al) phase to the (Al) + (Si) phase. For the following discussion, three important values of the C_{Si} within the Al are defined:

1. The *saturation concentration* C_{s} is the equilibrium concentration of Si in Al at a certain temperature (solubility). When Al is in contact with crystalline Si, there is a flow of Si atoms into the Al as long as the C_{Si} within the Al is lower than the C_{s}. For example, the C_{s} at 750 K is about 0.6at.%.
2. The difference in Gibbs energy G between the amorphous Si and the crystalline Si is the driving force for the ALILE process (about 0.1 eV per atom [5]). The higher chemical potential of amorphous Si compared to crystalline Si leads to diffusion of Si from the amorphous Si into the Al, even if the Al is already supersaturated with Si. While the C_{s} describes the equilibrium with crystalline Si, the *maximum concentration* $C_{\max}$ describes the equilibrium with amorphous Si. Due to the higher chemical potential of the amorphous Si, $C_{\max}$ is larger than C_{s}. During the ALILE process, $C_{\max}$

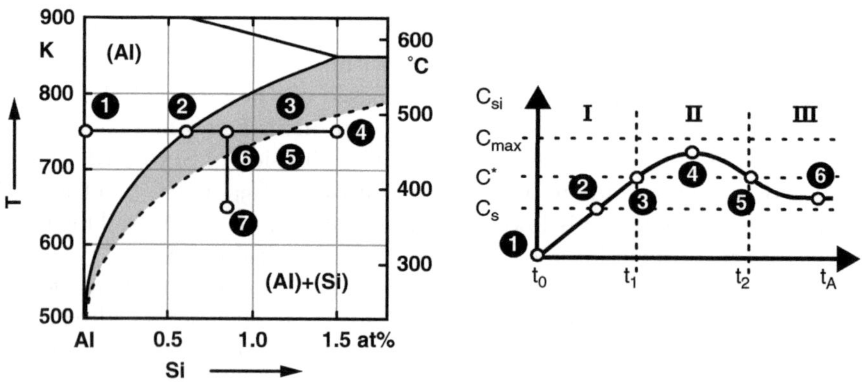

Fig. 12.12. *Left*: Al-rich part of the Al/Si phase diagram. The *dotted line* indicates schematically the critical concentration C^*. Within the *shaded* area no nucleation takes place but existing grains go on growing; *Right*: Si concentration C_{Si} within the Al layer versus annealing time t_{A} (C_{s}: saturation concentration; C^*: critical concentration; $C_{\max}$: maximum concentration); In both parts the different steps of the ALILE process are indicated by numbers (see main text). (from [52])

represents the upper limit of the C_{Si} within the Al layer. Due to the fact that C_{Si} is smaller than C_{max}, there is a steady flux of Si atoms from the amorphous Si into the Al layer. C_{max} is not shown in the phase diagram.

3. For both nucleation and growth supersaturation of the Al with Si is required ($S = C_{Si}/C_s > 1$). For diffusion-limited growth of existing grains, the growth rate J_G is proportional to S ($J_G \propto C_{Si} - C_s = C_s\ (S - 1)$). In the case of homogeneous nucleation, a stable nucleus is formed when the energy gain by the volume of the nucleus exceeds the energy, which is necessary to form the surface of the nucleus. This means that a stable nucleus cannot be formed at $C_{Si} = C_s$. The formation of a stable nucleus requires a C_{Si} within the Al layer, which exceeds the *critical concentration* C^*, i.e., the supersaturation S has to be above the critical supersaturation S^* ($S > S^* = C^*/C_s$). To form stable nuclei C_{max} has to be larger than C^*. In the phase diagram the critical concentration C^* is schematically indicated by a dotted line (the critical concentration C^* was assumed to be twice the C_s. This is an artificial definition, which was used to make C^* visible in the phase diagram and, therefore, to simplify the qualitative explanation of the ALILE process.

The time evolution of the C_{Si} within the Al layer, which is shown schematically in Fig. 12.12, plays a key role for the understanding of the ALILE process. The left part of Fig. 12.12 shows the time evolution of C_{Si} in the phase diagram (the T_A of 750 K was used as an example) and the right part of Fig. 12.12 shows C_{Si} as a function of the t_A. At the beginning of the ALILE process there is no Si within the Al layer. If the heating from room temperature to the T_A is fast compared to the diffusion of Si from the a-Si layer to the Al layer, C_{Si} remains zero when the T_A (in this example 750 K) is reached (1). With increasing t_A, C_{Si} increases due to the diffusion of Si into the Al layer. At certain time, the C_s is reached (2). Due to the higher chemical potential of amorphous Si compared to crystalline Si, C_{Si} increases further. Still no nuclei are formed although the Al is already supersaturated ($S = C_{Si}/C_s > 1$). Nucleation starts and the newly formed grains begin to grow when the critical concentration C^* is exceeded (3). The increasing number of nuclei and the corresponding grain growth reduces the increase of C_{Si} in the Al layer. At certain time, maximum of C_{Si} is reached (4). The maximum of C_{Si} corresponds to a maximum of the nucleation rate. At this point, the consumption of Si by nucleation and growth equals the supply of Si by diffusion from the a-Si layer. Due to the increasing Si consumption, C_{Si} is reduced. When the critical concentration C^* is reached again nucleation stops (5). Due to the fact that C_{Si} is still larger than C_s the growth of existing grains continues. The growing grains are not able to decrease C_{Si} below C_s because the grain growth requires supersaturation ($S > 1$). Therefore, C_{Si} stays between C_s and C^* for the rest of the process (6). In summary, three concentration regimes can be defined: (a) $C_{Si} \leq C_s$ (no nucleation, no growth), (b) $C_s < C_{Si} \leq C^*$ (no nucleation, but growth of existing grains – shaded area in the left part of Fig. 12.12), and (c) $C^* < C_{Si} < C_{max}$ (nucleation and growth). With this model, the origin of the above mentioned three stages of the process can be explained (the three stages I, II, and III are indicated in the right part of Fig. 12.12). The nucleation phase between t_1 and t_2 (stage II) is usually only a short part of the total ALILE process.

This model is supported by experiments where samples in stage III were cooled down in a very short time and subsequently reheated [52–55]. In addition to the already existing grains, many new grains appeared because upon cool down both the C_s and the critical concentration C^* are reduced while the C_{Si} stays constant (indicated by (7) in the left part of Fig. 12.12). Thus, the Al is strongly supersaturated ($C_{Si} > C^*$), which leads to the formation of numerous new nuclei. With such cool down and reheating experiments it was possible to make the depletion regions around existing grains visible [53–55]. Figure 12.13a shows an optical micrograph of the initial glass/Al interface of an ALILE sample after cool down and reheating. Many newly formed (small) grains can be seen between the few already existing (large) grains. However, all already existing grains are surrounded by a region where no additional grains were formed. These regions are referred to as depletion regions. In these depletion regions nucleation is strongly suppressed. In this example, the width of the depletion regions L was about 9 μm. This experimentally observed behavior can be understood by the schematically shown profile of the C_{Si} in the vicinity of an exiting grain (Fig. 12.13b). The existing grain is shown on the left and x marks the distance to the grain. The C_{Si} within the Al at the Si/Al interface is determined by the C_s before cool down $C_{s,a}$. Far away from the existing grain, C_{Si} is given by a certain value below the critical concentration before cool down C_a^* because the sample was in stage III (growth of existing grains but no new nucleation). The values of $C_{s,a}$ and C_a^* (a marks the

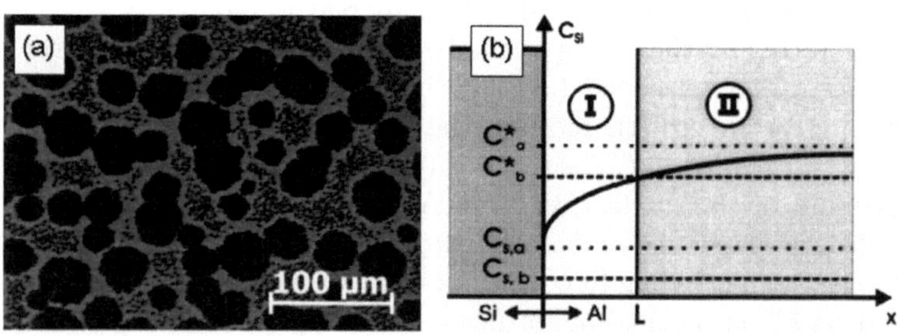

Fig. 12.13. (**a**) Optical micrograph of the initial glass/Al interface of an ALILE sample after cool down and reheating. *Dark* and *bright areas* correspond to Si and Al, respectively; (**b**) Schematic illustration of the Si concentration C_{Si} within the Al layer in the vicinity of an existing grain (x marks the distance from the grain). Saturation concentration C_s and critical concentration C^* are indicated for both before (index a) and after (index b) cool down. The width of the depletion region L separates region I (no nucleation) and region II (nucleation). (from [55])

situation before cool down) are indicated by dotted lines. The schematically shown profile of C_{Si} leads to lateral diffusion of Si towards the existing grain. Due to the fact that the sample was cooled down in a very short time, C_{Si} stays constant. Because of the lower temperature, both saturation concentration $C_{\mathrm{s,b}}$ and critical concentration C_{b}^{*} are reduced (see the phase diagram in the left part of Fig. 12.12). The values of $C_{\mathrm{s,b}}$ and C_{b}^{*} (b marks the situation after cool down) are indicated by dashed lines. After cool down, the C_{Si} far away from the existing grain exceeds the saturation concentration C_{b}^{*} (region II in Fig. 12.13b). Thus, many new nuclei are formed as observed in the experiment (Fig. 12.13a). Within a distance L from the existing grain C_{Si} is still below C_{b}^{*} and, therefore, no new nuclei are formed (region I in Fig. 12.13b). This is the already discussed depletion region, which is responsible for the suppression of nucleation. For the width of the depletion region L values between about 9 and 30 μm were experimentally observed [53–55]. From the schematic illustration in Fig. 12.13b, it is clear that the region where the C_{Si} is reduced (real depletion region) is larger than the experimentally observed width L.

The time evolution of the C_{Si} within the Al layer was also calculated to verify the above stated model of the ALILE process [56, 57]. All three experimentally observed stages of the ALILE process were investigated: stage I (no nucleation, no growth), stage II (nucleation and growth), and stage III (no nucleation, but growth). The results of the calculations are in good agreement with the model discussed above.

Based on experiments without a barrier layer between the initial layers Wang et al. suggested that Si diffusion within the Al layer takes place mainly along the Al grain boundaries [58, 59]. Thus, the Al grain boundaries are wetted by an amorphous Si layer. Nucleation occurs when the thickness of this amorphous Si layer reaches a critical value. Subsequent lateral growth of the Si nuclei is fed by Si diffusion along the interfaces between the Si nuclei and the surrounding Al.

After the description of the general model of the ALILE process, a model explaining the experimentally observed preferential (100) orientation of the poly-Si surface is discussed. The main experimental results regarding the preferential (100) orientation were presented in Sect. 12.4. In-situ investigations using an optical microscope showed no difference in the growth velocity of different grains [37]. Therefore, preferential growth was ruled out to be the origin of preferential orientation. Hence, a model based on preferential nucleation was suggested, which is able to explain the origin of the experimentally observed preferential (100) orientation [60,61]. With this model also the experimentally observed influence of both the T_{A} and the preparation conditions of the permeable membrane on the crystallographic surface orientation can be described.

The model is based on the following considerations: Due to the fact that {111} planes of crystalline Si feature the lowest specific surface energy these planes are preferentially formed [62,63]. This leads, for example, in the case of solid phase crystallization of a-Si, to a double-pyramid (octahedral) structure

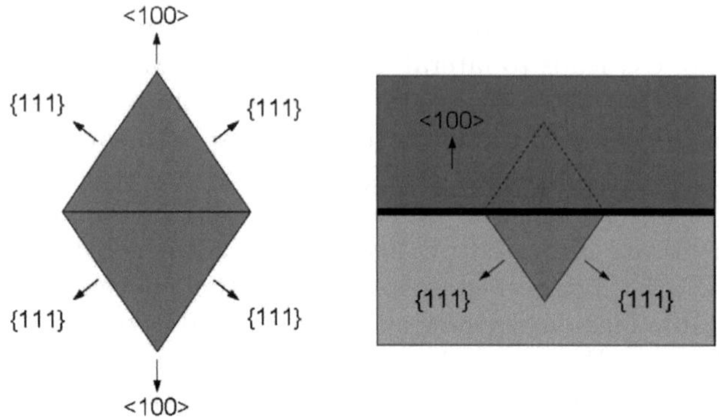

Fig. 12.14. *Left*: Side view of a Si nucleus with a double-pyramid (octahedral) structure; *Right*: Si nucleus formed within the initial Al layer at the interface between the initial Al layer (*bottom*) and the initial a-Si layer (*top*). The barrier layer is indicated by a *black line*. The nucleus is shown in the energetically most favorable alignment with respect to the interface. This alignment leads to a (100) surface orientation of the resulting grain. The Si nucleus is formed in the initial Al layer only (*the dashed line* in the initial a-Si layer is just a guide to the eye)

of the Si nuclei (see Fig. 12.14 (left)) [5]. Such a double-pyramid features eight {111} planes and two <100> tips. But just with the formation of such double-pyramids, the experimentally observed preferential (100) orientation cannot be explained. Therefore, it was additionally assumed that the nuclei are formed at the interface between the initial Al layer and the barrier layer. This assumption is based on experimental results, e.g., [18]. Calculations of the change of the Gibbs energy ΔG and hence, the activation energy for nucleation ΔG^* (maximum of ΔG with respect to the cluster size) as a function of the tilt of the nucleus at the interface showed that the alignment of a nucleus with the <100> tip perpendicular to this interface is energetically favorable (see Fig. 12.14 (right)) [60, 61]. This explains the origin of the experimentally observed preferential (100) orientation $\mathrm{R}_{(100)}$ of the poly-Si surface. The influence of the T_A and the preparation conditions of the barrier layer on $\mathrm{R}_{(100)}$ can be explained by the model as well. From the model point of view, it would be possible to reach a preferential (100) orientation $\mathrm{R}_{(100)}$ of almost 100% by the selection of a suitable barrier layer during sample preparation and a subsequent annealing step at very low temperatures.

12.7 Other Aspects of the ALILE Process

Usually the poly-Si films on foreign substrates, prepared by the ALILE process, are used as a template (seed layer) for subsequent homo-epitaxial growth of the absorber layer of a poly-Si thin-film solar cell. The highest efficiency obtained so far with an ALILE seed layer on a foreign substrate is 8% [64]. The corresponding structure of the poly-Si thin-film solar cell is: alumina substrate/spin-on oxide/p^+-type ALILE seed layer/p^+-type Si BSF (epitaxially grown)/p-type Si absorber layer (epitaxially grown)/i/n^+-type a-Si:H emitter (grown by PECVD)/indium tin oxide (ITO). The spin-on oxide was used to reduce the roughness of the alumina substrate and, therefore, to increase the grain size of the poly-Si seed layer formed by the ALILE process [65, 66]. In this case, the epitaxial thickening of the ALILE seed layer took place at 1,130°C using thermal CVD. This high-temperature process is not compatible with the utilization of glass substrates. At temperatures, which are compatible with the utilization of glass substrates (up to about 600°C), the efficiencies obtained so far are much lower. For example, the epitaxial thickening of an ALILE seed layer on glass at about 600°C using high-rate electron-beam evaporation has led to an efficiency of 3.2% [67]. So far, the poly-Si thin-film solar cells on ALILE seed layers are not limited by the grain size but by the intragrain defects. Therefore, future research will focus on these intragrain defects.

The poly-Si films obtained by ALILE process always show p-type behavior. On high temperature resistant foreign substrates, the p-type poly-Si can be transferred to n-type poly-Si by overdoping, e.g., by phosphorous diffusion at 950°C [68]. This allows for other solar cell configurations (e.g., substrate/n^+-type ALILE seed layer/n-type absorber/p^+-type emitter).

So far, we discussed the ALILE process only on insulating foreign substrates (e.g., glass). For applications, the possibility to form poly-Si films also on conductive foreign substrates is of high importance. It was shown that the ALILE process works also on some metal-coated glass substrates [28, 69] and on Al-doped zinc oxide (ZnO:Al)-coated glass [39, 70, 71]. Especially, the results on ZnO:Al, which is a transparent conductive oxide (TCO), are significant because ZnO:Al is a well established material for thin-film solar cells. On ZnO:Al-coated glass the grain size of the poly-Si film is slightly reduced and its preferential (100) orientation remains about the same (compared to bare glass) [39]. It was found that the ZnO:Al layer does not degrade during the ALILE process but the conductivity is even improved due to an increased carrier density [71]. This opens up attractive options for the preparation of poly-Si thin-film solar cells.

The ALILE process can not only be used to form poly-Si films but also to prepare polycrystalline silicon–germanium (poly-$Si_{1-x}Ge_x$) films in the entire composition range ($x = 0 \ldots 1$) [72–75]. For this purpose, an amorphous silicon–germanium (a-$Si_{1-x}Ge_x$) alloy is deposited on the oxidized Al layer to form the initial layer stack of the ALILE process. In order to prevent melting, the annealing should take place at relatively low temperatures, i.e., below the eutectic temperature of the Al/Ge system ($T_{eu} = 420°C$). Due to the low T_A, the process times are relatively long. With increasing Ge concentration, the nucleation density is reduced and the crystal growth gets more dentritic. No significant phase separation was found. By alloying with Ge, the optical absorption is increased. Thus, the utilization of poly-$Si_{1-x}Ge_x$ films could be an appealing option for the preparation of thin-film solar cells.

This chapter deals with the layer exchange process based on aluminum-induced crystallization of amorphous Si. Recently, a layer exchange process based on silver-induced crystallization was obtained [76]. The related process is called silver-induced layer exchange (AgILE). Apart from the fact that Ag is used instead of Al, the AgILE process is quite similar to the ALILE process. The eutectic temperature of the Ag/Si system is 836°C. Using the AgILE process, continuous intrinsic poly-Si films were formed. The related process times were quite long. For example at an T_A of 530°C, it took about 1,350 min to form a continuous poly-Si layer. In this case, the average diameter of the Si crystallites was about 30 μm.

The last two paragraphs show clearly that variations of the standard ALILE process can lead to interesting new results, which offer new promising perspectives. There are probably a lot of other options to vary the standard ALILE process. Therefore, further research is needed to completely utilize the possibilities of this process.

12.8 Summary and Conclusions

In this chapter, the formation of poly-Si films by the ALILE process has been discussed. During the ALILE process, a substrate/Al/a-Si stack is transformed into a substrate/poly-Si/Al(+Si) stack by a simple annealing step below the eutectic temperature of the Al/Si system (577°C). After the layer exchange process, the Al(+Si) layer on top of the poly-Si film is removed. The higher the T_A, the shorter is the time to form a continuous poly-Si film on the substrate. During the ALILE process, Si nuclei are formed within the initial Al layer. The lower the T_A, the lower is the density of Si nuclei (i.e., the larger is the final grain size). Nucleation takes place usually only in a very short time period. Afterwards, the formation of new nuclei is suppressed until the end of the process. This allows for the growth of large grains. The grain size of the resulting poly-Si films is typically about 10 μm, which is much larger than the film thickness of about a few hundred nm. Therefore, this material is called large-grained poly-Si. Due to the grain size and the preferential (100) orientation of the poly-Si surface, these layers are used as template (seed layer) for subsequent epitaxial growth. Due to the incorporated Al, the poly-Si films formed by the ALILE process feature a high doping concentration (p^+-type). The permeable membrane (barrier layer) between the initial Al layer and the initial a-Si layer plays a key role for the ALILE process. The barrier layer, which is usually formed by oxidation of the Al surface prior to the a-Si deposition, stays in place during the whole process. Important properties like process time, grain size, and preferential (100) orientation are influenced by this barrier layer. A general model of the ALILE process has been discussed. The different stages of the process can be explained using the phase diagram of the Al/Si system. In particular, the self-regulated suppression of nucleation, which is a characteristic feature of the ALILE process, can be explained by overlapping Si depletion regions around existing grains. In addition to the general model, a model explaining the origin of the preferential (100) orientation has been discussed. The related model is based on preferential nucleation and can explain the experimentally observed behavior.

The poly-Si films prepared by the ALILE process feature very attractive properties (e.g., large grains and preferential (100) orientation of the poly-Si surface). Therefore, this process has gained a lot of attention during the last years. The process is very sensitive with respect to the process parameters (e.g., the preparation conditions of the initial layer stack). The consequential advantage is that there are a lot of possibilities to influence the process and, therefore, the properties of the resulting poly-Si film. But the consequential disadvantage is that the process is not very robust. This makes it difficult to define standard process parameter, which can be used by different groups around the world leading always to the same poly-Si properties. This has to be solved to utilize the ALILE process not only on a research level but also in the production of Si thin-film solar cells. Another important point is that the research has so far mainly focused on the grain size and the crystallographic surface orientation of the poly-Si films. Future research should focus more on the intragrain defects, which still dominate the material quality. To benefit from the large grains the intragrain defect density has to be reduced significantly. For this purpose, a fundamental understanding of the origin of these intragrain defects is necessary.

There is already an excellent fundamental understanding of the ALILE process. However, many important questions remain open. Continuous research is needed to give adequate answers to these questions and to fully exploit the potential of the ALILE process. The upcoming years will show whether this fascinating process can be used in the industrial production of Si thin-film solar cells.

Acknowledgements

I like to thank my (partly former) colleagues at the Helmholtz-Zentrum Berlin für Materialien und Energie (formerly Hahn-Meitner-Institut Berlin) for their contributions during the last years in the context of our research on the ALILE process, in particular J. Berghold (formerly J. Klein), W. Fuhs, O. Hartley (formerly O. Nast), K.Y. Lee, M. Muske, B. Rech, A. Sarikov, J. Schneider and I. Sieber. Especially, the intensive collaboration with J. Schneider and W. Fuhs has provided the basis for this chapter. The research on the ALILE process has been financially supported by the European Commission in the framework of the projects METEOR and ATHLET. I have highly appreciated the collaborations with the project partners and also with other external colleagues.

References

1. J. Zhao, A. Wang, S.R. Wenham, M.A. Green, in *Proceedings of the 13th European Photovoltaic Solar Energy Conference*, Nice, France, 23–27 October 1995 p. 1566
2. M.A. Green, K. Emery, Y. Hishikawa, W. Warta, Progr. Photovoltaics Res. Appl. **17**, 85 (2009)
3. J. Meier, J. Spitznagel, U. Kroll, C. Bucher, S. Faÿ, T. Moriarty, A. Shah, Thin Solid Films **451–452**, 518 (2004)
4. B. Rech, T. Repmann, M.N. van den Donker, M. Berginski, T. Kilper, J. Hüpkes, S. Calnan, H. Stiebig, S. Wieder, Thin Solid Films **511–512**, 548 (2006)
5. C. Spinella, S. Lombardo, F. Priolo, J. Appl. Phys. **84**, 5383 (1998)
6. M.J. Keevers, T.L. Young, U. Schubert, M.A. Green, in *Proceedings of the 22nd European Photovoltaic Solar Energy Conference*, Milan, Italy, 3–7 September 2007, p. 1783
7. N.H. Nickel (ed.), *Laser Crystallization of Silicon* (Semiconductors and Semimetals **75**), (Elsevier, Amsterdam, 2003)
8. S.R. Herd, P. Chaudhari, M.H. Brodsky, J. Noncryst. Solids **7**, 309 (1972)
9. G. Ottaviani, D. Sigurd, V. Marrello, J.W. Mayer, J.O. McCaldin, J. Appl. Phys. **45**, 1730 (1974)
10. S.M. Sze, *Physics of Semiconductor Devices*, 2nd edn. (Wiley, New York, 1981), p. 21
11. G. Majni, G. Ottaviani, Appl. Phys. Lett. **31**, 125 (1977)
12. G. Ottaviani, G. Majni, J. Appl. Phys. **50**, 6865 (1979)
13. B.-Y. Tsaur, G.W. Turner, J.C.C. Fan, Appl. Phys. Lett. **39**, 749 (1981)
14. L.M. Koschier, S.R. Wenham, M. Gross, T. Puzzer, A.B. Sproul, in *Proceedings of the 2nd World Conference on Photovoltaic Energy Conversion*, Vienna, Austria, 6–10 July 1998, p. 1539
15. O. Nast, T. Puzzer. L.M. Koschier, A.B. Sproul, S.R. Wenham, Appl. Phys. Lett. **73**, 3214 (1998)
16. O. Nast, S. Brehme, D.H. Neuhaus, S.R. Wenham, IEEE Trans. Electron Devices **46**, 2062 (1999)
17. O. Nast, T. Puzzer, C.T. Chou, M. Birkholz, in *Proceedings of the 16th European Photovoltaic Solar Energy Conference*, Glasgow, UK, 1–5 May 2000, p. 292
18. O. Nast, S.R. Wenham, J. Appl. Phys. **88**, 124 (2000)
19. O. Nast, A.J. Hartmann, J. Appl. Phys. **88**, 716 (2000)
20. O. Nast, in Proceedings of the 28th IEEE Photovoltaic Specialists Conference, Anchorage (AK), USA, 15–22 September 2000, p. 284
21. O. Nast, S. Brehme, S. Pritchard, A.G. Aberle, S.R. Wenham, Solar Energy Mater Solar Cells **65**, 385 (2001)
22. O. Nast, Doctoral thesis, Philipps-Universität Marburg, Germany, 2000
23. E. Pihan, A. Slaoui, P. Roca i Cabarrocas, A. Focsa, Thin Solid Films **451–452**, 328 (2004)
24. T. Antesberger, C. Jaeger, M. Scholz, M. Stutzmann, Appl. Phys. Lett. **91**, 201909 (2007)
25. P.I. Widenborg, A.G. Aberle, in *Proceedings of the 29th IEEE Photovoltaic Specialists Conference*, New Orleans (LA), USA, 20–24 May 2002, p. 1206
26. P.I. Widenborg, A.G. Aberle, J. Cryst. Growth **242**, 270 (2002)

27. S. Gall, I. Sieber, M. Muske, O. Nast, W. Fuhs, in *Proceedings of the 17th European Photovoltaic Solar Energy Conference*, Munich, Germany, 22–26 October 2001, 1846 (2002)
28. S. Gall, M. Muske, I. Sieber, J. Schneider, O. Nast, W. Fuhs, in *Proceedings of the 29th IEEE Photovoltaic Specialists Conference*, New Orleans (LA), USA, 20–24 May 2002, p. 1202
29. S. Gall, J. Schneider, J. Klein, M. Muske, B. Rau, E. Conrad, I. Sieber, W. Fuhs, D. Van Gestel, I. Gordon, K. Van Nieuwenhuysen, L. Carnel, G. Beaucarne, J. Poortmans, M. Stöger-Pollach, P. Schattschneider, in *Proceedings of the 31st IEEE Photovoltaic Specialists Conference*, Lake Buena Vista (FL), USA, 3–7 January 2005, p. 975
30. P.I. Widenborg, T. Puzzer, J. Stradal, D.H. Neuhaus, D. Inns, A. Straub, A.G. Aberle, in *Proceedings of the 31st IEEE Photovoltaic Specialists Conference*, Lake Buena Vista (FL), USA, 3–7 January 2005, p. 1031
31. D. Van Gestel, I. Gordon, A. Verbist, L. Carnel, G. Beaucarne, J. Poortmans, Thin Solid Films **516**, 6907 (2008)
32. G. Ekanayake, H.S. Reehal, Vacuum **81**, 272 (2006)
33. G. Ekanayake, T. Quinn, H.S. Reehal, J. Cryst. Growth **293**, 351 (2006)
34. J.H. Kim, J.Y. Lee, Jpn. J. Appl. Phys. **35**, 2052 (1996)
35. S. Gall, M. Muske, I. Sieber, O. Nast, W. Fuhs, J. Noncryst. Solids **299–302**, 741 (2002)
36. U. Köster, Phys. Stat. Sol. A **48**, 313 (1978)
37. J. Schneider, J. Klein, M. Muske, S. Gall, W. Fuhs, J. Noncryst. Solids **338–340**, 127 (2004)
38. E. Pihan, A. Focsa, A. Slaoui, C. Maurice, Thin Solid Films **511–512**, 15 (2006)
39. K.Y. Lee, M. Muske, I. Gordon, M. Berginski, J. D'Haen, J. Hüpkes, S. Gall, B. Rech, Thin Solid Films **516**, 6869 (2008)
40. Ö. Tüzün, J.M. Auger, I. Gordon, A. Focsa, P.C. Montgomery, C. Maurice, A. Slaoui, G. Beaucarne, J. Poortmans, Thin Solid Films **516**, 6882 (2008)
41. B. Rau, I. Sieber, B. Selle, S. Brehme, U. Knipper, S. Gall, W. Fuhs, Thin Solid Films **451–452**, 644 (2004)
42. D.H. Neuhaus, N.P. Harder, S. Oelting, R. Bardos, A.B. Sproul, P. Widenborg, A.G. Aberle, Solar Energy Mater. Solar Cells **74**, 225 (2002)
43. P. Dogan, E. Rudigier, F. Fenske, K.Y. Lee, B. Gorka, B. Rau, E. Conrad, S. Gall, Thin Solid Films **516**, 6989 (2008)
44. W. Fuhs, S. Gall, B. Rau, M. Schmidt, J. Schneider, Solar Energy **77**, 961 (2004)
45. H. Kim, D. Kim, G. Lee, D. Kim, S.H. Lee, Solar Energy Mater. Solar Cells **74**, 323 (2002)
46. S. Gall, J. Schneider, J. Klein, K. Hübener, M. Muske, B. Rau, E. Conrad, I. Sieber, K. Petter, K. Lips, M. Stöger-Pollach, P. Schattschneider, W. Fuhs, Thin Solid Films **511**, 7 (2006)
47. E. Pihan, A. Slaoui, C. Maurice, J. Cryst. Growth **305**, 88 (2007)
48. D. Van Gestel, M.J. Romero, I. Gordon, L. Carnel, J. D'Haen, G. Beaucarne, M. Al-Jassim, J. Poortmans, Appl. Phys. Lett. **90**, 092103 (2007)
49. F. Liu, M.J. Romero, K.M. Jones, A.G. Norman, M. Al-Jassim, D. Inns, A.G. Aberle, Thin Solid Films **516**, 6409 (2008)
50. S. Gall, J. Schneider, M. Muske, I. Sieber, O. Nast, W. Fuhs, in *Proceedings of PV in Europe – From PV Technology to Energy Solutions*, Rome, Italy, 7–11 October 2002, p. 87

51. J. Schneider, J. Klein, M. Muske, A. Schöpke, S. Gall, W. Fuhs, in *Proceedings of the 3rd World Conference on Photovoltaic Energy Conversion*, Osaka, Japan, 11–18 May 2003, p. 106
52. J. Schneider, J. Klein, A. Sarikov, M. Muske, S. Gall, W. Fuhs, Mater. Res. Soc. Symp. Proc. **862**, A2.2.1 (2005)
53. J. Schneider, A. Schneider, A. Sarikov, J. Klein, M. Muske, S. Gall, W. Fuhs, J. Noncryst. Solids **352**, 972 (2006)
54. J. Schneider, Doctoral thesis, Technische Universität Berlin, Germany (2005)
55. J. Schneider, J. Klein, M. Muske, S. Gall, W. Fuhs, Appl. Phys. Lett. **87**, 031905 (2005)
56. A. Sarikov, J. Schneider, J. Klein, M. Muske, S. Gall, J. Cryst. Growth **287**, 442 (2006)
57. A. Sarikov, J. Schneider, M. Muske, S. Gall, W. Fuhs, J. Noncryst. Solids **352**, 980 (2006)
58. D. He, J.Y. Wang, E.J. Mittemeijer, J. Appl. Phys. **97**, 093524 (2005)
59. J.Y. Wang, Z.M. Wang, E.J. Mittemeijer, J. Appl. Phys. **102**, 113523 (2007)
60. J. Schneider, A. Sarikov, J. Klein, M. Muske, I. Sieber, T. Quinn, H.S. Reehal, S. Gall, W. Fuhs, J. Cryst. Growth **287**, 423 (2006)
61. A. Sarikov, J. Schneider, M. Muske, I. Sieber, S. Gall, Thin Solid Films **515**, 7465 (2007)
62. C. Messmer, J.C. Bilello, J. Appl. Phys. **52**, 4623 (1981)
63. D.J. Eaglesham, A.E. White, L.C. Feldman, N. Moriya, D.C. Jacobson, Phys. Rev. Lett. **70**, 1643 (1993)
64. I. Gordon, L. Carnel, D. Van Gestel, G. Beaucarne, J. Poortmans, Prog. Photovoltaics Res. Appl. **15**, 575 (2007)
65. I. Gordon, D. Van Gestel, K. Van Nieuwenhuysen, L. Carnel, G. Beaucarne, J. Poortmans, Thin Solid Films **487**, 113 (2005)
66. D. Van Gestel, I. Gordon, L. Carnel, K. Van Nieuwenhuysen, G. Beaucarne, J. Poortmans, in *Technical Digest of the 15th International Photovoltaic Science and Engineering Conference*, Shanghai, China, 10–15 October 2005, p. 853
67. S. Gall, C. Becker, E. Conrad, P. Dogan, F. Fenkse, B. Gorka, K.Y. Lee, B. Rau, F. Ruske, B. Rech, Solar Energy Mater. Solar Cells **93**, 1004 (2009)
68. Ö. Tüzün, A. Slaoui, I. Gordon, A. Focsa, D. Ballutaud, G. Beaucarne, J. Poortmans, Thin Solid Films **516**, 6892 (2008)
69. P.I. Widenborg, D.H. Neuhaus, P. Campbell, A.B. Sproul, A.G. Aberle, Solar Energy Mater. Solar Cells **74**, 305 (2002)
70. D. Dimova-Malinovska, O. Angelov, M. Kamenova, A. Vaseashta, J.C. Pivin, J. Optoelectron. Adv. Mater. **9**, 355 (2007)
71. K.Y. Lee, C. Becker, M. Muske, F. Ruske, S. Gall, B. Rech, Appl. Phys. Lett. **91**, 241911 (2007)
72. D. Dimova-Malinovska, O. Angelov, M. Sendova-Vassileva, M. Kamenova, J.C. Pivin, Thin Solid Films **451–452**, 303 (2004)
73. R. Lechner, M. Buschbeck, M. Gjukic, M. Stutzmann, Phys. Stat. Sol. C **1**, 1131 (2004)
74. M. Gjukic, M. Buschbeck, R. Lechner, M. Stutzmann, Appl. Phys. Lett. **85**, 2134 (2004)
75. M. Gjukic, R. Lechner, M. Buschbeck, M. Stutzmann, Appl. Phys. Lett. **86**, 062115 (2005)
76. M. Scholz, M. Gjukic, M. Stutzmann, Appl. Phys. Lett. **94**, 012108 (2009)

13

Thermochemical and Kinetic Databases for the Solar Cell Silicon Materials

Kai Tang*, Eivind J. Øvrelid, Gabriella Tranell, and Merete Tangstad

Abstract. The fabrication of solar cell grade silicon (SOG-Si) feedstock involves processes that require direct contact between solid and a fluid phase at near equilibrium conditions. Knowledge of the phase diagram and thermochemical properties of the Si-based system is, hence, important for providing boundary conditions in the analysis of processes. A self-consistent thermodynamic description of the Si–Ag–Al–As–Au–B–Bi–C–Ca–Co–Cr–Cu–Fe–Ga–Ge–In–Li–Mg–Mn–Mo–N–Na–Ni–O–P–Pb–S–Sb–Sn–Te–Ti–V–W–Zn–Zr system has recently been developed by SINTEF Materials and Chemistry. The assessed database has been designed for use within the composition space associated with the SoG-Si materials. An assessed kinetic database covers the same system as in the thermochemical database. The impurity diffusivities of Ag, Al, As, Au, B, Bi, C, Co, Cu, Fe, Ga, Ge, In, Li, Mg, Mn, N, Na, Ni, O, P, Sb, Te, Ti, Zn and the self diffusivity of Si in both solid and liquid silicon have extensively been investigated. The databases can be regarded as the state-of-art equilibrium relations in the Si-based multicomponent system. The thermochemical database has further been extended to simulate the surface tensions of liquid Si-based melts. Many surface-related properties, e.g., temperature and composition gradients, surface excess quantity, and even the driving force due to the surface segregation are possible to obtain directly from the database. By coupling the Langmuir–McLean segregation model, the grain boundary segregations of the nondoping elements in polycrystalline silicon are also possible to estimate from the assessed thermochemical properties.

13.1 Introduction

Metallurgical grade silicon (MG-Si) (>98%Si) is used as an alloying element in the aluminum industry, for silicones production and as a raw material for production of high-purity Si, targeted at the electronics and solar cell industries. All these markets have very different quality requirements. Silicon used in solar cells (SoG-Si) must have a total contamination level less than 1 ppm (6N) and silicon produced in ordinary reduction furnaces cannot meet this requirement without extensive refining processes. Supply of high-purity Si for the solar cell market has hence to date been relying on the energy intensive "Siemens" production process, originally targeted at producing electronic grade Si (EG-Si) with purity requirements of several orders of magnitude higher than that of SoG-Si. In order to expand the market for solar cells, new cheaper metallurgical and fluid bed routes for the production of SoG-Si are hence being developed.

The success of producing and refining Si feedstock materials to ultra high-purity depends heavily on the availability and reliability of thermodynamic, kinetic, and other physical data for the most common and important SoG-Si trace elements. For example, knowledge of the phase diagram and thermochemical properties of the Si-based system is important for providing boundary conditions in the analysis of processes. However, thermodynamic and kinetic data for these elements in the ppm and ppb ranges are scarce and often unreliable. More importantly, to our knowledge, no dedicated SoG-Si database exists, which has gathered and optimized existing data on high-purity Si alloys.

In this chapter, a self-consistent thermodynamic description of the Si–Ag–Al–As–Au–B–Bi–C–Ca–Co–Cr–Cu–Fe–Ga–Ge–In–Li–Mg–Mn–Mo–N–Na–Ni–O–P–Pb–S–Sb–Sn–Te–Ti–V–W–Zn–Zr system is introduced. The assessed database has been designed for use within the composition space associated with SoG-Si materials. The assessed kinetic database covers the same system as the thermochemical database. The impurity diffusivities of Ag, Al, As, Au, B, Bi, C, Co, Cu, Fe, Ga, Ge, In, Li, Mg, Mn, N, Na, Ni, O, P, Sb, Te, Ti, Zn and the self diffusivity of Si in both solid and liquid silicon have been extensively investigated. The thermochemical database has further been extended to simulate the surface tensions of liquid Si-based melts. By coupling the Langmuir–McLean segregation model, the grain boundary segregations of the nondoping elements in polycrystalline silicon have further been estimated from the assessed thermochemical properties.

13.2 The Assessed Thermochemical Database

Aluminum, boron, carbon, iron, nitrogen, oxygen, phosphorus, sulfur and titanium are the common impurities in the SoG-Si feedstock. Arsenic and antimony are frequently used as doping agents. Transition metals (Co, Cu, Cr, Fe, Mn, Mo, Ni, V, W, and Zr), alkali and alkali-earth impurities (Li, Mg, and Na), as well as Bi, Ga, Ge, In, Pb, Sn, Te, and Zn may appear in the SoG-Si feedstock. A thermochemical database that covers these elements has recently been developed at SINTEF Materials and Chemistry, which has been designed for use within the composition space associated with the SoG-Si materials. All the binary and several critical ternary subsystems have been assessed and calculated results have been validated with the reliable experimental data in the literature. The database can be regarded as the state-of-art equilibrium relations in the Si-based multicomponent system.

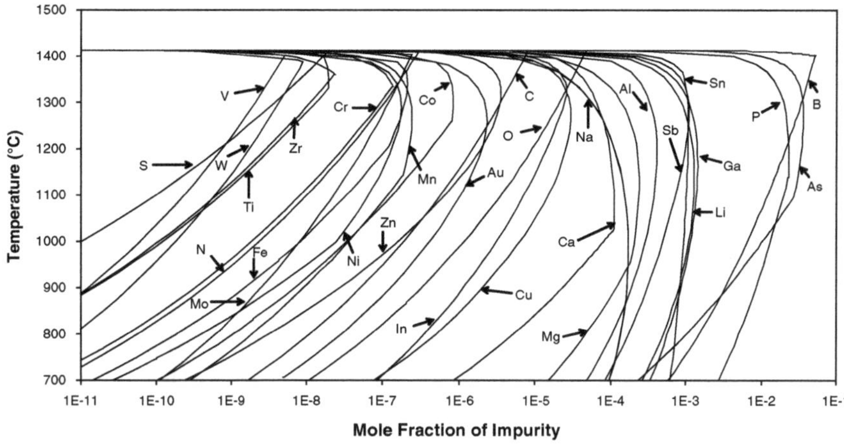

Fig. 13.1. Calculated the solubilities of the impurities in solid silicon

As illustrated in Fig. 13.1, the composition space of impurity in the SoG-Si materials generally ranges from ppb to a few percent. The database covers all the 33 silicon-impurity binary systems. Among the 33 binary silicon-containing systems, the Si–Al, Si–As, Si–B, Si–C, Si–Fe, Si–N, Si–O, Si–P, Si–S, Si–Sb, and Si–Te systems have thermodynamically been "reoptimized" primarily based on the assessed experimental information. Since thermodynamic calculations for the impurities in SoG-Si feedstock are normally multidimensional in nature, Gibbs energies of 36 other binary systems have also been included in the database. In this way, the effect of other impurities on the phase equilibria of principle impurity in SoG-Si materials can be reliably evaluated. Systematic validation of the database has been carried out using the experimental data for Si-based multicomponent systems. Examples of the validation will be given in the following section.

13.2.1 Thermodynamic Description

13.2.1.1 Element and Stoichiometric Compound

The Gibbs energy of the pure element "i," ${}^0G_{\mathrm{i}}^{\phi}(T)$, is described as a function of temperature by the following equation:

$$\begin{aligned} {}^0G_{\mathrm{i}}^{\phi}(T) =& a + bT + cT\ln(T) + dT^2 + eT^3 + fT^{-1} \\ &+ \cdots (\phi = \text{diamond, liquid, bcc, } \ldots). \end{aligned} \tag{13.1}$$

The coefficients of the above expression for each element are given by Dinsdale [1].

The Gibbs energy of stoichiometric compound, A_pB_q, is calculated using the equation:

$$G_{\mathrm{A}p\mathrm{B}q}(T) = p^0G_{\mathrm{A}}^{\phi} + q^0G_{\mathrm{B}}^{\phi} + \Delta_f G_{\mathrm{A}p\mathrm{B}q}^0(T). \qquad (13.2)$$

The expression for $\Delta_f G_{\mathrm{A}p\mathrm{B}q}^0(T)$, the Gibbs energy of formation of the compound, referring to the stable elements at T, is similar to that given by (13.1). Compounds with a narrow range of homogeneity, SiB_3 and SiB_6 for instance, are treated the same as the stoichiometric compounds in the database.

13.2.1.2 Solutions

The liquid Si-based solution, here abbreviated as "l", is described using a simple polynomial expression based on a substitutional solution with random mixing. The same model is employed for the diamond-structured Si-rich solid phase, denoted as "s." The Gibbs energies of the liquid and solid Si-based phases are given by the following equation:

$$G_{\mathrm{m}}^{\phi} = x_{\mathrm{Si}}^{\phi}\,{}^0G_{\mathrm{Si}}^{\phi} + \sum_{\mathrm{i}\neq\mathrm{Si}} x_{\mathrm{i}}^{\phi}\,{}^0G_{\mathrm{i}}^{\phi} + RT\sum_{\mathrm{i}} x_{\mathrm{i}}^{\phi}\ln x_{\mathrm{i}}^{\phi} + {}^{\mathrm{Ex}}G_{\mathrm{m}}^{\phi} \quad (\phi = \mathrm{l}, \mathrm{s}), \qquad (13.3)$$

where ${}^0G_{\mathrm{Si}}^{\phi}$ and ${}^0G_{\mathrm{i}}^{\phi}$ are, respectively, the molar Gibbs energy of Si and element "i" with the phase ϕ in a nonmagnetic state. The second term is the contribution of ideal mixing. The excess Gibbs energy, ${}^{\mathrm{Ex}}G_{\mathrm{i}}^{\phi}$, is expressed in the Redlich–Kister polynomial:

$$L_{\mathrm{ij}}^{\phi} = \sum_{k=0}^{n} {}^kL_{\mathrm{ij}}^{\phi}(x_{\mathrm{i}}^{\phi} - x_{\mathrm{j}}^{\phi})^k \qquad (13.4)$$

Since the concentrations of impurities in solar cell grade silicon are in the range from ppb to a few percent, it is not necessary to take ternary interaction parameters into account. The activity coefficient of impurity, "i", in a n-th multicomponent system is given by differentiating (13.3):

$$\begin{aligned} RT\ln\gamma_{\mathrm{i}}^{\phi} =\; & {}^0L_{\mathrm{Si-i}}^{\phi}x_{\mathrm{Si}}^{\phi}(1 - x_{\mathrm{i}}^{\phi}) + \sum_{k=1}^{n} {}^kL_{\mathrm{Si-i}}^{\phi}x_{\mathrm{Si}}^{\phi}(x_{\mathrm{i}}^{\phi} - x_{\mathrm{Si}}^{\phi})^{(k-1)} \\ & \times \left[(k+1)(1 - x_{\mathrm{i}}^{\phi})(x_{\mathrm{i}}^{\phi} - x_{\mathrm{Si}}^{\phi}) + kx_{\mathrm{Si}}^{\phi}\right] \\ & + \sum_{\mathrm{j}\neq\mathrm{i},\mathrm{Si}} \Bigg\{ {}^0L_{\mathrm{ij}}^{\phi}x_{\mathrm{j}}^{\phi}(1 - x_{\mathrm{i}}^{\phi}) + \sum_{k=1}^{n} {}^kL_{\mathrm{ij}}^{\phi}x_{\mathrm{i}}^{\phi}(x_{\mathrm{i}}^{\phi} - x_{\mathrm{j}}^{\phi})^{(k-1)} \\ & \qquad \times \left[(k+1)(1 - x_{\mathrm{i}}^{\phi})(x_{\mathrm{i}}^{\phi} - x_{\mathrm{j}}^{\phi}) + kx_{\mathrm{j}}^{\phi}\right] \Bigg\} \\ & - \sum_{\mathrm{j}\neq\mathrm{i}}\sum_{\mathrm{l}\neq\mathrm{i}} x_{\mathrm{j}}^{\phi}x_{\mathrm{l}}^{\phi}\left[{}^0L_{\mathrm{jl}}^{\phi} + \sum_{k=1}^{n} {}^kL_{\mathrm{jl}}^{\phi}(x_{\mathrm{j}}^{\phi} - x_{\mathrm{l}}^{\phi})^k(k+1)\right] \quad (\phi = \mathrm{l}, \mathrm{s}) \end{aligned} \qquad (13.5)$$

For the SoG-Si materials, $x_{\mathrm{i}}^{\phi} \to 0$ and $x_{\mathrm{Si}}^{\phi} \to 1$, the activity coefficient of "i" can be approximately expressed:

$$\gamma_{\mathrm{i}}^{\phi} \approx \exp\left(\sum_{k=0}^{n} {}^{k}L_{\mathrm{Si-i}}^{\phi} \Big/ RT\right) \quad (\phi = \mathrm{l,\ s}). \tag{13.6}$$

This is often called Henry's coefficient.

13.2.1.3 Solubility

The solubility is defined with respect to a second precipitated phase. The solubility of an impurity is the maximum concentration, which can be incorporated in the liquid or solid phase without precipitating a second phase. For most impurities in solid silicon at high-temperatures, equilibrium is achieved with the liquid phase governed by the liquidus in the phase diagram. Solid solubility is temperature-dependent as represented by the solidus or solvent curves in the phase diagram. At lower temperatures, the reference phase is usually a compound or an impurity-rich alloy. When the impurity is volatile, the saturated crystal is in equilibrium with the vapor, and the impurity solubility also depends on its vapor pressure.

For solid–solid and solid–vapor equilibria, which often occur at temperatures below the eutectic temperature, T_{eu}, the solubility can be described by an Arrhenius-type equation. For solid–liquid equilibria at temperatures above T_{eu}, the temperature dependence is more complex and will be discussed in detail in the following sections.

13.2.1.4 Equilibrium Distribution Coefficient

Segregation effects at the liquid–solid interface are controlled by the equilibrium distribution coefficient, $k_{\mathrm{i}}^{\mathrm{eq}}$, which is defined as the ratio of the solidus and the liquidus concentrations in atomic fractions:

$$k_{\mathrm{i}}^{\mathrm{eq}} = x_{\mathrm{i}}^{\mathrm{s}} \,/\, x_{\mathrm{i}}^{\mathrm{l}}. \tag{13.7}$$

The equilibrium distribution coefficient close to the melting point is also known as the partition coefficient. Since the partition coefficient controls the incorporation of impurities in the crystal during crystal growth and zone refining, it is one of the most important parameters that can be obtained from the thermochemical database. It is worth noting that the distribution coefficient determined by the ratio of volume concentrations, cm^{-3}, can be related to the distribution coefficient by introducing the density ratio of liquid and solid silicon:

$$k_{\mathrm{eq}}^{V} = \left(d_{\mathrm{Si}}^{\mathrm{liq}} \Big/ d_{\mathrm{Si}}^{\mathrm{sol}}\right) k_{\mathrm{eq}} \approx 1.1\, k_{\mathrm{eq}}. \tag{13.8}$$

Applying the phase equilibrium rule to this case, results in the formula for the determination of equilibrium distribution coefficient:

$$k_{eq} = \frac{\gamma_i^l}{\gamma_i^s} \exp\left(\frac{\Delta^0 G_i^{fus}}{RT}\right), \tag{13.9}$$

where $\Delta^0 G_i^{fus}$ refers the Gibbs energy of fusion of impurity "i". Activity coefficients of impurity "i" in liquid and solid phases can be determined using (13.6).

13.2.1.5 Retrograde Solubility

Retrograde solubility describes the change in impurity concentration in a solid above T_{eu}, i.e., a maximum solubility is observed at a temperature T_{max} lower than melting temperature of silicon T_m, but above T_{eu}. In this frequently encountered case, impurities tend to precipitate upon cooling.

Weber [2] proposed a formula to determine the maximum retrograde temperature, assuming that the impurity behaves ideally in liquid solution and regularly in solid phase:

$$T_{max} = \frac{\Delta H_{Si}^m / k_i^{eq}}{\Delta H_{Si}^m / k_i^{eq} - \ln\left[\frac{\Delta H_i^{s,l}}{\Delta H_i^{s,l} + \Delta H_{Si}^m}\right]}. \tag{13.10}$$

Thermodynamically, retrograde solubility requires a large positive value for $\Delta H_i^{s,l}$, the solid solution enthalpy of mixing.

Retrograde solubility can be usefully applied in solidification refining of impurities in SoG-Si materials. Yoshikawa and Morita proposed the methods for the removal of boron [3, 4] and phosphorus [5] by addition of Al and Ti. Shimpo et al. [6] and Inoue et al. [7] reported the Ca addition method to remove boron and phosphorus in silicon.

13.2.2 Typical Examples

The assessments for the Si–Al, Si–As, Si–B, Si–C, Si–Fe, Si–N, Si–O, Si–P, Si–S and Si–Sb systems will briefly be introduced in this section. Typical examples of the database calculation results are presented as diagrams.

The reassessment of the Si–Al binary system has been carried out based mainly on the experimental solubility data reported by Miller and Savage [8], Navon and Chernyshov [9], Lozovskii and Udyanskaya [10], and Yoshikawa and Morita [11]. Activities of Al in liquid Si measured by Miki et al. [12, 13] and Ottem [14] were also taken into account in the thermodynamic assessment. Figure 13.2 shows the calculated phase equilibria in the Si-rich part of the Si–Al system.

The Si–As system has been reassessed based on the phase equilibrium data reported by Klemm and Pirscher [15], Ugay and Miroshnichenko [16], and Ugay et al. [17]. Arsenic solubility data measured by Trumbore [18], Sandhu and Reuter [19], Fair and Weber [20], Ohkawa et al. [21], Fair and Tsai [22], Miyamoto et al. [23] and the activity data given by Reuter [19], Ohkawa

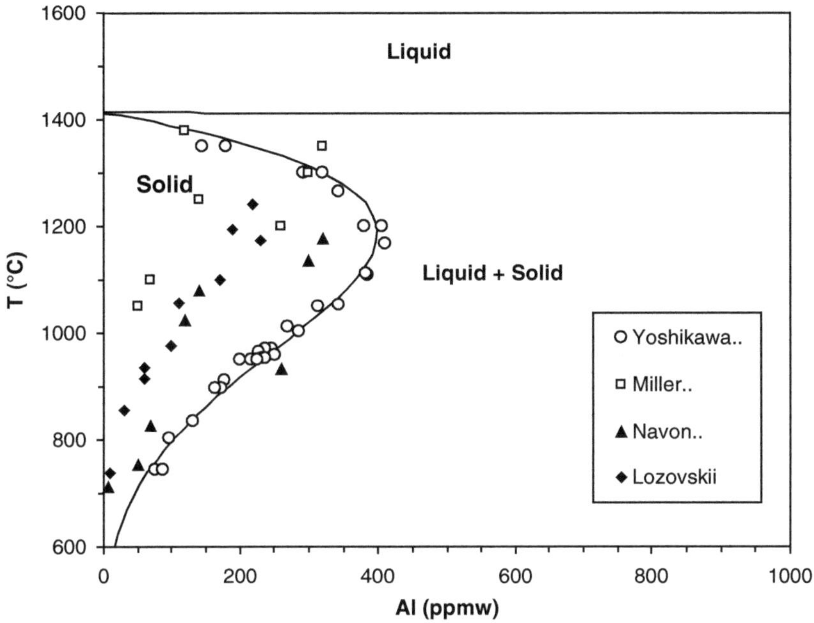

Fig. 13.2. The assessed Si–Al system in the Si-rich domain with the experimental data [8–14]

et al. [21] and Belousov [24] were also used in the thermodynamic "optimization." Figure 13.3 shows the calculated Si–As phase diagram in Si-rich domain.

Reassessment of the Si–B system was based primarily on the model parameters given by Fries and Lukas [25]. Modifications have been made on the thermodynamic properties of the liquid and solid diamond phases: Experimental liquidus data reported by Brosset and Magunsson [26], Armas et al. [27], and Male and Salanoubat [28], solid solubility data reported by Trumbore [18], Hesse [29], Samsonov and Sleptsov [30], and Taishi et al. [52], as well as the boron activities in liquid phase measured by Zaitsev et al. [32], Yoshikawa and Morita [33], Inoue et al. [7], and Noguchi et al. [31] were all used to determine the model parameters. Figure 13.4 shows the new assessed phase equilibria in the Si-rich Si–B system.

The new assessment for the Si–C system was primarily based on experimental SiC solubility data in liquid solution given by Scace and Slack [34], Hall [35], Iguchi [36], Kleykamp and Schumacher [37], Oden and McCune [38], and Ottem [14]. Solid solubility data given by Nozaki et al. [39], Bean [40], and Newman [41] were used to determine the properties of solid solution. The eutectic composition reported by Nozaki et al. [39] and Hall [35] and peritectic transformation temperature determined by Scace [34] and Kleykamp [37] were also used in the thermodynamic optimization. Thermodynamic description of the SiC compound was taken from an early assessment [42]. The

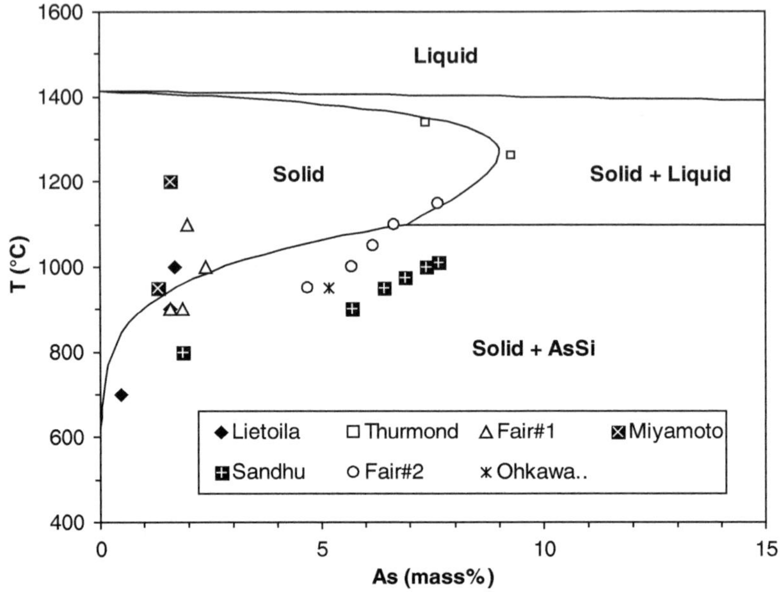

Fig. 13.3. The assessed Si–As system in the Si-rich domain with the experimental data [16–23]

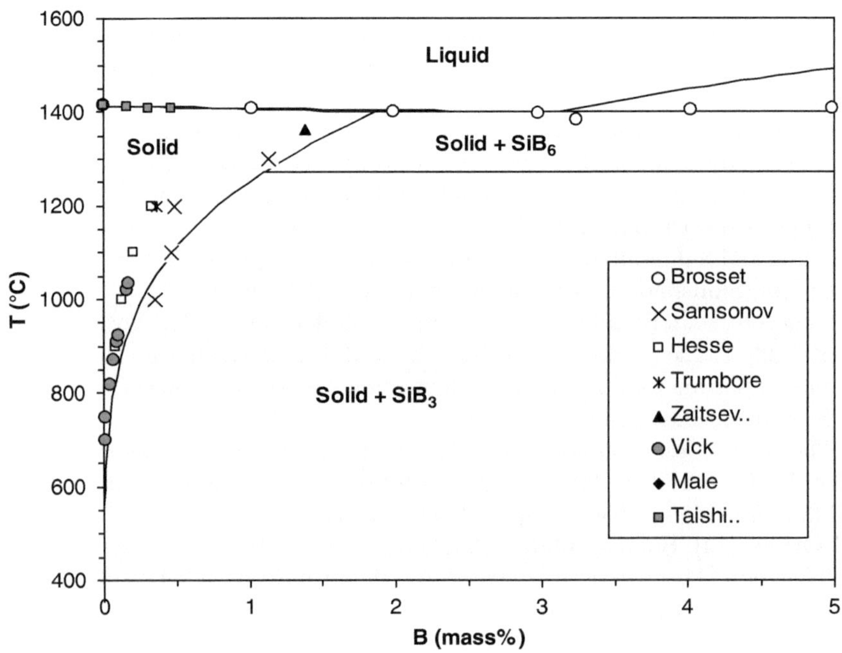

Fig. 13.4. The assessed Si–B system in the Si-rich domain with the experimental data [7, 18, 26–33, 52]

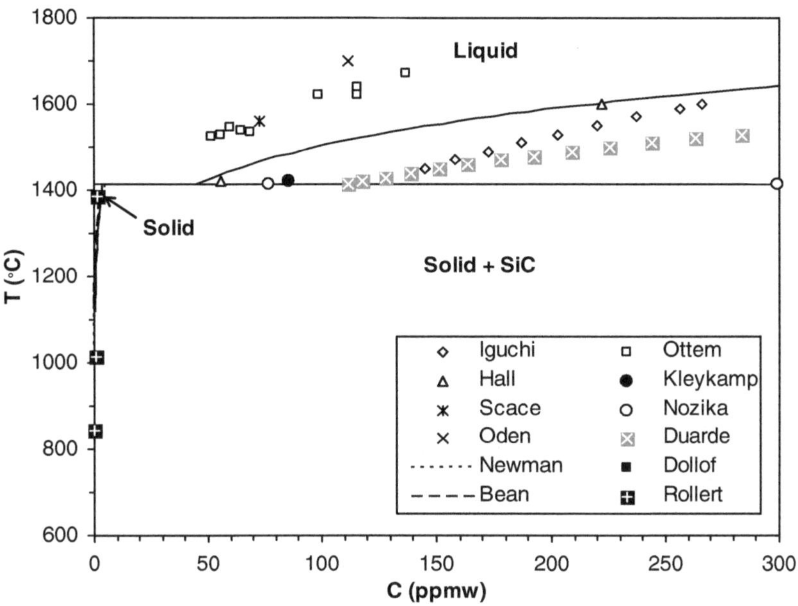

Fig. 13.5. The assessed Si–C system in the Si-rich domain with the experimental data [14,34–41,43]

calculated SiC solubilities in liquid and solid Si–C solutions are compared with the experimental values and shown in Fig. 13.5. Calculated SiC solubilities in liquid silicon have been confirmed by the most recently measurements carried out by Dakaler and Tangstad [43].

Experimental information on the solubility of Fe in solid silicon was reviewed by Istratov et al. [44] The retrograde solubility of iron above the eutectic temperature was reported by Trumbore [18], Feichtinger [45], and Lee et al. [46]. The solubility of iron below the eutectic temperature was studied by Mchugo et al. [47], Colas and Weber [48], Weber [49], Struthers [50], Nakashima et al. [51], Lee et al. [46], and Gills et al. [52] Thermodynamic description of the solid diamond phase was optimized using the above experimental phase equilibrium information. Parameters for the excess Gibbs energy of the liquid Si–Fe phase were optimized by Lacaze and Sundman [53], primarily based on the assessment by Chart [54]. The measured thermodynamic properties of Fe in molten Si, given by Miki et al. [55] and Hsu et al. [56], are reproducible using the assessed model parameters. Figure 13.6 is the new assessed phase equilibria in the Si-rich Si–Fe system in the temperature range of interest.

Figure 13.7 shows the reassessed Si–N phase diagram. The assessment has been carried out using the liquid solubility data reported by Yatsurugi et al. [58], Narushima et al. [59], Kaiser and Thurmond [60], and Iguchi et al. [36] Solubility of Si_3N_4 in solid silicon was determined by Yatsurugi et al. [58]

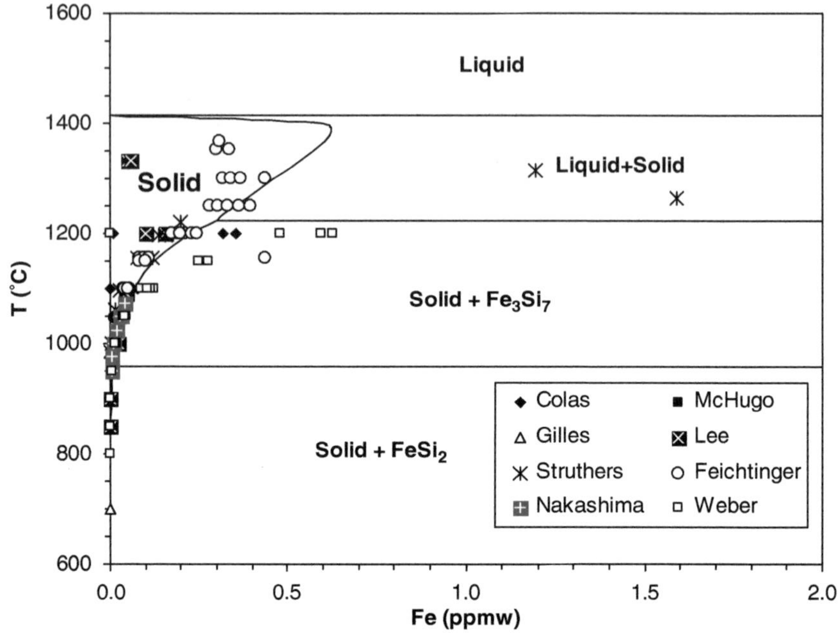

Fig. 13.6. The assessed Si–Fe system in the Si-rich domain with the experimental data [18, 33, 45–55, 57]

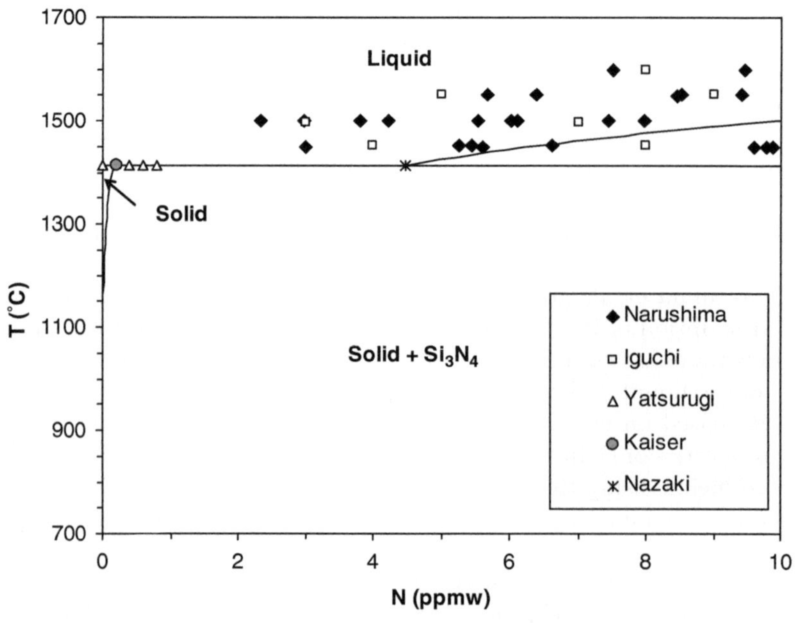

Fig. 13.7. The assessed Si–N system in the Si-rich domain with the experimental data [36, 58–60]

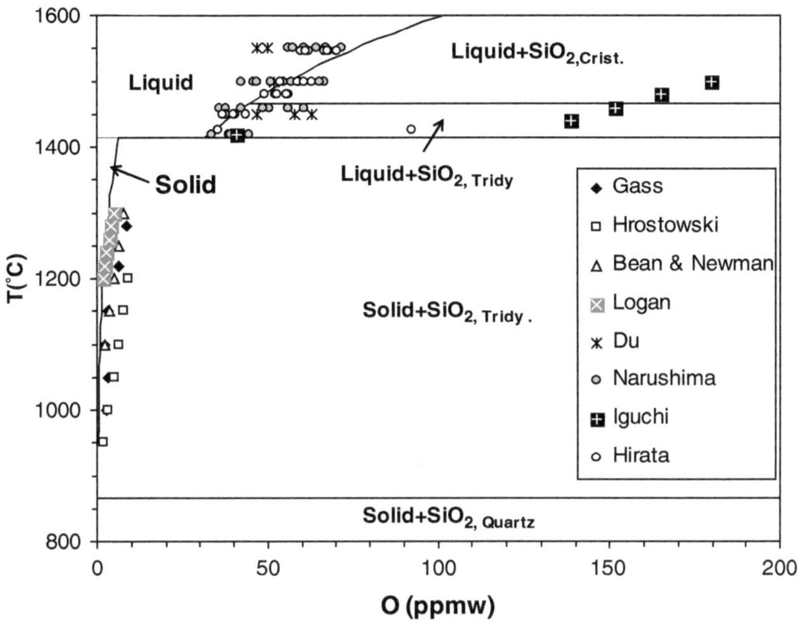

Fig. 13.8. The assessed Si–O system in the Si-rich domain with the experimental data [58, 63–68]

Thermodynamic properties of the Si_3N_4 compound were also reassessed based on the JANAF thermochemical tables [61].

The recent assessment for the Si–O system by Schnurre et al. [62] was accepted in the present database. The stable phase equilibria in the Si–O system, together with the experimental data [58, 63–68], is shown in Fig. 13.8.

Solubility of phosphorus in liquid and solid silicon were reported by Zaitsev et al. [69], Carlsson et al. [70], Giessen and Vogel [71], Korb and Hein [72], Miki et al. [73], Anusionwu et al. [74], Ugai et al. [75], Uda and Kamoshida [76], Kooi [77], Abrikosov et al. [78], Solmi et al. [79], Nobili et al. and Tamura [80]. Figure 13.9 is the partial Si–P phase diagram calculated using the present database.

The Si–S system has been assessed primarily based on the solid solubility reported by Carlson et al. [81] The solubility limits reported by Rollert et al. [82] and Migliorato et al. [83] were not taken into account.

The assessed Si–Sb system is primarily based on the experimental work of Rohan et al. [84], Thurmond and Kowalchik [85], Song et al. [86], Nobili et al. [87] and Sato et al. [88]. The equilibrium distribution coefficients as a function of Sb content were also given by Trumbore et al. [89]. The assessed phase diagram was thermodynamically modeled, primarily based on the work mentioned above, see Fig. 13.10.

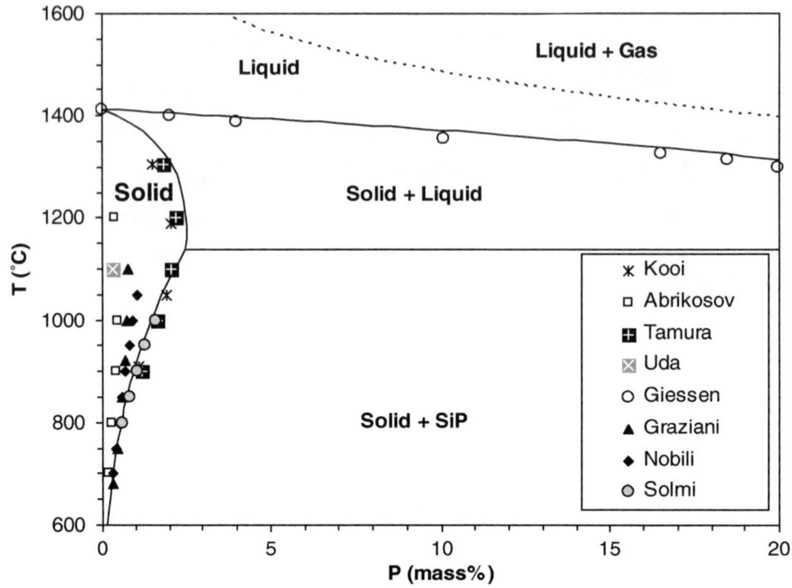

Fig. 13.9. The assessed Si–P system in the Si-rich domain with the experimental data [69–78, 80]

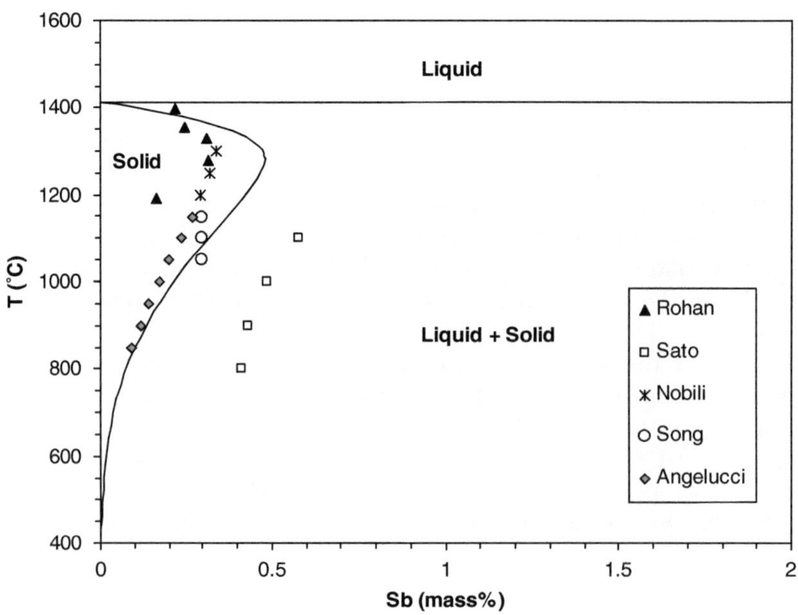

Fig. 13.10. The assessed Si–Sb system in the Si-rich domain with the experimental data [84–88]

13.3 The Kinetic Database

13.3.1 Impurity Diffusivity

Mechanisms for the solid diffusions are well-defined. The impurity diffusivity in solid silicon is usually described by the Arrhenius-type equation:

$$D = D_0 \exp\left(-\frac{E}{kT}\right), \tag{13.11}$$

where D_0 is the preexponential factor and E is the activation energy, usually given in eV. T is the absolute temperature and k is the Boltzmann constant. Equation (13.11) may be rewritten using the SI energy unit:

$$D = D_0 \exp\left(-\frac{Q}{RT}\right), \tag{13.12}$$

where Q is the activation energy in Joules and R is the gas constant. The parameters D_0 and Q can usually be evaluated from the measurements of diffusion coefficients at a series of temperatures.

The mechanism(s) for liquid diffusion are not well established yet. The Arrhenius equation is still the standard description for the self or impurity diffusivities in liquid [90, 91]. The preexponential factor and activation energy can be either fitted from the experimental data, or based solely on the first principle simulation [92–94] and theoretical estimation [95, 96] when no experimental value is available.

In the theoretical treatment of diffusive reactions, one usually works with diffusion coefficients, which are evaluated from experimental measurements. In a multicomponent system, a large number of diffusion coefficients must be evaluated, and are generally interrelated functions of alloy composition. A database would, thus, be very complex. A superior alternative is to store atomic mobilities in the database, rather than diffusion coefficients. The number of parameters which need to be stored in a multicomponent system will then be substantially reduced, as the parameters are independent. The diffusion coefficients, which are used in the simulations, can then be obtained as a product of a thermodynamic and a kinetic factor. The thermodynamic factor is essentially the second derivative of the molar Gibbs energy with respect to the concentrations, and is known if the system has been assessed thermodynamically. The kinetic factor contains the atomic mobilities, which are stored in the kinetic database.

From the absolute-reaction rate theory arguments, the mobility coefficient for an element B, M_{B}, may be divided into a frequency factor and an activation enthalpy Q_{B}, i.e.,

$$M_{\mathrm{B}} = \exp\left(\frac{RT \ln M_{\mathrm{B}}^0}{RT}\right) \exp\left(-\frac{Q_{\mathrm{B}}}{RT}\right) \frac{1}{RT} {}^{\mathrm{mg}}\Gamma. \tag{13.13}$$

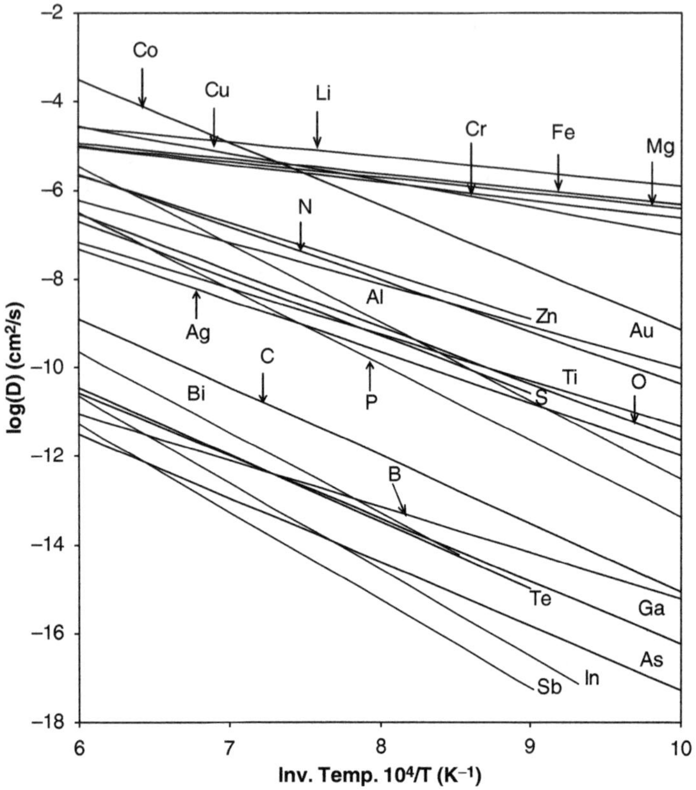

Fig. 13.11. Diffusivities of the impurities in solid silicon

Both $RT \ln M_{\rm B}^{0}$ and $Q_{\rm B}$ will, in general, depend upon the composition, temperature, and pressure.

The assessed kinetic database covers the same system as in the thermochemical database and is schematically shown in Figs. 13.11 and 13.12, respectively. Diffusivities of Al, As, B, C, Fe, N, O, P, and Sb in both solid and liquid silicon have been extensively investigated. The assessment of the impurity diffusivity is basically the same as for the thermodynamic properties. Experimental data were first collected from the literature. Then, each piece of selected experimental information was given a certain weight factor by the assessor. The weight factor could be changed until a satisfactory description of the majority of the selected experimental data was reproduced.

13.3.2 Typical Examples

The assessed diffusivities of Al, As, B, C, Fe, N, O, P, and Sb in both the solid and liquid silicon will be briefly introduced in this section. The measured

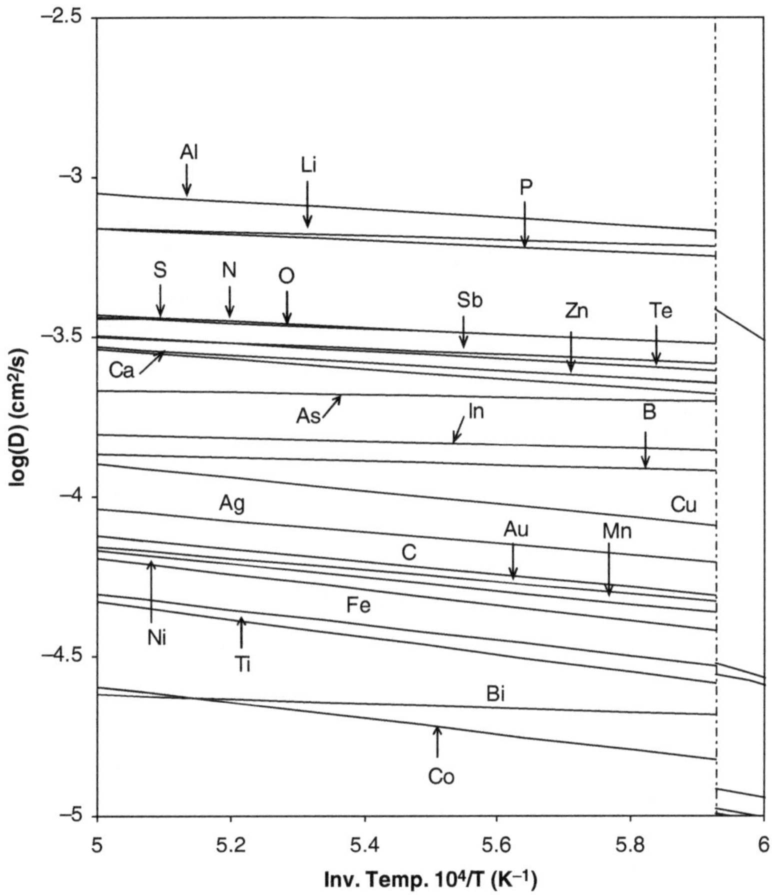

Fig. 13.12. Diffusivities of the impurities in liquid silicon

diffusivities in liquid silicon were also identified by the error bar given by the author(s).

The assessed diffusivity of Al in solid silicon was mainly based on the experimental data available in the literature [8, 97–107]. For the Al diffusivity in liquid silicon, the data estimated by Kodera [107] were later refined by Garandet [106]. The effect of temperature on the liquid diffusivity has then estimated using the theoretical approach proposed by Iida et al. [96]. Figure 13.13 shows the assessed Al diffusivities in both the solid and liquid silicon.

Figures 13.14 and 13.15 show the assessed diffusivities of As and B in both solid and liquid silicon, with the superimposition of experimental values reported in the literature [98, 106–125].

Carbon diffuses in solid silicon both substitutionally and interstitially. It has been experimentally confirmed that carbon diffuses substitutionally in

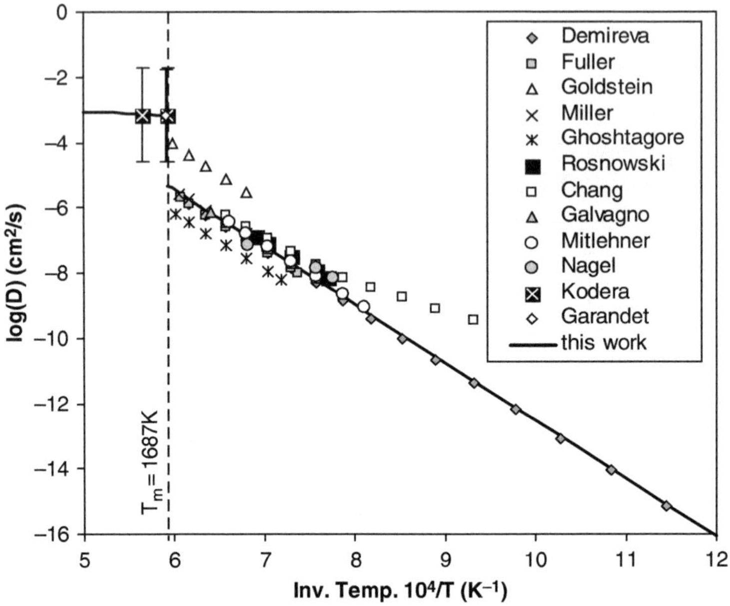

Fig. 13.13. Diffusivity of Al in solid and liquid silicon with the experimental data [8, 97–107]

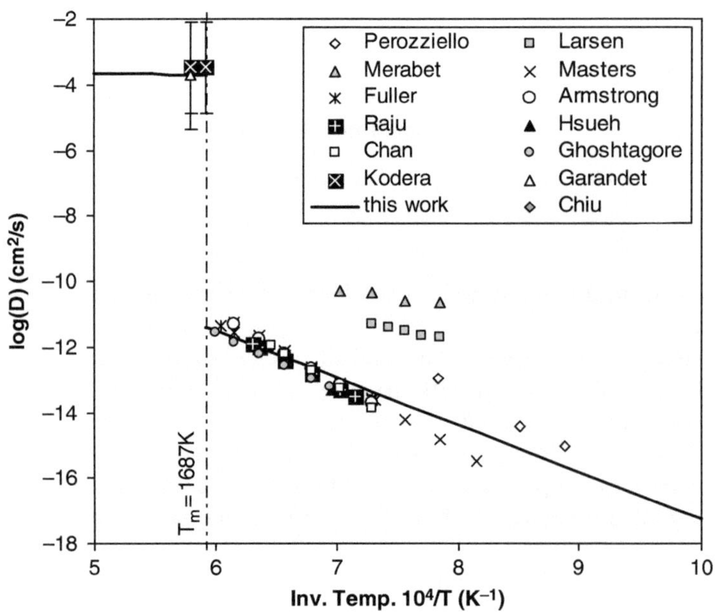

Fig. 13.14. Diffusivity of As in solid and liquid silicon with the experimental data [98, 106–116]

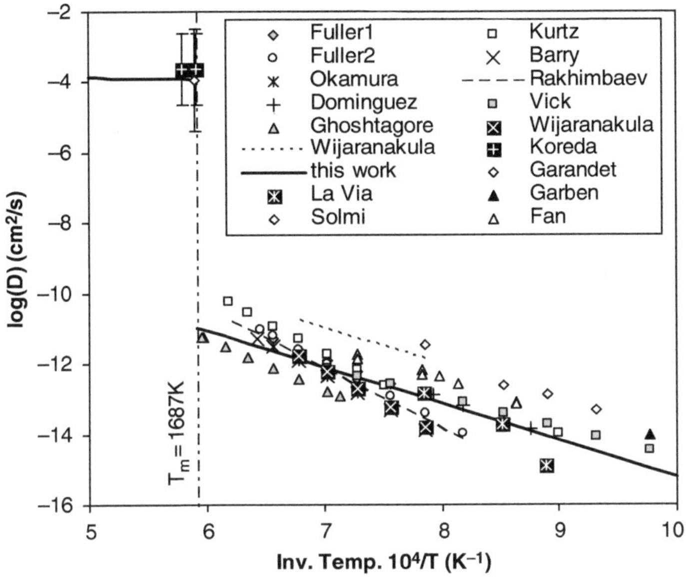

Fig. 13.15. Diffusivity of B in solid and liquid silicon with the experimental data [98, 106, 107, 117–125]

solid silicon at high-temperatures [41]. The carbon diffusivities in liquid silicon measured by Pampuch et al. [126] were used in the present kinetic database. Figure 13.16 shows the assessed carbon diffusivity in liquid and solid silicon.

The Fe diffusivities at high-temperatures were measured using either the diffusion profile of Fe, the radiotracer method [50, 52, 129, 130], or deep level transient spectroscopy (DLTS) [131]. The measurements for iron diffusivity obtained by Struthers [50] were done in the high temperature range (1,100–1,270°C) and featured noticeable scatter. Uskov [132] evaluated iron diffusivity from the kinetics of iron diffusion from the single crystal sample at 1,200°C. The measurement carried out by Antonova et al. [133] was of the diffusion depth profiles of Fe after annealing the samples at temperatures between 1,000 and 1,200°C. Isobe et al. [131] determined the in-diffusion depth profiles of iron in silicon at 800, 900, 1,000, and 1,070°C by the DLTS measurements. Gilles [52] used the tracer method to determine the dependence of the solubilities and diffusion coefficients of Mn, Fe, and Co on doping with B, P, and As in solid Si. The diffusivity of Fe obtained from their measurement at 920°C was in good agreement with the assessed value given by Weber [2]. Figure 13.17 shows the measured and assessed diffusivities of Fe in the solid silicon.

Few experimental data are available about nitrogen diffusivity in solid and liquid silicon in the literature [136–142]. The liquid diffusivities reported by Mukerji and Biswas [142] were several orders of magnitude lower than the solid

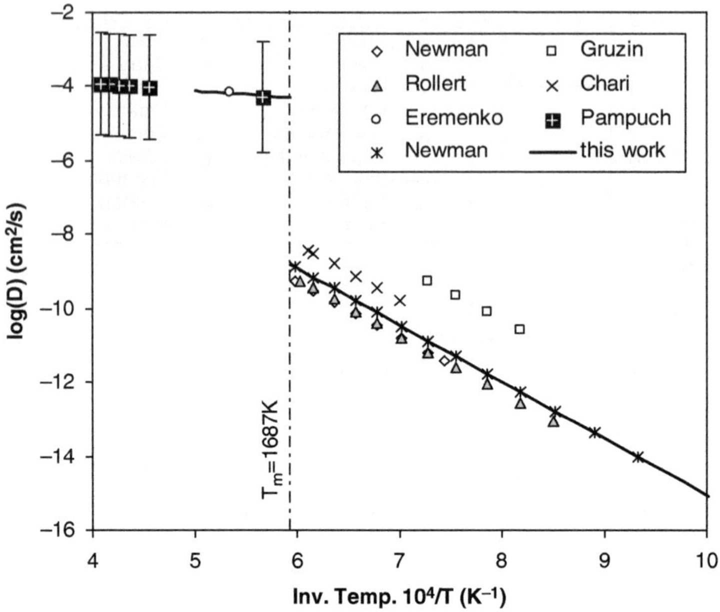

Fig. 13.16. Diffusivity of C in solid and liquid silicon with the experimental data [41, 126–128]

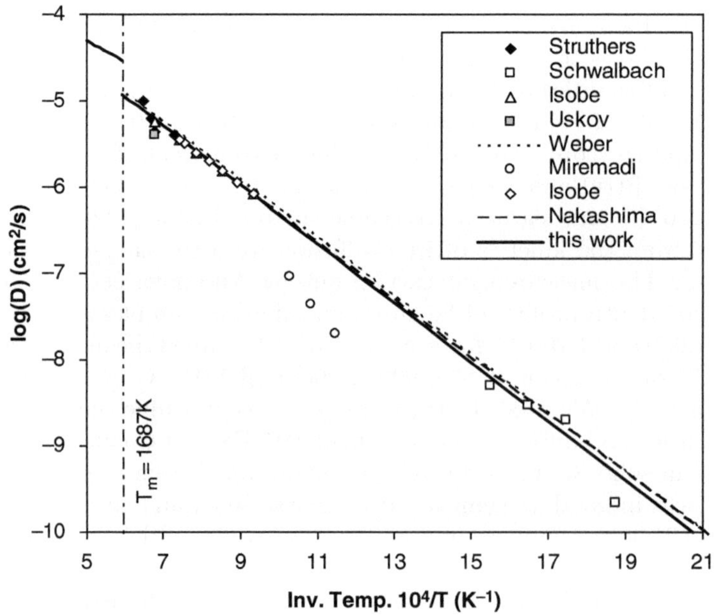

Fig. 13.17. Diffusivity of Fe in solid and liquid silicon with the experimental data [49, 50, 129, 131, 132, 134, 135]

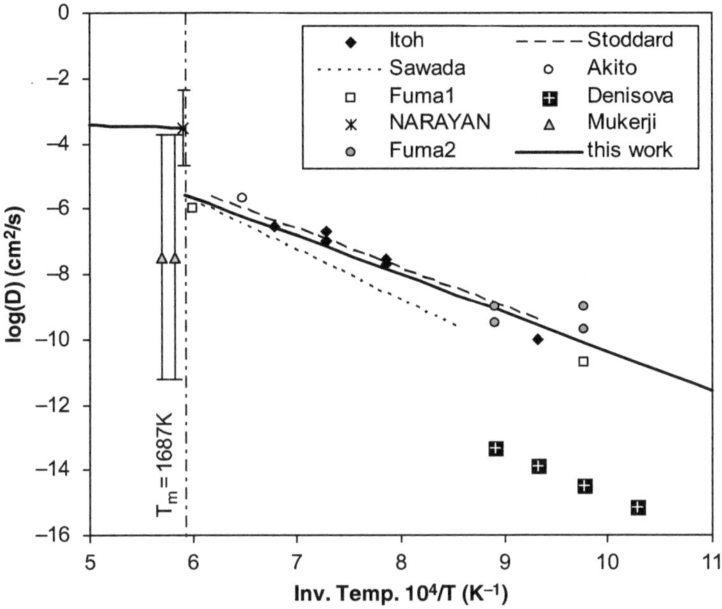

Fig. 13.18. Diffusivity of N in solid and liquid silicon assessed with the experimental data [136–142]

diffusivities. The estimated values given by Narayan et al. were used in this work. Figure 13.18 summarizes the experimental and assessed N diffusivities in solid and liquid silicon.

Oxygen diffusion in solid silicon was extensively studied in the past few decades. Figure 13.19 shows the oxygen diffusivities in solid and liquid silicon. Finally, the diffusivities of P and Sb in both the solid and liquid silicon have been evaluated and are shown in Figs. 13.20 and 13.21, respectively.

13.4 Application of the Thermochemical and Kinetic Databases

13.4.1 Effect of Solubility, Distribution Coefficient, and Stable Precipitates in Solar Cell Grade Silicon

The assessed thermodynamic properties of liquid and solid Si-based solution can be directly applied to evaluate the influence of third element on the solubility of the main impurity in silicon melt. For example, the effect of the impurity element on the solubility of C in pure Si melt can be evaluated by the following equation:

$$\ln x_{\mathrm{C}}^{\mathrm{l}} \cong -\ln K_{\mathrm{SiC}}^{0} - \ln \gamma_{\mathrm{C}}^{\mathrm{l}} + \sum_{\mathrm{i}} x_{\mathrm{i}} \left(\ln \gamma_{\mathrm{Si-C}}^{\mathrm{l}} - \ln \gamma_{\mathrm{i-C}}^{\mathrm{l}} + \ln \gamma_{\mathrm{Si-i}}^{\mathrm{l}}\right), \quad (13.14)$$

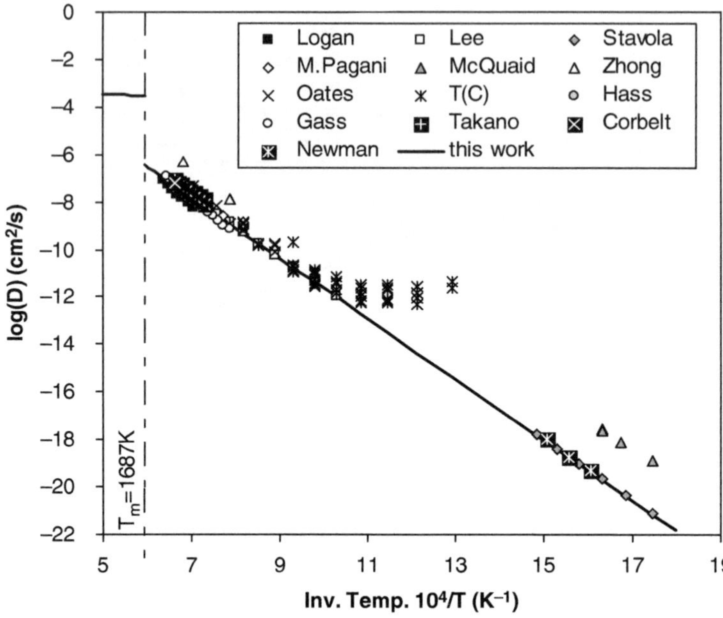

Fig. 13.19. Diffusivity of O in solid and liquid silicon assessed with the experimental data [67, 143–152]

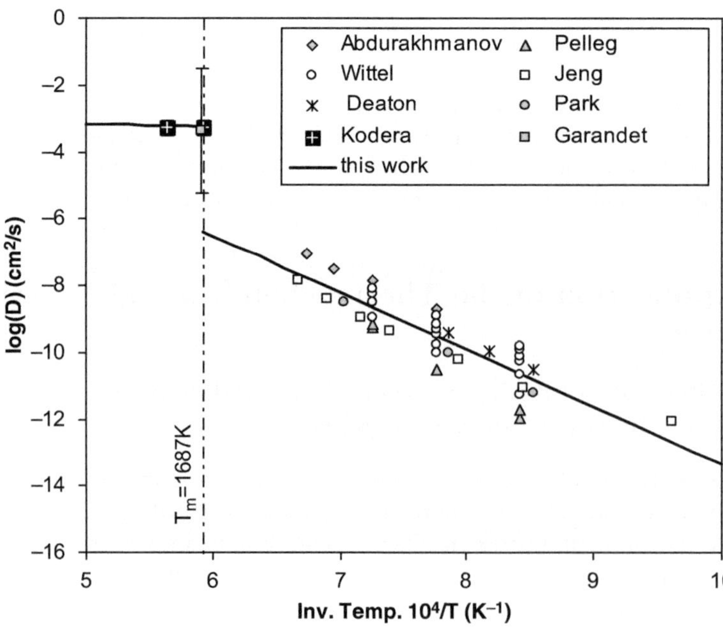

Fig. 13.20. Diffusivity of P in solid and liquid silicon assessed with the experimental data [106, 107, 143–148]

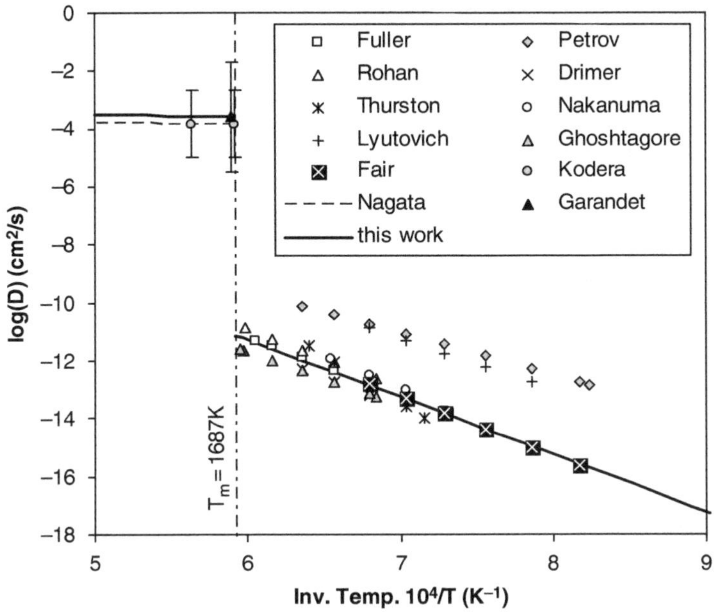

Fig. 13.21. Diffusivity of Sb in solid and liquid silicon with the experimental data [84, 106, 107, 116, 149–154]

where K^0_{SiC} is the equilibrium constant for the reaction: $\underline{\mathrm{C}} + \underline{\mathrm{Si}} = \mathrm{SiC}_\beta$. Figure 13.22 shows the evaluated effect of impurities elements on the solubility of carbon in pure Si. Addition of Zr, P, B, Zn, As, Mn, and Al leads to increased carbon solubility while addition of O, Cr, Cu, Ca, Fe, Ni, S, and N has the opposite influence. Comparing these results with the experimental results reported by Yanaba et al. [155], the model calculations are satisfactory.

Since the impurities in SoG-Si feedstock are normally multidimensional in nature, it is important to estimate the influence of other impurities on the distribution coefficient of one impurity in pure Si. A simple relation has been derived for the effects of secondary impurities, "j," on the distribution coefficient of "i" in pure Si, if the Henry's activity coefficients, $\gamma^{\phi}_{\mathrm{i-j}}$, are available:

$$\ln {}^3k^{\mathrm{i}}_0 \cong \ln {}^2k^{\mathrm{i}}_0 + x^{\mathrm{l}}_{\mathrm{j}} \left({}^2k^{\mathrm{j}}_0 \ln \gamma^{\mathrm{s}}_{\mathrm{Si-j}} - \ln \gamma^{\mathrm{l}}_{\mathrm{Si-j}} \right) = \ln {}^2k^{\mathrm{i}}_0 + x^{\mathrm{l}}_{\mathrm{j}} \Delta. \qquad (13.15)$$

Here ${}^2k^{\mathrm{i}}_0$ and ${}^3k^{\mathrm{i}}_0$ are, respectively, the distribution coefficient of "i" in Si–i binary and Si–i–j ternary systems. Calculated Δ values for different impurities in pure Si are shown in Fig. 13.23. Most of the common impurities in pure Si have a positive contribution to the impurity distribution coefficient, i.e., the appearance of secondary impurities will increase the distribution coefficient of the primary impurity in silicon.

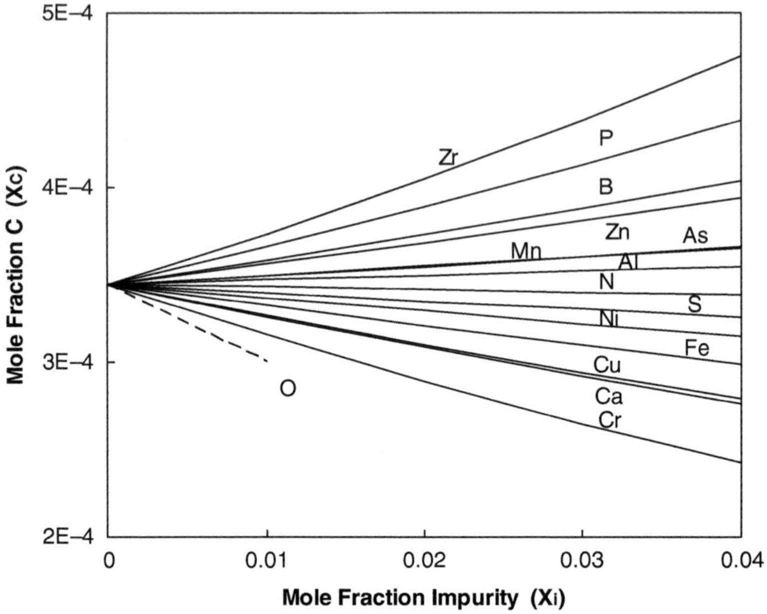

Fig. 13.22. Effect of element on the carbon solubility in liquid silicon

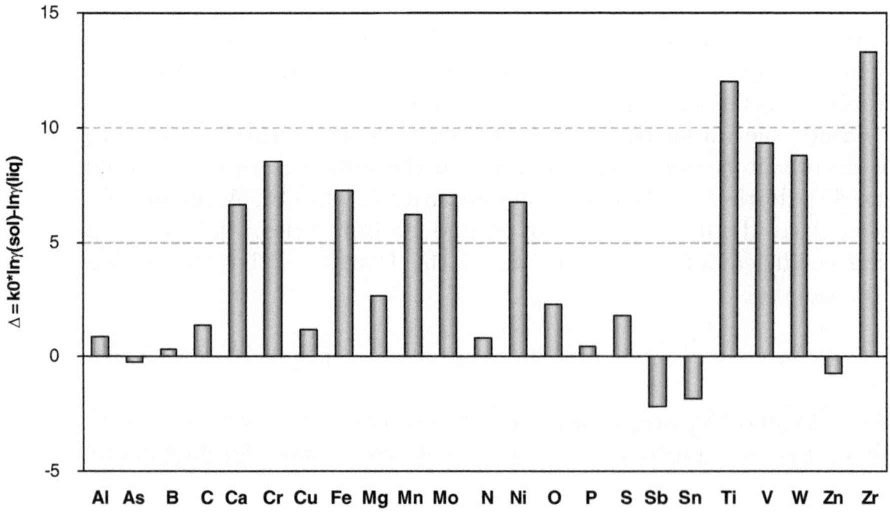

Fig. 13.23. The effect of third element on the equilibrium distribution coefficient of the primary impurity

13.4.2 Surface/Interfacial Tensions

Single crystal silicon is one of the important fundamental materials for the modern photovoltaic industry. The Czochralski method of growing single crystal silicon is affected by the thermocapillary convection. Temperature and concentration gradients at the free surface of the melt give rise to surface tension-driven Marangoni flow, which can lead to crystal defects, if it is sufficiently large.

Based on the statistic thermodynamic treatment [156, 157], equilibria between the bulk phase and a monolayer thickness surface phase is assumed. Following the Koukkari and Pajarre [158] proposal, a fictitious species, "Area," is introduced in the imaginary "surface" phase. The components of the surface phase are the fictitious components: $\mathrm{SiA_m}$, $\mathrm{AlA_n}$, $\mathrm{BA_p}$, $\mathrm{CA_q}$, etc. The stoichiometric coefficients of the components in the surface phase can be determined by the molar surface area of the pure element. The chemical potentials of the components in surface phase are determined by the relation:

$$\mu_{\mathrm{i}}^{\mathrm{S}} = \mu_{\mathrm{i}}^{\mathrm{B}} + A_{\mathrm{i}}\sigma_{\mathrm{i}}, \tag{13.16}$$

where the superscript S and B denote surface and bulk phases; A_{i} and σ_{i} are the molar surface area ($\mathrm{m^3\,mol^{-1}}$) and surface tension of pure element "i," respectively.

The molar surface areas and surface tensions of the metastable liquid B, C, O, and N have been estimated from the experimental values available in the literature using code written for this purpose. The chemical potential of the fictitious component, μ_{Area}, is equivalent to the surface tension of liquid melt, σ, with the unit, $\mathrm{mN\,m^{-1}}$. In this way, surface tension of a multicomponent melt can be directly determined using commercial thermodynamic software, ChemSheet, for example.

Figure 13.24 shows the calculated surface tension of Si–B (left) and Si–C (right) melts. The modeled surface tension and temperature gradient for the Si–O melts at different oxygen partial pressures are shown in Fig. 13.25. The calculation results can reproduce the experimental data [159–163] within their uncertainties.

Using the surface tension implemented Si-based thermochemical database, many surface-related properties, e.g., temperature and composition gradients, surface excess quantity, and even the driving force due to the surface segregation are readily obtained. Figures 13.26 and 13.27 show, respectively, the calculated temperature and composition gradients of impurities in a silicon melt. The composition gradient values estimated by Keene [164] are also given in the diagram for comparison. In addition to the surface tension, phase equilibria and thermochemical properties of the corresponding system can simultaneously also be obtained in the calculation. This may provide more efficient and accurate ways to simulate practical problems.

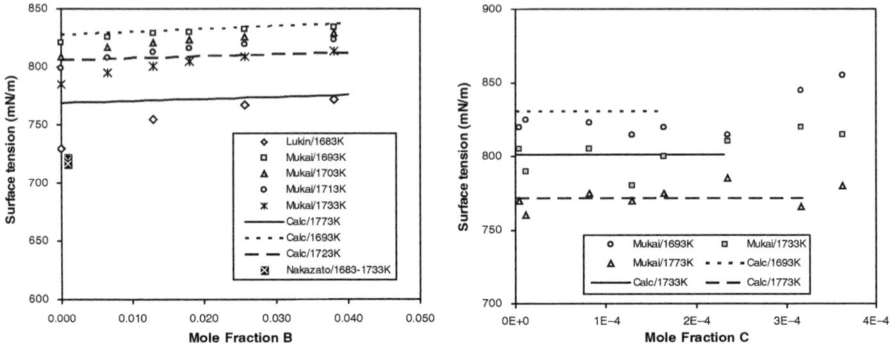

Fig. 13.24. Calculated surface tension of Si–B (*left*) and Si–C (*right*) melts

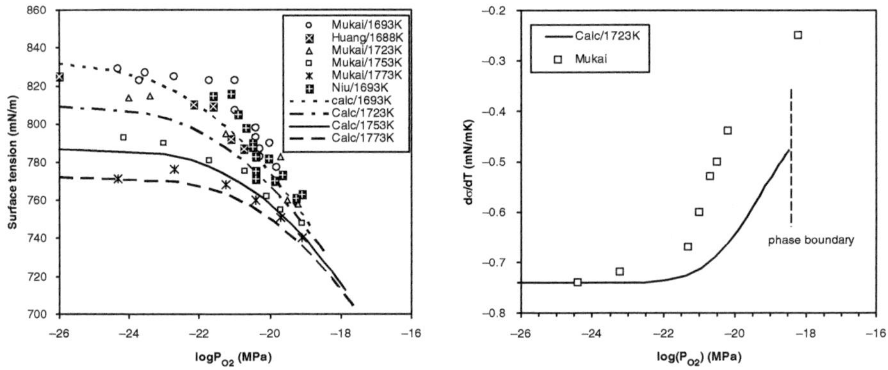

Fig. 13.25. Calculated surface tension of Si–O melt (*left*) and temperature coefficient (*right*)

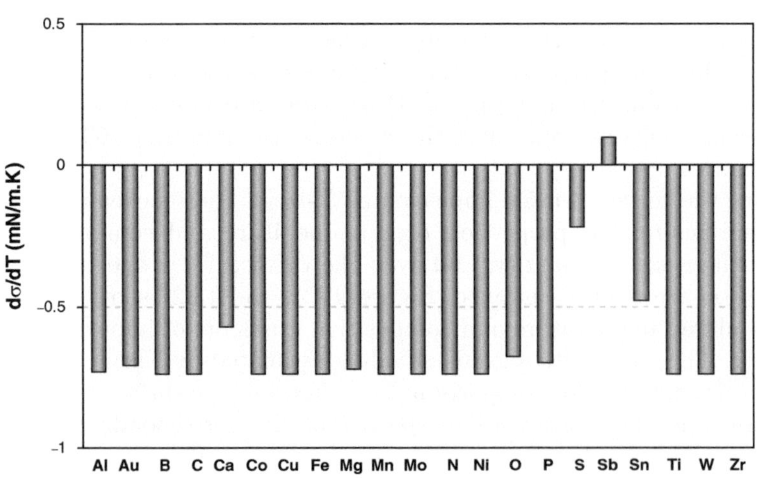

Fig. 13.26. Calculated temperature gradients of impurities in silicon

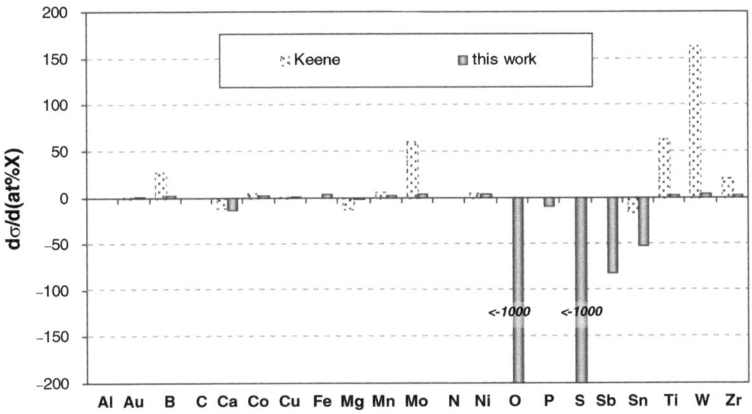

Fig. 13.27. Calculated concentration gradients of impurities in silicon melt

13.4.3 Grain Boundary Segregation of Impurity in Polycrystalline Silicon

The grain boundary segregations of impurities in polycrystalline silicon are of great importance for the intrinsic gettering of oxygen and other nondoping impurities, trapping of transition metal impurities, and annealing treatments. In principle, there exist two main approaches to the thermodynamic description of grain boundary segregation: the Gibbs adsorption isotherm and Langmuir–McLean types of segregation isotherm. The Gibbs adsorption isotherm is based on the changes of the interfacial energy with the bulk activity of the solute [165]. The Langmuir–McLean isotherm describes the segregation equilibrium from the view point of the minimum Gibbs energy [166].

The segregation of a solute I at an interface ϕ of a matrix, M, in a binary M–I system can be represented by the exchange of the components M and I between grain boundary (GB) and the crystal interior. This exchange can be represented by the following "equilibrium reaction":

$$\mathrm{M^{GB}} + \mathrm{I} = \mathrm{M} + \mathrm{I^{GB}}. \tag{13.17}$$

At equilibrium, the chemical potentials of the final and initial states should be the same:

$$\Delta G = \left(\mu_{\mathrm{I}}^{0,\mathrm{GB}} + \mu_{\mathrm{M}}^{0} - \mu_{\mathrm{M}}^{0,\mathrm{GB}} + \mu_{\mathrm{I}}^{0}\right) + RT \ln\left(\frac{a_{\mathrm{I}}^{\mathrm{GB}} a_{\mathrm{M}}}{a_{\mathrm{I}} a_{\mathrm{M}}^{\mathrm{GB}}}\right) = 0. \tag{13.18}$$

If the first term in the right side of (13.18) is defined as the standard molar Gibbs energy of segregation, $\Delta G_{\mathrm{I}}^{0}$, the general form of the segregation equation can then be expressed as:

$$\frac{a_{\mathrm{I}}^{\mathrm{GB}}}{a_{\mathrm{M}}^{\mathrm{GB}}} = \frac{a_{\mathrm{I}}}{a_{\mathrm{M}}} \exp\left(\frac{\Delta G_{\mathrm{I}}^{0}}{RT}\right). \tag{13.19}$$

In the M–I binary system, $X_M = 1 - X_I$, we may rewrite (13.19) as: pcpc

$$\frac{X_I^{GB}}{1 - X_I^{GB}} = \frac{X_I}{1 - X_I} \exp\left(-\frac{\Delta G_I^0 + \Delta G_I^E}{RT}\right), \tag{13.20}$$

where ΔG_I^E is the so-called the excess molar Gibbs energy of interfacial segregation and can be written as: pcpc

$$\Delta G_I^E = RT \ln\left(\frac{\gamma_I^{GB}\gamma_M}{\gamma_I \gamma_M^{GB}}\right). \tag{13.21}$$

For the SoG-Si polycrystalline silicon, both the X_I^{GB} and X_I are far less than unity. Furthermore, if the impurities follow Henry's law, e.g., (13.6), the effect of the excess Gibbs energy on the grain boundary segregation can be neglected. Equation (13.20) may, thus, be simplified to: pcpc

$$\beta_I = \frac{X_I^{GB}}{X_I} \cong \exp\left(-\frac{\Delta G_I^0}{RT}\right) = \exp\left(\frac{\Delta S_I^0}{R} - \frac{\Delta H_I^0}{RT}\right), \tag{13.22}$$

where β_I is the segregation ratio; ΔH_I^0 and ΔS_I^0 are, respectively, the enthalpy and entropy of segregation.

The compositional profiles of C, O, and Si in polycrystalline silicon were reported by Pizzlni et al. [167], who used the secondary ion mass spectrometry (SIMS) to determine the interfacial segregations. The segregation enthalpies and entropies for C and O in polycrystalline silicon have been estimated and listed in Table 13.1. The calculated C and O segregations (solid lines) are compared with the measured values and shown in Fig. 13.28.

A more sophisticated approach for determination of the grain boundary segregation is similar to the determination of the surface tension of silicon melt. The novel approach of surface tension simulation has been successfully implemented in the thermochemical database. Hence, the assessment of the parameters for impurity segregation in solid silicon phase may greatly extend the application of the thermochemical database. The calculation results for C and O segregation are shown as dashed lines in Fig. 13.28. The McLean segregation isotherm can be reproduced using the approach similar to the surface tension simulation.

Table 13.1. Estimated segregation enthalpies and entropies for C and O in polycrystalline Si

Impurity	ΔH_I^0 (J mol^{-1})	ΔS_I^0 (J mol K^{-1})
C	1,500	0.5
O	46,500	52.0

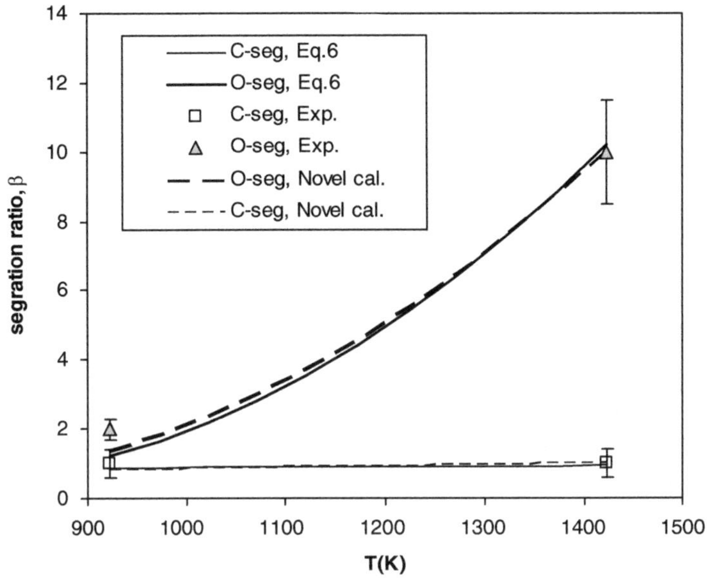

Fig. 13.28. Grain boundary segregations of C and O in polycrystalline silicon

13.4.4 Determination of the Denuded Zone Width

For the defect, engineering of semiconductor industry can be introduced to the solar cell wafer processing. As an example, the width of the denuded zone (DZ) for the intrinsic gettering annealing depends mainly on the annealing conditions at high-temperatures and the microdefect densities. From a practical point of view, it is especially interesting to examine depth profiles of interstitial oxygen concentration in relation to out-diffusion and precipitation. Isomae et al. [168] proposed the following equation to estimate the DZ width:

$$C(x,t) = C_{\mathrm{s}} + (C_{\mathrm{i}} - C_{\mathrm{s}})\,\mathrm{erf}\left(\frac{x}{2\sqrt{Dt}}\right), \tag{13.23}$$

where C_{i} is the initial oxygen concentration, which is assumed to be independent of the depth; C_{s}, D, and t are the oxygen solubility, the oxygen diffusivity, and the annealing time at temperature T, respectively. The depth profile after the annealing is shown in Fig. 13.29. The oxygen solubility and impurity diffusivity were taken from the current assessed databases. The calculated DZ widths at different annealing conditions are in rather good agreement with experimental values [168].

13.5 Conclusions

A self-consistent thermodynamic description of the Si–Ag–Al–As–Au–B–Bi–C–Ca–Co–Cr–Cu–Fe–Ga–Ge–In–Li–Mg–Mn–Mo–N–Na–Ni–O–P–Pb–S–Sb–Sn–Te–Ti–V–W–Zn–Zr system has recently been developed by SINTEF

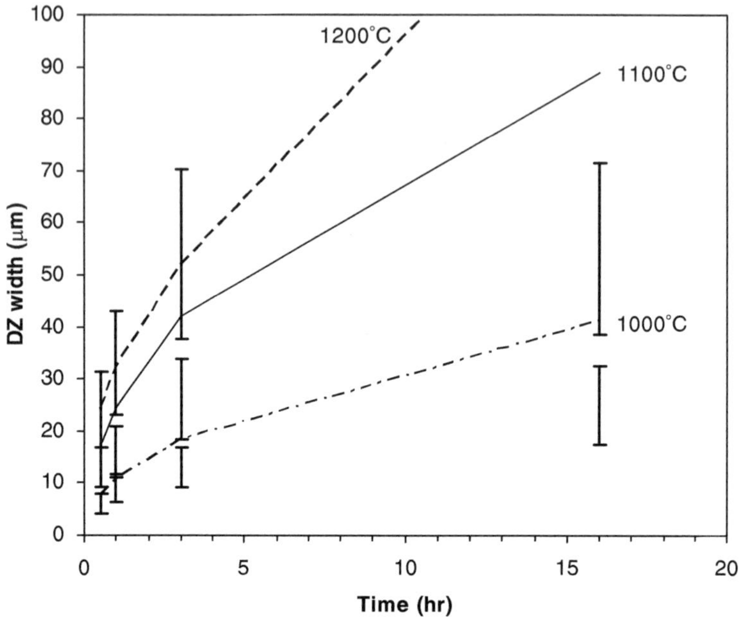

Fig. 13.29. Dependence of denuded zone width on the thermal conditions

Materials and Chemistry. The assessed database has been designed for use within the composition space associated with the SoG-Si materials. Among the 35 binary silicon-containing systems, the Si–Al, Si–As, Si–B, Si–C, Si–Fe, Si–N, Si–O, Si–P, Si–S, Si–Sb, and Si–Te systems have thermodynamically been "reoptimized," primarily based on the assessed experimental information. Since thermodynamic calculations for the impurities in SoG-Si feedstock are normally multidimensional in nature, Gibbs energies of other 36 binary systems have also included in the database.

The assessed kinetic database covers the same system as in the thermochemical database. The impurity diffusivities of Ag, Al, As, Au, B, Bi, C, Co, Cu, Fe, Ga, Ge, In, Li, Mg, Mn, N, Na, Ni, O, P, Sb, Te, Ti, Zn and the self diffusivity of Si in both solid and liquid silicon have extensively investigated. The databases can be regarded as state-of-the-art equilibrium relations in the Si-based multicomponent system.

The thermochemical database has further been extended to simulate the surface tension of liquid Si-based melts. Many surface-related properties, e.g., temperature and composition gradients, surface excess quantity, and even the driving force due to the surface segregation are able to directly obtain from the database. By coupling the Langmuir–McLean isotherm formulism, the grain boundary segregations of the nondoping elements in polycrystalline silicon are now able to estimate from the assessed thermochemical properties.

Finally, application of the thermochemical and kinetic databases to determine the denuded zone width of the intrinsic gettering annealing has been demonstrated.

Acknowledgments

Financial support from the Norwegian Research Council is gratefully acknowledged.

References

1. A.T. Dinsdale, CALPHAD **15**, 317 (1991)
2. E.R. Weber, Appl. Phys. A **30**, 1 (1983)
3. T. Yoshikawa, K. Arimura, K. Morita, Metall. Mater. Trans. B **36**, 837 (2005)
4. T. Yoshikawa, K. Morita, Metall. Mater. Trans. B **36**, 731 (2005)
5. T. Yoshikawa, K. Morita, Sci. Technol. Adv. Mater. **4**, 531 (2003)
6. T. Shimpo, T. Yoshikawa, K. Morita, Metall. Mater. Trans. B **35**, 277 (2004)
7. G. Inoue, T. Yoshikawa, K. Morita, High Temp. Mater. Processes **22**, 221 (2003)
8. R.C. Miller, A. Savage, J. Appl. Phys. **27**, 1430 (1956)
9. D. Navon, V. Chernyshov, J. Appl. Phys. **28**, 823 (1957)
10. V.N. Lozovskii, A.I. Udyanskaya, Izvestiya Akademii Nauk SSSR-Neorganicheskie Materialy (Translate: Inorg. Mater.) **4**, 1174 (1968)
11. T. Yoshikawa, K. Morita, J. Electrochem. Soc. **150,** G465 (2003)
12. T. Miki, K. Morita, N. Sano, Mater. Trans. JIM **40**, 1108 (1999)
13. T. Miki, K. Morita, N. Sano, Metall. Mater. Trans. B, 1998. **29**(5): p. 1043–1049
14. L. Ottem, *Løselighet og termodynamiske data for oksygen og karbon i flytende legeringer av silisium og ferrosilisium* (SINTEF Metallurgy, Trondheim, Norway, 1993), p. 1
15. W. Klemm, P. Pirscher, Z. Anorg. Chem. **247**, 211 (1941)
16. Y.A. Ugay, S.N. Miroshnichenko, Izn. Akad. Nauk SSSR Neorg. Mat. **9**, 2051 (1973)
17. Y.A. Ugay et al., Fiz.Kh.im.Prots.Polyprouod Pouerk.h, 138 (1981)
18. F.A. Trumbore, Bell Syst. Tech. J. **39**, 205 (1960)
19. J.S. Sandhu, J.L. Reuter, IBM J. Res. Dev. **15**, 464 (1971)
20. R.B. Fair, G.R. Weber, J. Appl. Phys. **44**, 273 (1973)
21. S. Ohkawa, Y. Nakajima, Y. Fukukawa, Jpn. J. Appl. Phys. **14**, 458 (1975)
22. F.B. Fair, J.C.C. Tsai, J. Electrochem. Soc. **122**, 1689 (1975)
23. N. Miyamoto, E. Kuroda, S. Yoshida, SAE Prepr. 408 (1973)
24. V.I. Belousov, Russian J. Phys. Chem. **53**, 1266 (1979)
25. S. Fries, H.L. Lukas, in *COST507: Thermodynamic Assessment of the Si-B system*, 1998, p. 77
26. C. Brosset, B. Magnusson, Nature **187**, 54 (1960)
27. B. Armas et al., J. Less Common Metals **82**, 245 (1981)
28. G. Male, D. Salanoubat, Revue Inter. Haut. Temp. Refrac. fractaires **18**, 109 (1981)
29. J. Hesse, Zeitschrift fur Metallkunde **59**, 499 (1968)
30. G. V. Samsonov, V. M. Sleptsov: Russ. J. Inorg. Chem., **8** **1047**(1963)

31. R. Noguchi et al., Metall. Mater. Trans. B **25**, 903 (1994)
32. A.I. Zaitsev, A.A. Kodentsov, J. Phase Equilib. **22**, 126 (2001)
33. T. Yoshikawa, K. Morita, Mater. Trans. **46**, 1335 (2005)
34. R.I. Scace, G.A. Slack, J. Chem. Phys. **30**, 1551 (1959)
35. R.N. Hall, J. Appl. Phys. **29**, 914 (1958)
36. Y. Iguchi, T. Narushima, in *First International Conference on Processing Materials for Properties*, 1991
37. H. Kleykamp, G. Schumacher, Berichte der Bunsen-Gesellschaft-Phys. Chem. Chem. Phys. **97**, 799 (1993)
38. L.L. Oden, R.A. McCune, Metall. Trans. A **18**, 2005 (1987)
39. T. Nozaki, Y. Yatsurugi, N. Akiyama, J. Electrochem. Soc. **117**, 1566 (1970)
40. A.R. Bean, R.C. Newman, J. Phys. Chem. Solids **32**, 1211 (1971)
41. R.C. Newman, Mater. Sci. Eng. B **36**, 1 (1996)
42. J. Grobner, H.L. Lukas, J.C. Anglezio, CALPHAD **20**, 247 (1996)
43. H. Dalaker, M. Tangstad, in *23rd European Photovoltaic Solar Energy Conference*, Valencia, Spain, 2008
44. A.A. Istratov, H. Hieslmair, E.R. Weber, Appl. Phys. A **69**, 13 (1999)
45. H. Feichtinger, Acta Phys. Aust. **51**, 161 (1979)
46. Y.H. Lee, R.L. Kleinhenz, J.W. Corbett, Appl. Phys. Lett. **31**, 142 (1977)
47. S.A. McHugo et al., Appl. Phys. Lett. **73**, 1424 (1998)
48. E.G. Colas, E.R. Weber, Appl. Phys. Lett. **48**, 1371 (1986)
49. E. Weber, H.G. Riotte, J. Appl. Phys. **51**, 1484 (1980)
50. J.D. Struthers, J. Appl. Phys. **27**, 1560 (1956)
51. H. Nakashima et al., Jpn. J. Appl. Phys. **27**, 1542 (1988)
52. D. Gilles, W. Schröter, W. Bergholz, Phys. Rev. B **41**, 5770 (1990)
53. J. Lacaze, B. Sundman, Metall. Trans. A **22**, 2211 (1991)
54. T.G. Chart, High Temp. High Pressures **2**, 461 (1970)
55. T. Miki, K. Morita, N. Sano, Metall. Mater. Trans. B **28**, 861 (1997)
56. C.C. Hsu, A.Y. Polyakov, A.M. Samarin, Izv. Vyssh. Uchebn. Zaved. Chern. Metall. 12 (1961)
57. T. Taishi et al., Jpn. J. Appl. Phys. Lett. **38**, L223 (1999)
58. Y. Yatsurugi et al., J. Electrochem. Soc. **120**, 975 (1973)
59. T. Narushima et al., Mater. Trans. Jim. **35**, 821 (1994)
60. W. Kaiser, C.D. Thurmond, J. Appl. Phys. **30**, 427 (1959)
61. M.W. Chase, *NIST-JANAF Thermochemical Tables,* vol. 2b (Published by the American Chemical Society and the American Institute of Physics for the National Institute of Standards and Technology, Woodbury, N.Y., 1998)
62. S.M. Schnurre, J. Grobner, R. Schmid-Fetzer, J. Noncryst. Solids **336**, 1 (2004)
63. H. Hirata, K. Hoshikawa, J. Cryst. Growth **1990**, 657 (1990)
64. H.J. Hrostowski, R.H. Kaiser, J. Phys. Chem. Solids **11**, 214 (1959)
65. X.M. Huang et al., Jpn. J. Appl. Phys. **32**, 3671 (1993)
66. R.A. Logan, A.J. Peters, J. Appl. Phys. **30**, 1627 (2007)
67. J. Gass, et al., J. Appl. Phys. **51**, 2030 (1980)
68. T. Narushima et al., Mater. Trans. Jim **35**, 522 (1994)
69. A.I. Zaitsev, A.D. Litvina, N.E. Shelkova, High Temp. High Pressure **39**, 227 (2001)
70. J.R.A. Carlsson et al., J. Vac. Sci. Technol. A Vac. Surf. Films **15**, 8 (1997)
71. V.B. Giessen, R. Vogel, Z. Metallkde. **50**, 274 (1959)
72. J. Korb, K. Hein, Z. Anorg. Allg. Chine., **425**, 281 (1976)

73. T. Miki, K. Morita, N. Sano, Metall. Mater. Trans. B **27,** 937 (1996)
74. B.C. Anusionwu, F.O. Ogundare, C.E. Orji, Indian J. Phys. Proc. Indian Assoc. Cultivation Sci. A **77A**, 275 (2003)
75. Y.A. Ugai et al., Izvestiya Akademii Nauk SSSR, Neorganicheskie Materialy **17**, 1150 (1981)
76. K. Uda, M. Kamoshida, J. Appl. Phys. **48**, 18 (1977)
77. E. Kooi, J. Electrochem. Soc. **111**, 1383 (1964)
78. N.K. Abrikosov, V.M. Glazov, L. Chen-yuan, Russ. J. Inorg. Chem. **7**, 429 (1962)
79. S. Solmi et al., Phys. Rev. B **53**, 7836 (1996)
80. M. Tamura, Philos. Mag. **35**, 663 (1977)
81. R.O. Carlson, R.N. Hall, E.M. Pell, J. Phys. Chem. Solids **8**, 81 (1959)
82. F. Rollert, N.A. Stolwijk, H. Mehrer, Appl. Phys. Lett. **63**, 506 (1993)
83. P. Migliorato, A.W. Vere, C.T. Elliott, Appl. Phys. **11**, 295 (1976)
84. J.J. Rohan, N.E. Pickering, J. Kennedy, J. Electrochem. Soc. **106**, 705 (1959)
85. C.D. Thurmond, M. Kowalchik, Bell Syst. Tech. J. **39**, 169 (1960)
86. S.H. Song et al., J. Electrochem. Soc. **129**, 841 (1982)
87. D. Nobili et al., J. Electrochem. Soc. **136**, 1142 (1989)
88. A. Sato et al., Roc. IEEE 1995 Int. Conference on Microelectronic Test Structures, Vol8, March 1995, Nara, Japan
89. F.A. Trumbore, P.E. Freeland, R.A. Logan, J. Electrochem. Soc. **108**, 458 (1961)
90. Y. Du et al., Mater. Sci. Eng. A **363**, 140 (2003)
91. J.O. Andersson et al., CALPHAD **26**, 273 (2002)
92. I. Stich, R. Car, M. Parrinello, Phys. Rev. B **44**, 4262 (1991)
93. C.Z. Wang, C.T. Chan, K.M. Ho, Phys. Rev. B **45**, 12227 (1992)
94. W. Yu, Z.Q. Wang, D. Stroud, Phys. Rev. B **54**, 13946 (1996)
95. Y. Liu et al., Scr. Mater. **55**, 367 (2006)
96. T. Iida, R. Guthrie, N. Tripathi, Metall. Mater. Trans. B **37**, 559 (2006)
97. D. Demireva, B. Lammel, J. Phys. D Appl. Phys. **30**, 1972 (1997)
98. G.S. Fuller, J.A. Ditzenberger, J. Appl. Phys. **27**, 544 (1956)
99. B. Goldstein, Bull. Am. Phys. Soc. Ser. II, 145 (1956)
100. R.N. Ghoshtagore, Phys. Rev. B **3**, 2507 (1971)
101. W. Rosnowski, J. Electrochem. Soc. **125**, 957 (1978)
102. M. Chang, J. Electrochem. Soc. **128**, 1987 (1981)
103. G. Galvagno et al., Semiconductor Sci. Technol. **8**, 488 (1993)
104. D. Nagel, C. Frohne, R. Sittig, Appl. Phys. A **60**, 61 (1995)
105. H. Mitlehner, H-J. Schulze, EPE J. **4**, 37 (1994)
106. J.P. Garandet, Int. J. Thermophys. **28**, 1285 (2007)
107. H. Kodera, Jpn. J. Appl. Phys. **2**, 212 (1963)
108. E.A. Perozziello, P.B. Griffin, J.D. Plummer, Appl. Phys. Lett. **61**, 303 (1992)
109. A.N. Larsen et al., Nucl. Instrum. Methods Phys. Res. B**,** 697 (1993)
110. A. Merabet, C. Gontrand, Physica Status Solidi a-Appl. Res. **145**, 77 (1994)
111. B.J. Masters, J.M. Fairfield, J. Appl. Phys. **40**, 2390 (1969)
112. W.J. Armstrong, J. Electrochem. Soc. **109**, 1065 (1962)
113. P.S. Raju, N.R.K. Rao, E.V.K. Rao, Indian J. Pure Appl. Phys. **2**, 353 (1964)
114. Y.W. Hsueh, Electrochem. Technol. **6**, 361 (1968)
115. T.C. Chan, C.C. Mai, Proc. Inst. Electr. Electron. Eng. **58**, 588 (1970)
116. R.N. Ghoshtagore, Phys. Rev. B **3**, 397 (1971)
117. C.S. Fuller, J.A. Ditzenberger, J. Appl. Phys. **25**, 1439 (1954)

118. E.L. Williams, J. Electrochem. Soc. **108** 795 (1961)
119. M.L. Barry, P. Olofsen, J. Electrochem. Soc. **116**, 854 (1969)
120. M. Okamura, Jpn. J. Appl. Phys. **8**, 1440 (1969)
121. D. Rakhimbaev, A. Avezmuradov, M.D. Rakhimbaeva, Inorg. Mater. **30**, 418 (1994)
122. E. Dominguez, M. Jaraiz, J. Electrochem. Soc. **133**, 1895 (1986)
123. G.L. Vick, K.M. Whittle, J. Electrochem. Soc. **116**, 1142 (1969)
124. R.N. Ghoshtag, Phys. Rev. B **3**, 389 (1971)
125. W. Wijaranakula, Jpn. J. Appl. Phys. **32** 3872 (1993)
126. R. Pampuch, E. Walasek, J. Bialoskórski, Ceramics Int. **12**, 99 (1986)
127. N.N. Eremenko, G.G. Gnesin, M.M. Churakov, Powder Metall. Metal Ceramics **11**, 471 (1972)
128. A. Chari, P. de Mierry, M. Aucouturier, Revue de Physique Appliquée **22**, 655 (1987)
129. P. Schwalbach et al., Phys. Rev. Lett. **64**, 1274 (1990)
130. L.C. Kimerling, J.L. Benton, J.J. RubIn, Inst. Phys. Conf. Ser. **59**, 217 (1981)
131. T. Isobe, H. Nakashima, K. Hashimoto, Jpn. J. Appl. Phys. **28**, 1282 (1989)
132. V.A. Uskov, Sov. Phys. Semiconductors USSR (English Transl.) **8**, 1573 (1975)
133. I. V. Antonova, K.B.K.E.V.N.L.S.S., Phys. Stat. Sol. (a) **76**, K213 (1983)
134. B.K. Miremadi, S.R. Morrison, J. Appl. Phys. **56**, 1728 (1984)
135. H. Nakashima et al., Jpn. J. Appl. Phys. **23**, 776 (1984)
136. Y. Itoh, T. Nozaki, Jpn. J. Appl. Phys. **24**, 279 (1985)
137. N. Stoddard et al., J. Appl. Phys. **97** (2005)
138. H. Sawada et al., Phys. Rev. B **65**, 075201 (2002)
139. H. Akito et al., Appl. Phys. Lett. **54**, 626 (1989)
140. N. Fuma et al., Mater. Sci. Forum **196–201**, 79 7 (1995)
141. N. Fuma et al., Jpn. J. Appl. Phys. **35**, 1993 (1996)
142. J. Mukerji, S.K. Brahmachari, J. Am. Ceramic Soc. **64**, 549 (1981)
143. K.P. Abdurakhmanov et al., Sov. Phys. Semiconductors-USSR, **22**, 1324 (1988)
144. J. Pelleg, M. Sinder, J. Appl. Phys. **76**, 1511 (1994)
145. F. Wittel, S. Dunham, Appl. Phys. Lett. **66**, 1415 (1995)
146. N. Jeng, S.T. Dunham, J. Appl. Phys. **72**, 2049 (1992)
147. K. Park et al., J. Electron. Mater. **20**, 261 (1991)
148. R. Deaton, U. Gosele, P. Smith, J. Appl. Phys. **67**, 1793 (1990)
149. C.S. Fuller, J.C. Severiens, Phys. Rev. **96**, 21 (1954)
150. D.A. Petrov, Y.M. Shaskov, I.P. Akimchenko, Chem. Abstr. (1960), p. 17190c.
151. M.O. Thurston, J.C.C. Tsai, Ohio State University Research Foundation, 1962
152. S. Nakanuma, S. Yamagishi, J. Electrochem. Soc. Jpn. **36**, 3 (1968)
153. R.B. Fair, M.L. Manda, J.J. Wortman, J. Mater. Res. **1**, 705 (1986)
154. K. Nagata et al., in *Handbook of Physico-Chemical Properties at High Temperatures* ed. by Y. Kawai, Y. Shiraishi, (Iron and Steel Institute of Japan, Tokyo, Japan, 1988)
155. K. Yanaba et al., Mater. Trans. Jim **39**, 819 (1998)
156. E.A. Guggenheim, Trans. Faraday Soc. **41**, 150 (1945)
157. J.A.V. Bulter, Proc. R. Soc. **A135**, 348 (1932)
158. P. Koukkari, R. Pajarre, CALPHAD **30**, 18 (2006)
159. S.V. Lukin, V.I. Zhuchkov, N.A. Vatolin, J. Less Common Metals **67**, 399 (1979)
160. K. Mukai et al., ISIJ Int. **40** (Supplement), S148 (2000)

161. K. Mukai, Z.F. Yuan, Mater. Trans. Jim **41**, 331 (2000)
162. X. Huang et al., J. Cryst. Growth **156**, 52 (1995)
163. Z. Niu et al., J. Jpn. Assoc. Cryst. Growth **23**, 374 (1996)
164. B.J. Keene, Surf. Interf. Anal. **10**, 367 (1987)
165. P. Wynblatt, D. Chatain, Metall. Mater. Trans. A **37**, 2595 (2006)
166. D. McLean, *Grain Boundaries in Metals*, vol. 10 (Clarendon, Oxford, 1957), 346s
167. S. Pizzini et al., Appl. Phys. Lett. **51**, 676 (1987)
168. S. Isomae, S. Aoki, K. Watanabe, J. Appl. Phys. **55**, 817 (1984)

Index

MIX
Papier aus verantwortungsvollen Quellen
Paper from responsible sources
FSC® C105338

If you have any concerns about our products,
you can contact us on
ProductSafety@springernature.com

In case Publisher is established outside the EU,
the EU authorized representative is:
Springer Nature Customer Service Center GmbH
Europaplatz 3, 69115 Heidelberg, Germany

Printed by Libri Plureos GmbH
in Hamburg, Germany